*Charles Schneider, La Jolla, California*

## CECIL HOWARD GREEN

# ADVANCES IN EARTH SCIENCE

# TIC DATA RECEIPT

**Space Division** — Rockwell International

|  |  |
|---|---|
| ACCESSION NUMBER | 054146 |
| DOCUMENT / DEWEY NUMBER | 0003 |
| COPY NUMBER | |
| TICKET NUMBER | U 06131 |
| TITLE (25) | 550A.1964 |
| TITLE (45) | ADV IN |
| CORPORATE SOURCE / PERSONAL AUTHOR (60) | EENICE EARTH SC |
| DOCUMENT DATE | |
| ENCLOSURES | |
| PAGES | |
| REGISTRY NUMBER | |
| SECURITY CLASSIFICATION | |
| L R D (17) | / / |
| REQUESTER (LAST NAME FIRST) (33) | |
| DUE DATE MO/DAY/YR | |
| CLEARANCE NUMBER | |
| $ SECRET / C CONFIDENTIAL | C |
| DATE RECEIVED MO/DAY/YR | 02 / 23 / 81 |
| DEPARTMENT (29) | DONAVAN R&D |
| GROUP (57) | 743305K86 |
| LOCATION (60) | 4187-448 |
| EXTENSION / EXTERNAL LINE (69) | 3273-2482 |
| ID SERIAL NUMBER | |
| RECEIPT SIGNATURES | M.D. Donavan SR |

ENT. PROPERTY OF ROCKWELL INTERNATIONAL HAS BEEN RE-
ERSIGNED, AND IF CLASSIFIED, WILL BE USED UNDER THE
ENT SECRECY AGREEMENT. FAILURE TO EXHIBIT DURING
K. RETURN THE DOCUMENT AT THE EXPIRATION OF THE
OF AN EMPLOYEE'S TERMINATION, WILL RESULT IN A-
AMOUNT TO THE REPLACEMENT COST OF THE DOCU-
STRIAL SECURITY FOR APPROPRIATE ACTION.

# ADVANCES IN EARTH SCIENCE

*Contributions to the International Conference on the Earth Sciences*
*Massachusetts Institute of Technology*
*September 1964*

*Edited by* P. M. Hurley

THE M.I.T. PRESS
*Massachusetts Institute of Technology*
*Cambridge, Massachusetts, and London, England*

*Copyright © 1966 by
The Massachusetts Institute of Technology*

*Second printing, September 1967*

*All rights reserved. This book may not be reproduced,
in whole or in part, in any form (except by reviewers for the public press),
without written permission from the publishers.*

*Library of Congress Catalog Card Number: 65-25438
Printed in the United States of America*

This volume is dedicated to
CECIL HOWARD GREEN
*Geophysicist*

# PREFACE

The Conference recorded in this volume was held on the occasion of the dedication of the Cecil and Ida Green Building. This new building houses the Center for the Earth Sciences of the Massachusetts Institute of Technology, comprising the Departments of Meteorology and of Geology and Geophysics.

The annals of science do not frequently honor a man of industry, as the faculty in earth sciences at M.I.T. have done in choosing to dedicate this volume to Cecil H. Green, geophysicist and alumnus of the Institute. In so doing, they wished to recognize his contributions to the development of geophysics in the service of mankind, his deep concern for education in the earth sciences, and his warm friendship with many scientists, including a large number of the conference participants. By constant demand for basic insight, Dr. Green has developed and maintained an organization of scientists which has an exceptional record of achievement and leadership in applied geophysics, as well as in certain other areas. Dr. Green's interest in geophysics education has had an impact in many parts of the world. His concern has been the extent of basic science in curricula, emphasis on the strong connections between earth sciences and other disciplines of science and engineering, the opportunity for field training, and the stimulation of keen interest in the subject of exploration geophysics.

The occasion of the dedication of the new building also marks an important step in the Institute's plans for fostering an integrated development of the earth sciences and for expanding activities in these fields, plans which were precipitated by the interest and actions of Cecil and Ida Green. Substantial increases in staff, students, research, and facilities are accompanying the move to the new Center. Meteorology, oceanography, and some aspects of planetary science have been brought into closer contact with geochemistry, geology, and geophysics, and students may select paths in instruction and research throughout this range of subjects.

The Conference itself demonstrated the increasing integration of the earth sciences, the introduction of many new techniques, and the growing activity and sphere of earth scientists. This Conference, with the participation of many colleagues from many institutions, has given inspiration to the new Center, and has also produced a fascinating and thoughtful series of discussions on many of the most lively topics in the earth sciences. It is hoped that they will be an inspiration to the readers as well and a base for many more interesting developments.

CHARLES H. TOWNES

*Cambridge, Massachusetts*
*November 1, 1965*

# PROGRAM

*International Conference on the Earth Sciences*

*Conference Chairman*, Charles H. Townes, *M.I.T.*

WEDNESDAY, SEPTEMBER 30
*Morning Session, 9:30–12:00*

*Chairman*, Bengt Strömgren, *Princeton University*
*Co-Chairman*, John V. Harrington, *M.I.T.*

**THE EARTH'S ENVIRONMENT**

**The Sun and Solar Physics**
Leo Goldberg, *Harvard University*

**The Moon, Planets, and Their Origin**
Gerard P. Kuiper, *University of Arizona*

**The Interplanetary Medium and Solar-Planetary Relations**
Ludwig F. Biermann, *Max-Planck-Institut für Physik und Astrophysik*

*Afternoon Session, 2:00–5:00*

*Chairman*, Robert M. White, *United States Weather Bureau*
*Co-Chairman*, Henry G. Houghton, *M.I.T.*

## ATMOSPHERIC MOTIONS

**Large-Scale Motions of the Atmosphere: Circulation**
Edward N. Lorenz, *M.I.T.*

**Motions of Intermediate Scale: Fronts and Cyclones**
Arnt Eliassen, *University of Oslo*

**Atmospheric Turbulence**
A. M. Obukhov, *Institute of Atmospheric Physics, Moscow*

THURSDAY, OCTOBER 1
*Morning Session, 9:00–12:00*

*Chairman*, W. Maurice Ewing, *Columbia University*
*Co-Chairman*, Columbus O. Iselin
*Woods Hole Oceanographic Institution and M.I.T.*

## DYNAMICS OF THE OCEANS

**Long-Period Phenomena of the Oceans Revealed by Chemistry**
Gustaf Arrhenius, *University of California, San Diego*

**Large-Scale Circulation of the Oceans**
Henry M. Stommel, *M.I.T.*

**The Spectrum of Waves**
Walter H. Munk, *University of California, San Diego*

*Afternoon Session, 2:00–5:00*

*Chairman*, J. Tuzo Wilson, *University of Toronto*
*Co-Chairman*, Raymond Hide, *M.I.T.*

## THE "SOLID" EARTH I

**Long-Term Mechanical Properties of the Earth and Internal Motions**
Gordon J. F. MacDonald, *University of California, Los Angeles*

**Seismological Information and Advances**
Frank Press, *California Institute of Technology*

**Composition and Phases of the Mantle**
A. E. Ringwood, *Australian National University*

FRIDAY, OCTOBER 2
*Morning Session, 9:00–12:00*

*Chairman*, Sir Edward C. Bullard
*Churchill College, University of Cambridge*
*Co-Chairman*, Patrick M. Hurley, *M.I.T.*

## THE SOLID EARTH II

**Temperature, Heat Production, and Thermal History of the Earth**
Francis Birch, *Harvard University*

**Geochronology, and Isotope Data Bearing on the Development of the Continental Crust**
G. J. Wasserburg, *California Institute of Technology*

**Mechanics of the Upper Mantle**
Walter M. Elsasser, *Princeton University*

# CONTENTS

## *The Earth's Environment*   1

**The Sun and Solar Physics**   3
Leo Goldberg, *Harvard University*

**The Moon and the Planet Mars**   21
Gerard P. Kuiper, *University of Arizona*

**The Interplanetary Medium and Solar Planetary Relations**   71
Ludwig F. B. Biermann, *Max-Planck-Institut für Physik und Astrophysik*

## *Atmospheric Motions*   93

**Large-Scale Motions of the Atmosphere: Circulation**   95
Edward N. Lorenz, *Massachusetts Institute of Technology*

**Motions of Intermediate Scale: Fronts and Cyclones**   111
Arnt Eliassen, *University of Oslo*

**Atmospheric Turbulence**   139
A. M. Obukhov, *Institute of Atmospheric Physics, Moscow*

## *Dynamics of the Oceans*   153

**Sedimentary Record of Long-Period Phenomena**   155
Gustaf Arrhenius, *University of California, La Jolla*

**The Large-Scale Oceanic Circulation**   175
Henry M. Stommel, *Massachusetts Institute of Technology*

**Coupling of Waves, Tides, and the Climatic Fluctuation of Sea Level**   185
Walter H. Munk, *University of California, La Jolla*

## The "Solid" Earth I  197

**The Figure and Long-Term Mechanical Properties of the Earth**  199
Gordon J. F. MacDonald, *University of California, Los Angeles*

**Seismological Information and Advances**  247
Frank Press, *California Institute of Technology*

**The Chemical Composition and Origin of the Earth**  287
A. E. Ringwood, *Australian National University*

**Mineralogy of the Mantle**  357
A. E. Ringwood, *Australian National University*

## The "Solid" Earth II  401

**Earth Heat Flow Measurements in the Last Decade**  403
Francis Birch, *Harvard University*

**Geochronology, and Isotopic Data Bearing on Development of the Continental Crust**  431
G. J. Wasserburg, *California Institute of Technology*

**Thermal Structure of the Upper Mantle and Convection**  461
Walter M. Elsasser, *Princeton University*

# PART I

# THE EARTH'S ENVIRONMENT

# THE SUN AND SOLAR PHYSICS

*Leo Goldberg*

*Harvard and Smithsonian Observatories, Cambridge, Massachusetts*

**Introduction**

The ultimate goal of solar physics is the construction of a solar model that agrees with observation and is based on the known laws of physics. This is not an impossibly difficult task. Over forty years ago, Sir Arthur Eddington pointed out that a hypothetical physicist on a cloudbound planet could have "discovered" the stars merely by calculating the equilibrium of globes of gas of various sizes. He would have found, in fact, that equilibrium exists only for a rather small range of mass between $10^{33}$ and $10^{35}$ grams, and it is no accident that this is just the range in which most actual stars are found to occur. Moreover, he could have demonstrated that the entire internal structure of the star, as well as its luminosity and radius, is an unambiguous consequence of its mass and chemical composition and the well-known physical laws governing hydrostatic equilibrium, thermal equilibrium, and the transport of energy.

The solar interior cannot be directly observed, which may explain why it appears to be well understood. But the solar atmosphere is quite another matter, and even after four centuries of observation, it has not been explained theoretically with any degree of certainty or completeness. Thus, most of the problems of solar physics are located in the outer atmospheric fringe, which includes about 1 part in $10^{11}$ of the sun's mass.

Greatly oversimplified, there are two reasons, one general and the other specific, why the solar atmosphere seems to have eluded understanding. The first is that we are prone to human error in predicting the consequences of known physical laws, and the second is that we do not yet fully comprehend the hydrodynamics of gases, especially in the presence of magnetic fields. As a result, solar physics has acquired a strong observational bias. The selection of critical observations or experiments is not always obvious, since even an apparently minor phenomenon may turn out to be of major significance. It is becoming increasingly evident, however, that although most solar phenomena are exceedingly complex, they are interconnected and are governed by the laws of magnetohydrodynamics. Furthermore, all of the phenomena can in one way or another be traced back to two rather simple physical properties, namely, an unstable convection zone below the visible surface and a weak but widespread magnetic field. Much of the effort in solar physics today is being devoted to the investigation of these properties and their consequences.

**Convective Motions**

Let us first glance at the observational evidence for convective motions in the solar atmosphere. As shown in Figure 1, photographs of the solar disk made in white light display brightness fluctuations, the so-called granulation, on a scale of about 1 second of arc ($1'' = 750$ km on the sun). It now seems quite certain that the solar granulation is caused by convection from the layers below, the bright regions representing hot gases moving upward and the darker areas cooler material on the way down. The brightness fluctuations have an rms value of about 7 per cent. The granulation shows a pattern similar to that of nonstationary convection. Thus, it is quasi-cellular with a mean lifetime of about 6 minutes.

There is little doubt that the solar granulation originates in a convectively unstable zone a few hundred kilometers below the visible surface, the *photosphere*. Throughout most of its interior the sun is in radiative equilibrium. This is tantamount to saying that the actual temperature gradient is less than the adiabatic gradient. Near the surface, however, we encounter a zone of transition between ionized hydrogen and neutral hydrogen. As the hydrogen becomes neutral, the opacity increases, which makes the radiative temperature gradient steeper. At the same time, the low ratio of specific heats associated with the ionization of hydrogen reduces the adiabatic temperature gradient. This leads to a situation in which an upward

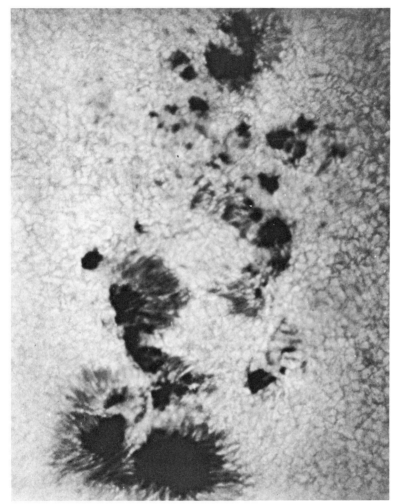

Fig. 1. A portion of the solar disk in the neighborhood of a sunspot group, showing characteristic cellular pattern of convection. 14 April 1963. Sacramento Peak Observatory, Air Force Cambridge Research Laboratories.

rising bubble of gas, expanding to match the drop in the surrounding pressure, finds itself at a higher temperature than its surroundings and therefore continues to rise. However, by the time the visible surface is reached, the convective motions have died out and the gases are again in radiative equilibrium. But the effects of the convective motions do penetrate into the visible layers and are in fact probably responsible for the heating of the chromosphere and corona.

In 1950, Schwarzschild and Richardson conjectured that high-resolution solar spectra might reveal the radial motions associated with

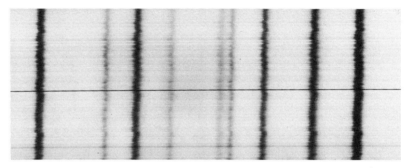

FIG. 2. Local Doppler shifts in the solar spectrum between 5194.3 Å (left) and 5199.2 Å (right). The zigzag or wiggly pattern is caused by adjacent columns or bubbles of gas alternately ascending and descending. The displacements are quasi-periodic with a period of about 5 minutes. Sacramento Peak Observatory, Air Force Cambridge Research Laboratories.

the granulation. Thus, the absorption lines formed when a spectrograph slit is placed across the center of the disk should not be perfectly straight but should exhibit a kind of zigzag or wiggly pattern of Doppler shifts. Although the displacements were not obvious on the best spectra available at that time, careful measurements by Schwarzschild and Richardson at the Mount Wilson Observatory did reveal a pattern of velocity fluctuations on a scale of about 2 seconds of arc, with an rms value of about 0.4 km/sec. Thanks to some rather major improvements in spectrographs, these shifts can now be easily observed and studied in very great detail (Figure 2).

It was also found many years ago that the widths of the solar absorption lines were much greater than those to be expected purely on the basis of thermal Doppler broadening at a temperature of 6000°K. In explanation of these excessive line widths, it was assumed that the photospheric gases were turbulent, with rms velocities of

1 to 2 km/sec. Finally, the systematic increase of line widths toward the solar limb suggested that the motions also had a large horizontal component. Until recently, it was assumed that the motions deduced from the spectral lines were directly and intimately connected with the granulation and were explained by a circulatory flow pattern in which hot bubbles rise to the surface and then flow horizontally before cooling and sinking.

This interpretation has been greatly modified by new observational studies at the Sacramento Peak Observatory and at the Mount Wilson and Palomar Observatories, which have resulted in the discovery of two separate and distinct velocity fields in the photosphere. One of these is oscillatory and vertical while the other is a pattern of horizontal motions which is called the supergranulation because its linear scale is about 20 times larger than that of the oscillating elements.

At any given instant, the vertical velocity field near the center of the solar disk shows a random pattern of up and down motions on a linear scale somewhat larger than that of the brightness fluctuations, that is to say, about 1000 km. However, the pattern varies sinusoidally with a period of almost exactly 5 minutes. This is now well established as a characteristic physical property of the solar photosphere. The observed properties of the fine-scale photospheric velocity field have recently been summarized by Leighton and by Evans and Michard. Most of the kinetic energy of macroscopic motions of regions in the size range 1000 to 5000 km is contained in quasi-oscillatory motions with a period of about 5 minutes, the oscillation damping out after 3 or 4 cycles. These oscillating regions do not represent the granulation itself but lie above it. However, they seem to be initiated by the impact of hot rising granules on the lower boundary of the stable layer. Thus, bright granules are followed by oscillations, beginning with an upward motion, the time delay being about 40 seconds. An important point is whether the oscillations are progressive waves moving vertically with phase lags between the motion at different heights, or stationary waves with the same phase at different heights. The evidence on this point is rather mixed, but there are indications that a given disturbance starts out as a running wave but then develops into a standing wave. As the height in the atmosphere increases, the characteristic period decreases slowly but the velocity amplitude increases rapidly from about 0.4 to about 1.6 km/sec. A substantial fraction of the energy seems to be propagated as acoustic or pressure waves, which travel into the chromosphere where they are transformed into shock waves which are eventually dissipated into heat.

Horizontal motions are best studied near the solar limb, where they produce Doppler shifts. They can be seen readily on so-called Doppler spectroheliograms (Figure 3) according to an ingenious

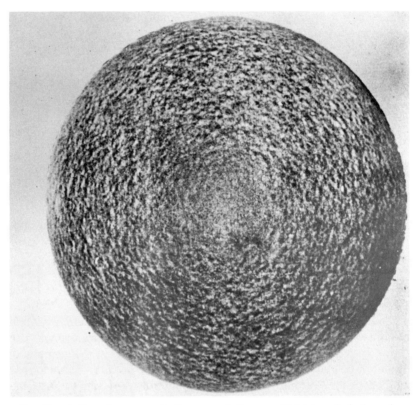

FIG. 3. A Doppler spectroheliogram of the sun. Bright and dark regions denote velocities of approach and recession, respectively. Mount Wilson and Palomar Observatories.

scheme devised by Leighton, in which line-of-sight velocities appear as intensity modulations. Thus bright regions represent motion outward and dark regions motion inward. The oscillatory velocity field is suppressed by combining two Doppler spectroheliograms taken half a period or about 150 seconds apart. It can be seen first of all that there is almost no large-scale motion at the center of the disk, which means that the flow is chiefly horizontal. Next, one sees a large number of patches of motion which are typically lighter on the side toward the center of the disk, implying a velocity of approach on one side and of recession on the other. Detailed study reveals a cellular system in which the flow within each cell is from the center toward the edge as in cellular convection. The supergranulation pattern has a lifetime of at least several hours. There seems to be no correlation between the horizontal and vertical oscillatory motions. Finally, the supergranulation is best observed at the centers of very strong lines and must therefore be located in the chromosphere.

At the recent meetings of the International Astronomical Union in Hamburg a month ago, there was general agreement that the basic hydrodynamics of stellar convection zones is still very far from being understood. A number of questions were asked but not answered. For example, how are convective motions in the deep layers transformed to oscillatory motion in the higher layers of the photosphere? Why do the observed oscillations have such a narrow frequency spectrum? Calculations suggest that the convective modes may overshoot into higher layers of the atmosphere. But if the supergranulation is an overshooting phenomenon of purely convective modes, why is it observed only in the chromosphere and not in the photosphere?

**Magnetic Fields**

Motions in the solar atmosphere are greatly complicated by the presence of magnetic fields, which were first discovered by George Ellery Hale in 1908. The original discovery was made by measurement of the Zeeman splitting of lines in the spectra of sunspots, where typical fields lie in the range two to three thousand gauss. Such fields are easily detectable because they produce displacements of about 0.1 Å. Most of the effort in recent years, however, has been devoted to the search for and the measurement of a so-called general magnetic field, which can be inferred from the presence of coronal streamers. This effort has led to the development of techniques for the measurement of fields as small as a few tenths of one gauss, for which the Zeeman splitting is about $10^{-5}$ Å. The major advances in this area were made first by H. W. Babcock at Mount Wilson Observatory about 10 years ago and more recently by A. B. Severny at the Crimean Astrophysical Observatory in the Soviet Union. Measurements made over a period of years show that the sun's magnetic field has a very patchy distribution, but there is a very close coincidence between the occurrence of centers of activity and local enhancements of the magnetic field (Figure 4). In the low latitudes where active regions occur, the magnetic maps show concentrations of localized bipolar magnetic fields. At these low latitudes, the magnetic field is rarely less than 50

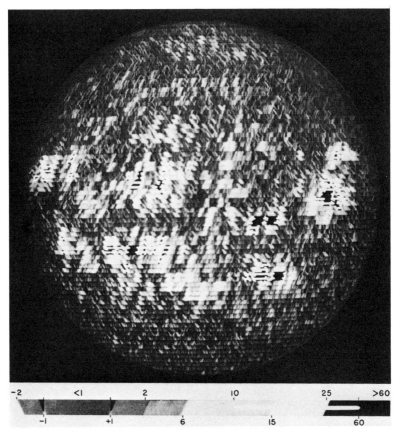

Fig. 4. Magnetic map of the sun's disk, 21 July 1961, showing the location, field intensity, and polarity of weak magnetic fields. The records are made automatically by a scanning system that employs a polarizing analyzer, a powerful spectrograph, and a sensitive photoelectric detector for measuring the line-of-sight component of the magnetic field by means of the Zeeman effect. The calibration strip at the bottom of the picture shows how the recording line slants to right or left to indicate magnetic polarity, and how it changes brightness and form to indicate seven different levels of magnetic field intensity in gauss. The extended magnetic areas on the solar disk are characteristically bipolar and usually produce sunspots as well as other solar activity. Mount Wilson and Palomar Observatories.

gauss, and therefore the presence of a general field must be sought near the poles, away from the zones of activity. The existence of such a general field, with opposite polarity at the two poles, seems to have been established. Its intensity is on the order of 1 gauss, and it appears to have reversed its polarity near the time of the last sunspot maximum, although the reversal of polarity at the South Pole occurred nearly 18 months after the reversal at the North Pole.

There is now a great abundance of observational evidence showing that sunspots and other observable features of solar activity such as coronal loops, flares, and the chromospheric network are somehow caused by the sun's magnetic field and its variations. For many years

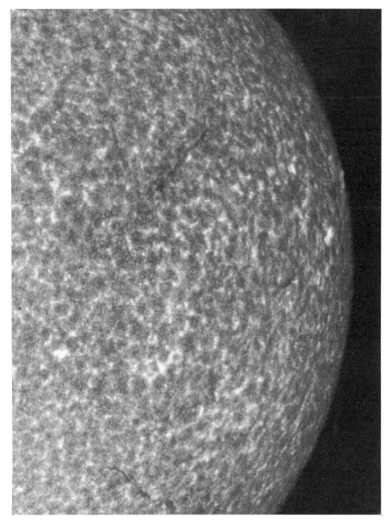

Fig. 5. Spectroheliogram of a portion of the solar disk in the monochromatic light of the emission line of Ca+ at 3933 Å. Mount Wilson and Palomar Observatories.

the idea has been developing that the sunspot cycle and its attendant phenomena are caused by the internal amplification of the sun's general dipole field that results from the systematic increase in the angular rate of the sun's axial rotation from the higher latitudes to the Equator. Very recently Leighton has shown that the process is reversible, namely, that the general field may be caused by the migration of disintegrating sunspots toward the poles.

The Mount Wilson measurements upon which the conclusions about the general field are based have an angular resolution of 23 seconds of arc. Recently, however, Severny has reported measurements made with much higher angular resolution, a few seconds of arc,

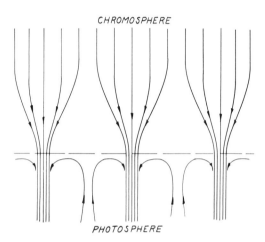

FIG. 6. Model of the supergranulation showing interaction between hydrodynamic flow and magnetic fields according to R. B. Leighton. *Annual Review of Astronomy and Astrophysics*, vol. 1, 1963.

which seem to show that on this much smaller scale even the polar regions seem to contain many small elements of different polarities. Severny also shows that the dependence of the solar polar field on latitude is quite unlike that of either a dipole field or a uniform magnetized sphere. But Severny's results do not contradict the models of Babcock and Leighton, which require only that the fields in polar latitudes be *predominantly* but not necessarily entirely of one sign.

In addition to the rather obvious connection between magnetic fields and solar activity, there also seems to be a rather striking relationship between the distribution of magnetic fields and the pattern of the supergranulation. Figure 5 shows a monochromatic spectroheliogram of the sun in the emission line of Ca+ at 3933 Å. This emission comes from the low chromosphere and its distribution in the form of a bright network has been familiar to solar observers for many years. In fact it was suggested by Deslandres in 1910 that the chromospheric network might correspond to the boundaries of a system of convection cells. This suggestion is proving to be correct. I referred earlier to the supergranulation as a pattern of cells in which the flow is always from the center toward the edge. It has also been found by Simon that the cells of the supergranulation are surrounded by the bright linear features of the chromospheric network and furthermore that the motion in these narrow bright cell boundaries is predominantly downward. Finally, it has been noted that there is a close geometrical correspondence between the intensity of weaker magnetic fields, with strength about 2 to 5 gauss, and the brightness of emission in the Ca+ line. This suggests an interaction between the supergranulation and the weak magnetic fields, as shown in Figure 6. Ionized material flows downward into the photosphere from the high chromosphere along lines of force of the magnetic field. The predominantly sideways motion of material in the supergranulation cells tends to push and compress the lines of force to the boundaries of the cells. The downward flowing ionized material is thus channeled downward into the space between the cells.

**Heating of the Chromosphere and Corona**

The explanation of the chromospheric network is a very good example of how the structure of the solar atmosphere is governed by magnetohydrodynamics, even though the subject can by no means as yet be treated as an exact science. Even more dramatic examples are provided by the high temperature of the corona and the occurrence of solar flares. One of the great surprises in the history of solar physics was the discovery by Edlén a little over 20 years ago that the corona is a very hot plasma with temperature on the order of a million degrees. As a corollary, the chromosphere must be a transition region in which the temperature rises quite steeply between the photosphere and the corona. It was shown in 1948, independently by Schwarzschild and Biermann, that the outward increase of temperature in the solar atmosphere is caused by the dissipation of shock waves, which originate as acoustic waves in the hydrogen convection zone and propagate outward. As the density decreases, the wave amplitude builds up and shock waves are formed which are then dissipated into heat. To a first approximation, the velocity of sound in the chromosphere is constant, but the velocity of magnetohydrodynamic Alfvén waves increases rapidly outward and eventually exceeds the sound velocity. Therefore, magnetohydrodynamic effects must also be important in the heating of the chromosphere and corona. It is believed that the spicules or jets, which are observed to emerge from the sun at supersonic velocities of about 30 km/sec, are visible examples of shock waves propagating into the corona (Figure 7). Since the spicules appear to move along

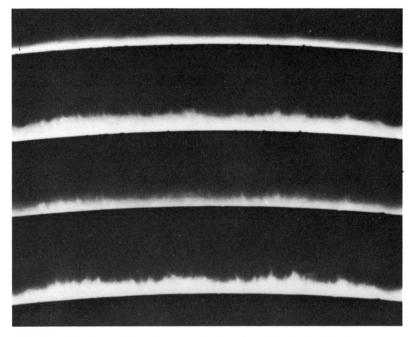

FIG. 7. Spicule activity photographed at the solar limb. The bottom two photographs were made in the H-$\alpha$ line and the top in the nearby continuum. Sacramento Peak Observatory, Air Force Cambridge Research Laboratories.

the direction of lines of force of the solar magnetic field, there is obviously a close connection between the solar field and the outward propagating disturbances.

The rough outlines of this theory of the heating of the solar corona seem unshakable, and indeed Moore and Spiegel have shown in some detail how oscillations in the convective zone can get through to the upper regions and provide a sufficient flux of mechanical energy to maintain the corona at its high temperature. On the other hand, the details of the theory are very incomplete, and it has not yet proved possible to calculate the structure of the chromosphere and corona with any kind of assurance. Here theory must look to observation for guidance, but unfortunately it is somewhat the case of the blind leading the blind. For example, there is general agreement that the temperature in the low chromosphere, in the height range 0 to 1000 km, increases slowly from 4500°K up to values in the neighborhood of 8000 to 10,000°, and furthermore that somewhere in the region of heights between 1000 and 5000 km the temperature rises very steeply to its value in the corona. But temperature determinations in this intermediate zone are widely discrepant, as may be seen in Figure 8.

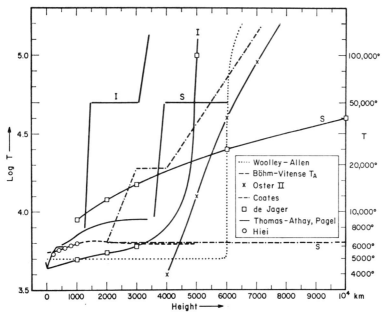

FIG. 8. Temperature gradients in the solar chromosphere according to several different investigators. The letter S refers to conditions within spicules, whereas I denotes the regions between or interspicular regions. Compiled by B. E. J. Pagel. *Annual Review of Astronomy and Astrophysics*, vol. 2, 1964.

The steep temperature rise provides the solar spectroscopist with some rather fascinating problems. Since the outer layers of the sun are essentially transparent down to the level of the photosphere, the solar spectrum exhibits emission lines formed over an enormous range of temperature from a few thousand to a few million degrees. These emission lines are best observed in the far ultraviolet region of the spectrum from rockets and satellites. One sees such low excitation lines as those of neutral oxygen and hydrogen at one extreme to lines of O VIII and Fe XVII at the other. The corona also radiates forbidden lines of highly ionized atoms of Fe, Ca, Ni, and others in the visible region of the spectrum.

### Ionization Equilibrium of the Chromosphere-Corona

The lines of different ionization potential originate from different levels in the atmosphere and hence offer the possibility of a fairly exact determination of the variation of temperature and density with height. Unfortunately, attempts to derive the structure of the outer atmosphere are hampered by inhomogeneities in the structure, by the complexity of the motions, and worst of all by gross departures from thermodynamic equilibrium. In the analysis of the visible absorption line spectrum, it is still fairly safe for most purposes to assume thermodynamic equilibrium and therefore to employ the Saha-Boltzmann equations for the calculation of the degree of ionization and excitation. But in the middle and upper chromosphere, and especially in the corona, there is not even a semblance of thermodynamic equilibrium, and therefore analyses of the spectrum must be based on consideration of the rates of the various detailed processes by which atoms exchange energy with their surroundings. The difficulty is that we have hardly begun to make the needed laboratory measurements on the cross sections for excitation, ionization, and recombination, particularly for highly ionized atoms, and therefore present-day analyses of the spectrum are based on very approximate calculations, which have often led to puzzling and contradictory results. For example, the temperature of the corona may be derived both from ionization theory, as applied to the relative intensities of forbidden iron lines observed in several different stages of ionization from Fe X to Fe XVI, and also from the thermal broadening of the profiles of the same emission lines. In the past, the ionization temperature has been consistently lower than the Doppler temperature by about a million degrees. A second anomaly has been that the greater the temperature of the corona, the greater is the ionization potential of the ion upon which the temperature determination is based. Another curious result has been that the abundances of heavy elements relative to hydrogen in the corona seem to be an order of magnitude greater than abundances inferred from absorption-line spectra of the sun and stars.

It has generally been assumed that, in the corona, the rate of ionization by electron impact is exactly balanced by the rate of electron capture by ordinary radiative recombination, all other processes being assumed negligible.

It now appears that a major process has been omitted from the conventional theory of ionization equilibrium, and as a result the cross sections for electron recombination have been underestimated by a factor of 20 or more. Following a suggestion by Unsöld, A. Burgess has shown that under certain conditions, frequently found in astrophysics, dielectronic recombination plays a dominant role in the ionization equilibrium. This is a process in which a radiationless capture of a free electron in a discrete state of energy greater than that required to ionize the outer electron is followed by a downward transition to a state below the first ionization limit. Burgess' discovery will require the reinvestigation of many astrophysical problems for which the rate of capture by atoms other than hydrogen needs to be known within a factor of 2. Burgess and Seaton have already recalculated the ionization equilibrium of iron in the solar corona. The higher rate of recombination raises the ionization temperature considerably and largely removes the former discrepancy.

**Solar Flares**

Finally, I should refer to the connection between motions and magnetic fields on the sun and the occurrence of solar flares (Figure 9). Solar flares always break out in active regions on the sun. Usually, sunspots are also present, but these are only visible evidence for strong magnetic fields which is the real precondition for a flare. Flares are accompanied by increased flux of radio, optical, and X-ray emission. Very often, the region of the flare is also the scene of generation of relativistic electrons as well as high-energy protons and heavier nuclei. Large flares also seem to be accompanied by an intensification of the solar wind. It has been estimated that the total expenditure of energy in a flare may exceed $10^{32}$ ergs.

Theories of the nature and origin of flares always start by considering the energy budget. With reasonable estimates of the emitting volume of a flare, the average expenditure is somewhere in the range 100 to 500 $ergs/cm^3/sec^{-1}$. On the other hand, the thermal energy density of the chromosphere is about 1 $erg/cm^3$ and of the corona about 0.01 $erg/cm^3$, but the energy density in a magnetic field of 50 gauss is about 100 $ergs/cm^3$, and therefore it is reasonable to suppose that a flare represents the conversion of magnetic into thermal energy. There is considerable observational support for this hypothesis, particularly from the magnetic mapping of active regions by Severny, who finds that magnetic fields in active regions are frequently diminished by about the right order of magnitude after the occurrence of major flares.

Despite the general attractiveness of the idea, there is no proof that the energy for solar flares is supplied by the dissipation of magnetic fields. As yet, nobody has developed a satisfactory theory for the

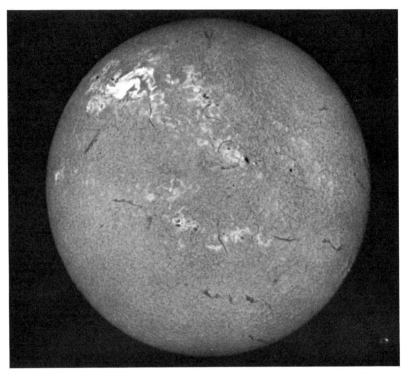

FIG. 9. A bright solar flare of 10 May 1959, importance 3+. North is at the top, east to the left. Lockheed Solar Observatory.

mechanism by which such an enormous amount of magnetic energy can be converted into heat in a time as short as a thousand seconds, although several authors have shown how magnetic fields may be annihilated to produce joule heating. Most of the suggested mechanisms require times that are too long by at least a factor of 10, although Petschek's recent proposal of standing magnetohydrodynamic waves as an annihilation mechanism is promising in this regard. It is interesting to note that one of the leading theorists working in the field, E. N. Parker, stresses that "theoretical progress is presently in the doldrums with no real understanding of the basic dynamical processes of a flare" and suggests that the need is for new types of observations without which "there is every prospect that the flare will always remain what it is today, a growing mass of puzzling data."

# THE MOON AND
# THE PLANET MARS

*Gerard P. Kuiper*

*Lunar and Planetary Laboratory, The University of Arizona, Tucson, Arizona*

## 1. Introduction

The earth belongs to a family of nine major and millions of minor planets. The origin of this complex system was apparently incidental to the formation of the central body, the sun, about $4.7 \times 10^9$ years ago. The multiplicity of our system is not an exception among the stars. Stellar formation, more often than not, leads to double and multiple rather than single stars or to stars attended by planetlike bodies. This multiplicity stems from the improbability of a contracting prestellar cloud possessing so small a total angular momentum that it is able to contract into a single stable body. Normally, the excess angular momentum causes the formation of a plural system.

Because of the common origin of earth and planets, intercomparisons between the members of the planet family are instructive and place in perspective the properties of the planet best known to us, the earth. It is then found that the principal distinguishing parameter among the planets is their *mass*; the distance from the sun is of lesser importance. Since the moon's distance from the sun averages the same as for the earth, comparison of earth and moon reveals clearly the effects of the large mass ratio, about 1:80. Mars is geometrically almost precisely midway between earth and moon, with a mass slightly in excess of 10 per cent of the earth. Since Mars' distance from the sun is 1.52 astronomical units, the absolute equilibrium temperature is only $\sqrt{1.52} = 1.234$ times lower than for a similar planet placed at unit distance; but since Mars has a lower reflectivity than the earth (reflects 15 per cent of the solar radiation versus 36 per cent by the earth), the ratio of the equilibrium radiation temperatures earth/Mars is even closer to unity than computed from their distances alone. The values are approximately 246°K and 217°K, differing only 13 per cent. The chief differences between Mars and the earth may therefore be attributed also to their different masses rather than to their somewhat different locations in the solar system.

Mars and the moon, therefore, extend geophysical experience downward in the scale of masses by one and two orders of magnitude, respectively. The solar system also contains planets ranging upward from the earth by one and by two orders of magnitude: Uranus with 15 and Neptune with 17 earth masses on the one hand, and Saturn with 95 and Jupiter with 318 earth masses on the other. These bodies are much more remote from the sun than is the earth, and the temperature effects on their structure must be considered. Even in these cases, however, the solar distance appears to play only a subsidiary role. The masses are all-important. It is found that Jupiter, though its mass is only 0.001 of the sun, nevertheless approaches it in composition, about 80 per cent being hydrogen and nearly all of the remainder helium [de Marcus, 1958; Wildt, 1961]. Saturn is still nearly of solar composition [de Marcus, 1959], but Uranus and Neptune deviate markedly from the sun in having a greatly reduced hydrogen content [de Marcus and Reynolds, 1963]. The loss of hydrogen and helium is nearly completed at the earth mass and, of course, persists down to the smallest independent bodies near the earth. The variations to be observed between earth, Mars, and the moon are thus less in bulk composition than in the important secondary aspects of atmosphere formation and crustal development, problems essentially geophysical in nature. Such differences in bulk composition as do exist may well be related to other secondary effects, caused by internal temperature differences (in turn, largely due to mass differences) during a period immediately following planet formation, with resulting differences in the degree of reduction, as suggested by A. E. Ringwood [1962].

The moon has a velocity of escape so low (2.4 km/sec) that a small but finite fraction of molecules of all but the heaviest gases maintained at earthlike temperatures will exceed this limit. Thus the moon is unable to retain gases exhaled from its crust. In one sense this makes lunar history simpler than geologic history, since no erosion by water, ice, or wind can have taken place. In another sense, it complicates its history, since the crust is unprotected against the ravages of cosmic weathering by impacting meteorites and micrometeorites, secondary debris spread over the lunar surface by major primary impacts, and proton bombardment by solar plasma, cosmic rays, solar ultraviolet light, etc. The crust of the moon therefore bears witness to phenomena not well observed on the earth, and can in principle do so for the entire history of the earth-moon system, terrestrial-type erosion being absent. Lunar and terrestrial studies therefore supplement each other in a most interesting manner. The additional fact that the earth-moon system has a joint history in its tidal relationship, which has caused the distance between these bodies to increase 10- to 20-fold since their formation, heightens the interest of lunar studies in providing clues to an early period effaced from the earth.

The planet Mars offers a rich mixture of lunar and terrestrial problems. Among these are all the geophysical and geochemical aspects of atmosphere formation and retention, its state of oxidation ($CO$—$CO_2$—$O_2$ balance), its water-vapor content (silicate-carbonate balance in crust), and, possibly foremost, its ability to promote and sustain life. Its crust, also, suggests problems analogous to those of the lunar maria and terrae, called maria and deserts on Mars. While neither the lunar nor the Martian maria are in nature true to their name, and probably never have been, the duality of the crustal pattern observed on all three planets invites comparison and analysis.

## 2. The Moon

The moon subtends an angle of 31' and measures 3476 km in diameter. To the unaided eye the resolution is therefore 100 to 200 km; in the best earth-based photography it is 0.3 km, but most photography does not achieve much better than 1 km. Ranger VII attained 0.25 and 0.4 meter on the two closest frames, but 10 to 100 meters is more typical of Ranger records. In the vertical resolution of broad structural features, such as the numerous lava flows on the maria, the earth-based resolution is 3 to 10 meters for observations made very close to the terminator (sun $\frac{1}{2}$ or $1°$ above the horizon). This high vertical resolution has important aspects for the analysis of mare structure.

As stated in the Introduction, the lunar surface shows principally two types of terrain, the dark *maria* and the much brighter *terrae*. It is shown in the following that the maria are lava-filled basins. The terrae appear older since they have suffered far greater dislocations by successive impacts. There is a third type of terrain on the lunar surface that cannot properly be classed as either mare or terra. Such terrain is found immediately adjacent to circular maria and appears to be

Fig. 1. Mare Crisium, rectified, showing surrounding walls. Plate Y369.

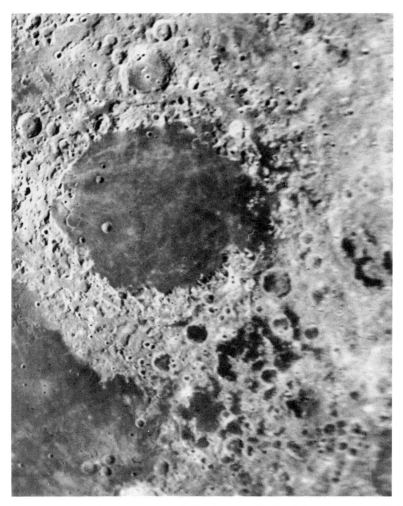

Fig. 2. Mare Crisium and neighboring small flooded craters, rectified. Plate Y738.

Fig. 3. Mare Nectaris, rectified, showing three surrounding walls. Plate Y1334.

either mare or terra terrain, very highly disturbed by an overburden and new mountain ranges, both evidently caused by the process of mare formation.

The lunar *maria* are of two main types: (*1*) those that are nearly circular and are surrounded by near-circular mountain arcs or fault scarps; these maria also show some circular symmetry in the peripheral rille systems and in the ridge systems of the mare floors; and (*2*) those that are irregular. Examples of class *1* are Mare Crisium (Figures 1 and 2), Mare Nectaris (Figure 3), and Mare Orientale (Figure 4). Mare Crisium is surrounded by one prominent mountain wall (diameter 450 km) and fragments of two weaker walls (diameters 670 and 1060 km); Mare Nectaris is surrounded by three rings (diameters 400, 600, and 840 km); Mare Orientale is surrounded by at least five rings (diameters 320, 480, 620, 930, and 1300 km). Examples

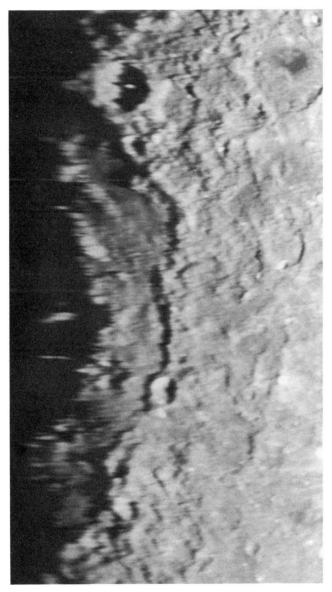

Fig. 4. Arcuate fault scarps surrounding Mare Orientale, rectified. Plate M372.

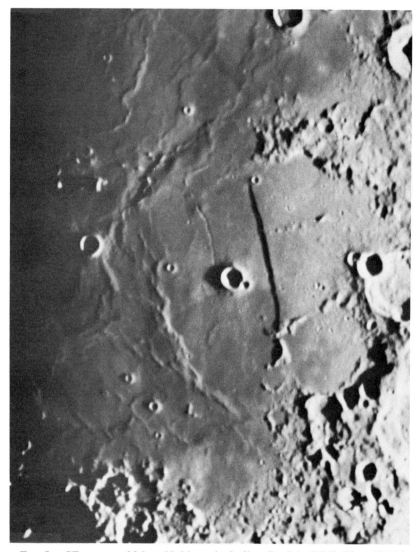

Fig. 5. SE corner of Mare Nubium, including Straight Wall. Plate Y1271.

of class *2* (irregular maria) are Mare Nubium (part of which is shown in Figure 5) and Mare Tranquillitatis (Figure 6). Here the ridge systems show patterns of a *regional* nature, with several such systems covering the entire mare. Further, the outlines of these maria are irregular, as if due to flooding onto terrain with previous irregular

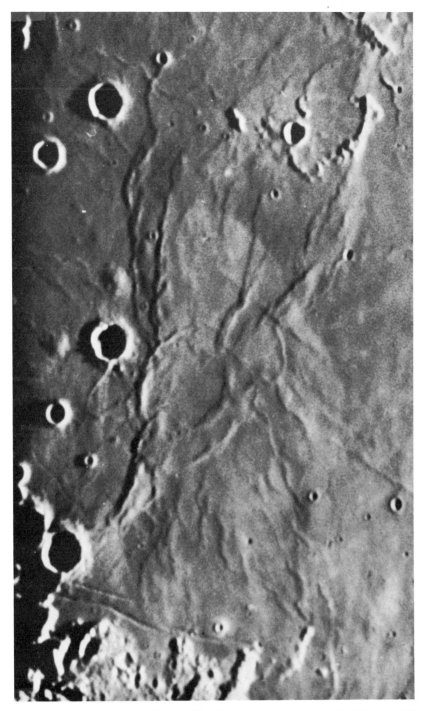

Fig. 6. Part of Mare Tranquillitatis near Arago, at sunrise (Ranger VI impact area). Plate Y1338.

contours and complex tectonic origin, and no peripheral mountain walls or scarps are observed.

Among the various *hypotheses* that have been advanced to account for the existence and general appearance of the maria, the following are mentioned.

*a.* The maria are lava fields, though not necessarily identical in composition or fine structure with terrestrial lava beds. It is usually assumed that the nearly circular maria were caused by *impact* as gigantic impact craters and *subsequent flooding*; the irregular maria, without surrounding walls and associated rille systems, by flooding only. The source of the lava requires a separate explanation, discussed later.

*b.* The same classification of maria is made, but the explanation of the dark mare material is not lava, but dust, or lava and dust. It is assumed that this dust resulted from the impacts that caused the mare basins.

*c.* It is assumed that the mare basins were filled with dust that accumulated in these low-lying areas during the long lunar history, the source of the dust being erosional and the transportation to the basins by electrical forces.

*d.* It is assumed that the maria are the oldest regions on the moon, with the dark surface caused by solar radiation effects.

Hypothesis (*d*) may be ruled out if it is assumed, as is done on excellent grounds, that the great majority of the lunar craters are caused by impacts. It then follows, from the crater densities in the maria and on the terrae, that the maria do not constitute the oldest parts of the lunar surface but the most recent. Hypothesis (*c*) would require that the oldest lunar craters would have their floors covered with the largest amount of dark material and the most recent craters with the smallest amount. This is contrary to fact. The craters showing dark floors such as shown in Figure 2 appear of *intermediate* age, with both the oldest and the most recent craters showing light-colored floors. Also, the mare basins would be expected to be essentially featureless. In reality, structural details related to the general basin geometry may be found all over the maria when observed under favorable conditions of illumination and resolution, and these structural units show the nature of the lunar maria to be basically plutonic.

A compelling argument against a general migration of dust across the surface of the moon is based on the sharp boundaries observed for the different color provinces. These boundaries are so sharp and often coincide with structural units such as flows (to be discussed later) that any migration must be within subtelescopic dimensions.

The plates selected for reproduction here were taken in part from the *Consolidated Lunar Atlas* now in preparation, which includes new earth-based lunar photography. For a more general description and systematic photographic coverage of the lunar maria, reference is made to a paper by Hartmann and Kuiper, entitled "Concentric Structures Surrounding the Lunar Basins" [1962]. This paper contains 77 photographic reproductions, many of them rectified, with a summary for each of the maria on the ridge and rille system, as well as on the presence or absence of concentric surrounding mountain walls or faults. A sequel to this study is contained in two papers by Hartmann on "Radial Structures Surrounding Lunar Basins" [1963, 1964], giving incidentally much additional photographic coverage. Earlier systematic lunar photography is reproduced in the *Photographic Lunar Atlas* and the *Rectified Lunar Atlas*.

Figure 5 shows the SE corner of Mare Nubium, including the region of the Straight Wall, about 20° from Mare Cognitum. It is noted that the mare floor exhibits a complex system of ridges, unlike Mare Crisium (Figure 1), Mare Nectaris (Figure 3), and other near-circular maria, such as Mare Serenitatis (Figure 7) and Mare Humorum (Figure 8).

Another region, showing local rather than mare-wide structural patterns, is portrayed in Figure 6. It is at once evident that the prominent ridge system reflects a complex deep-seated structure that is probably related to one or more major impacts and old craters, all preceding the filling of the mare basin. The complex structure is again in contrast with the rather simple concentric ridge patterns found in Mare Crisium, Mare Nectaris, and Mare Humorum; nevertheless, the pattern is not random and appears in principle interpretable in terms of one or two main events. Very close to the terminator, elevation differences of 3 to 10 meters can be easily detected on ridges sufficiently long to be shown on the photographs. The straight diagonal ridges seen in Figure 6 crossing the radial pattern emanating from Lamont agree in direction with lineaments found in the nearby terrae and indicate a basic crustal pattern continuing into the mare region. This fact and the detailed appearance of the ridge system, as well as the prominent mounds near Arago, are unmistakable signs of plutonic activity *extending right up to the visible surface*. The color differences within the mare, to be discussed later, reinforce this conclusion.

Figure 7 shows the east sector of Mare Serenitatis. This mare, because of its near-circular outline, its roughly symmetrical ridge pattern, and its surrounding mountain walls on the south and east sectors, is considered to have formed by impact early in the lunar history (prior to the formation of Mare Imbrium). We call attention to the *en échelon* structure of the Serpentine Ridge shown in Figure 7,

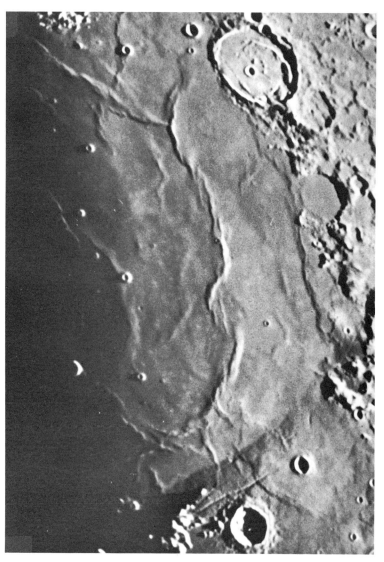

Fig. 7. Mare Serenitatis from Posidonius to Plinius. Note *en échelon* structure in ridge system and Tranquillitatis flow entering Serenitatis N of Plinius, shown to be bluish. Plate Y936.

Fig. 8. Mare Humorum, concentric rilles and ridges, and tilted craters on shoreline. Plate M630.

the dark threshold just north of Plinius which later is shown to have a bluer color than the remainder of the mare basin, and the very remarkable crossing and intertwining of ridges in the vicinity of Plinius.

Figure 8 illustrates the occurrence of several major craters near the shoreline of Mare Humorum whose rims are not level but dip toward the mare. Similar structures are observed on the shorelines of Mare Crisium, Mare Nectaris, and elsewhere. Apparently these craters were caused by impact *after* the basin had formed (since they interrupt and locally destroy the mountain wall surrounding the mare), but *prior* to the flooding of the basin, as follows from the details of the mare

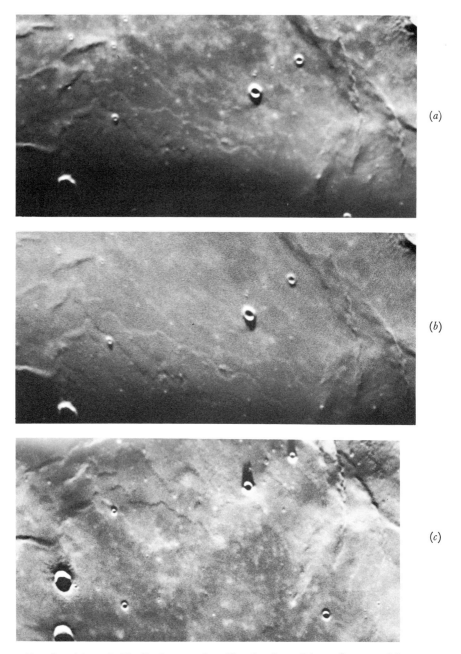

FIG. 9. (a) and (b). Early morning illumination of lava flows on Mare Imbrium shown with two different librations. (c). Same area shown with evening illumination.

level where the craters were invaded. The concentric ridge system of Mare Humorum is shown also, though not as well as on photographs taken with a low sun angle.

An important source of information is the *color of the lunar surface*. These color differences are not large but they are real and definite. The color boundaries between "bluish" and "reddish" areas are usually very sharp, and by comparison with direct photographs it is found that these boundaries often coincide with visible flows such as shown in Figure 9. In Figure 10 eight flows have been mapped in a limited region of Mare Imbrium; the boundaries of 5 or 6 of these are coincident with edges of color provinces. These color contrasts indicate, first, that *the lunar maria are not covered with even 1 millimeter of cosmic dust*, which would have obliterated the color differences; and that the color contrasts seen on the lunar surface are, in many instances, related to observable structural units, such as the flows shown in Figure 9. The *uppermost flows*, obviously the most recent ones, *are often the bluest*; the older deeper-lying flows, partly covered, are often but not always the reddest. The color differences may be due to differences in the state of oxidation possibly related to variations in the *water content* of the magmas. On Mauna Loa, Hawaii, one often observes a reddish coloration of vents which is attributed to escaping steam. Color differences might be due also to differences in age (cosmic weathering effects). The adjacent terrae and particularly the islands in the maria are often redder than the maria themselves. This holds also for much of the crater ray material deposited on the maria. Regardless of the causes of the observed colors, the existence of successive well-marked flows as part of a general tectonic pattern unmistakably supports hypothesis (*a*).

There is another important observation that bears on the choice between hypotheses (*a*) and (*b*). It is that the flooding of the circular maria did not occur immediately upon the formation of the basins but *after a considerable interval*, apparently different for the different basins, in some cases perhaps as long as a few million years. This follows from the study of the inclined and partly flooded craters near the borders of Mare Nectaris, Mare Humorum, Mare Crisium, and others. These craters are tipped, having clearly been formed on an *inclined* surface, apparently the floor of the basin before it was flooded. Since the number of such structures differs for different maria, it appears that the intervals between formation and flooding may not be constant. In any case, the intervals must have been many thousands of years. Hypothesis (*b*) does not explain this phenomenon, whereas it is compatible with hypothesis (*a*) provided the formation of the lavas was not the result

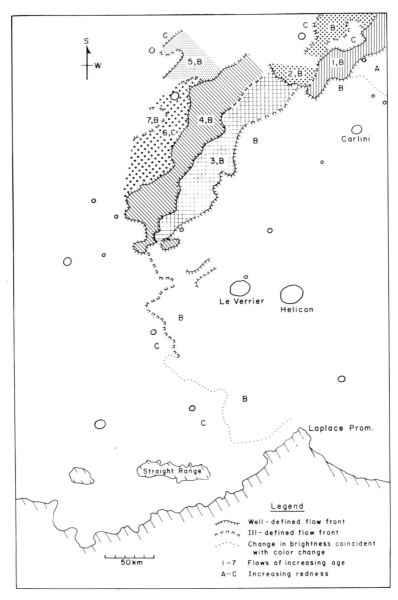

Fig. 10. Flows photographed in Mare Imbrium, mapped by R. Strom. Cf. Figure 9.

of the impact itself but *due to a general heating process of the moon* causing subsurface magmas to be available during a certain period of lunar history. The assumption of the moon having passed through such a period of considerable internal heat and partial melting is consistent with the observation that the parent bodies of the meteorites, which by all indications were even smaller than the moon, have gone through a similar process of melting, differentiation, and freezing. It is difficult, if not impossible, to explain various other lunar phenomena without the assumption of a period of maximum melting. Among these are the *isostatic adjustment* that has taken place in approximately twenty large premaria craters, such as Clavius and Hevelius, which show convex, not concave, floors; and also the flooding and filling of many craters with elevated flat, dark floors, such as Wargentin. The hypothesis that the flooding was indeed by available magmas, caused by previous lunar heating and not by the impact themselves, is further supported by the fact that several very large premaria craters were never flooded, whereas even small craters dating from the period of maria formation are flooded (Figure 2), and that distinctly postmare craters, such as Tycho and Copernicus, again show no sign of flooding.

The conclusion that hypothesis (a) is consistent with the observed features does not mean that the *structure* and *density* of the lunar lava, as exposed in the maria, are the same as of terrestrial lavas. The deposition in a vacuum, the lower surface gravity of the moon, and the subsequent erosion by particles and solar radiation will have caused departures which must be ascertained. For instance, the lunar surface lavas will upon deposition probably be extremely porous and vesicular, perhaps like pumice or reticulite, such as found on Hawaiian lava flows, but probably much more extreme. The pressures occurring in terrestrial lavas at a depth of 10 cm are encountered on the moon at a depth of some 120 meters if the lunar surface rock has a specific gravity of 0.5.

That the uppermost layers may have an even lower density is indicated by the experiments of Dobar *et al.* [1964] who found about 0.12 for the bulk density of solid silica extruded as a liquid into a vacuum. These authors introduced the name "simolivac" for such material (silica molten in vacuum). Almost certainly *the surface of the original lunar lava flows will have had a rock-froth structure* somewhat like simolivac. It will not be identical because the lunar rock will probably have no more than about 70 per cent $SiO_2$ content and gases dissolved in the subsurface magmas will cause upon extrusion further complexities. A rapid boiling-off of many substances will occur since the sublimation points of $SiO_2$, $MgO$, $BaO$, $ZnO$, $Al_2O_3$, and other oxides, as well as the metals Ca, Cr, Cu, Fe, Mg, Ni, Si, Zn, etc., are much below the melting points at such "high" lunar pressures as $10^{-5}$ mm Hg. We are indebted to Dr. Alvar Wilska for calling these facts to our attention. The substances that boil off are likely to precipitate on colder rock surfaces in the neighborhood and cause powdery deposits not unlike "fairy castles." Recent experiments with *extruded basalt* by Dr. S. Hoenig at this Laboratory yielded *0.3 for the bulk density of this material*.

Some estimate of the thickness of a rock-froth layer on each of the lava flows may be gained from the Mare Imbrium and Mare Serenitatis observations. The flows on Imbrium are about 50 to 100 meters thick and the most extensive are some 200 km long (Figure 9). These Imbrium flows appear to have issued not from the mare shoreline but from fissures well inside the mare, within a rather confined area. The slopes of the flows have not been measured (this could probably be done photometrically) and may not be the original ones at the time of deposition. They seem to be of the order of 1°. In Mare Serenitatis, near the S Shore (Figure 7), a distinctly steeper slope, perhaps 3°, occurs though other flows are found nearby with very small slopes.

The *composition of these flows* cannot, of course, be inferred with any certainty by inspection from the earth. O'Keefe and Cameron [1962] have drawn an analogy with terrestrial ash flows because of their great fluidity and suitable composition, if it be adopted that tektites are of lunar origin and their composition (about 70 per cent $SiO_2$) indicative of that of the lunar crust, and that lunar ash flows would lead to a material of similar composition as on earth (ignimbrite). On the other hand, as Mr. R. Strom has pointed out, terrestrial ash flows spread in all directions from the source, unlike flows such as shown in Figure 9, and have no appreciable terminal walls, unlike the observed lunar flows, which more resemble the Hawaiian flows observed on Mauna Loa.

It must be noted, however, that the observed average slopes of the terminal walls (or flow fronts) are not as steep as observed on terrestrial lava flows. Estimates indicate average slopes of about 3° to 7° for the fronts; the well-known flow between the Apennines and the Caucasus, toward Archimedes, has a front slope of about 17° (Strom and Whitaker). However, these low figures are not necessarily representative because the limited resolution used will reduce the values. It is further noted that ash flows tend to sweep over low hills and other obstacles in their paths, whereas the slower lava flows avoid and bypass them. On Figure 9 a number of hills are seen to have been bypassed, in curved arcs, by the principal flow there shown. On the earth ash flows are associated with explosive volcanism and basaltic lava flows are quiescent. The topography of the lunar maria appears to fit the second description better. It is true that some authors have assumed the craters on the maria to be of volcanic origin; however, there is nothing to indicate that the observed lava flows and the craters are related.

The retention of the carrier gas for an ash flow in the lunar vacuum would be difficult to understand unless it is assumed that the flow is promptly sealed off at the surface with a layer of sufficient strength to confine the gas. Also, ash flows are normally associated with lava flows. There thus appear to be reasons to assume that the observed flows are lavas, not ash, though allowance must be made for the different lunar conditions before one can be certain. If they are ash flows, the composition is likely to be rhyolitic; if they are lavas, they are likely to be basaltic.

Since the flows that have been recognized on the lunar maria cover only a fraction of the total surface, it may be assumed that either the flows in the other portions were too thin for recognition or were of somewhat different composition and texture. Conceivably a sequence of differentiated deposits, from basalt to rhyolite to ash, was involved as has been observed, e.g., in the Valles caldera of New Mexico [Ross et al., 1961]. It is probable that great progress may be made in the recognition of lunar surface types by the combination of refined telescopic observation, colorimetric studies from 0.3 to 2 $\mu$, and thermal mapping throughout a lunation at 5 $\mu$, 10 $\mu$, 20 $\mu$, and 1 mm.

The mare *ridges* cut through the color provinces without visible interference. *They are not the edges of lava flows.* Instead, they appear to have a *dynamical* cause, some being related to the symmetrical pattern of the near-circular maria, and others to the lunar grid system, of which some ridges are parts. This may be verified in part by inspection of Figures 5 through 9. Also the ridges are quite invisible at high sun, showing that the lunar surface is essentially continuous across them. There are no close analogs of the lunar ridges on the earth. Their non-interference with the color provinces and their invisibility at high sun indicate that their formation is by *uplift*. In some cases, the uplift appears to have been so large that the surface has split open; and often it may be observed that out of such fissures, a new dikelike rock mass has been extruded, usually whiter in color than the original mare floor. These whitish "lunar dikes" have been known [Kuiper, 1958] and some are clearly seen on several of the Ranger photographs.

The phenomenon of lunar dikes indicates that the ridges are surface manifestations of subcrustal structural planes along which upwelling of magmas has taken place. These structural planes resulted presumably from dynamical causes, most by tension. This model accounts for the presence of ridges as part of the grid system and also as structural units, roughly concentric and radial, of the impact maria, because, as the lava basins cooled to increasingly great depths, the deeper spherical layers of the moon shrank, causing subsurface tensional cracks and the admission of deeper magmas.

The formation of ridges by a laccolithic-type uplift of the lunar crust above dynamically determined fracture planes is facilitated by the layered structure of the lunar maria noted earlier (individual layers typically 50 to 100 meters thick), particularly if alternate layers are very vesicular and weak structurally. The low surface gravity will also facilitate the vertical displacements. Thus, it may not be surprising that the equivalent of lunar ridges has not been found on the earth.

The ridges are therefore assumed to date from an early postmare period and to have resulted from effects accompanying the cooling of the mare basins. Later in lunar history magmas will not have been available except possibly in isolated "volcanic" regions. The apparent absence of ridges from the terrae is consistent with their presumed nature as consolidated dust to great depth and the absence of near-surface magmas.

*In summary*, the maria are low-level areas flooded by lunar lavas. Some of these depressions were caused by large impacts that resulted also in extensive surrounding tectonic structures, both peripheral and radial. The lavas were not caused by the impacts themselves but resulted from extrusion of available magmas. The mare surfaces show clear evidence of several successive lava flows. Each flow is found to have a very homogeneous color, but different flows may have different colors. The most recent (upper) flows tend to be the bluest, but the age sequence is not everywhere the same as the color sequence. Composition differences rather than mere age differences are therefore held responsible. These composition differences may reflect the state of oxidation, possibly related to the water content of the magmas.

The flows are bounded by terminal walls, and some large flows (up to 200 km in length) are comparatively narrow. By terrestrial analogy this indicates that the flows are basaltic lavas rather than ash. Because of the small slopes of the flows and their great lengths, the lavas must have been very fluid. Yet their terminal walls show the presence of a substantial solid crust during the flow. On this basis and in view of recent laboratory experiments by Dobar and others, it is assumed that the lunar flows consist of two layers each, an upper layer of rock froth, possibly about 10 meters thick, and a lower layer of more compacted rock that resulted from the low-viscosity lava. Rhyolitic and ash flows may exist in mare regions not bounded by visible terminal walls.

Subsequent meteorite impacts and cosmic abrasion will have occurred in this multilayered structure. Small craters will have formed in the rock froth only, larger craters will have extended into the denser rock, but the succession of flows on the maria may locally have caused irregular alternations between very vesicular and denser rock material. Regional unevenness in crater shape is therefore to be expected.

Dobar *et al.* [1964] found the bearing strength of simolivac to be 1 to 4 tons per square foot for static loads and this may be regarded as a first approximation for the lunar maria wherever they are not disturbed by impacts.

A similar strength has been found by the writer for recticulite that was solidified in free fall from bits of liquid lava ejected by Laimana Crater, Hawaii, in 1960.

An actual measure of the bearing strength was obtained for the floor of Crater Alphonsus, observed on Ranger IX, from the partial penetration into the moon of some 50 rocks each about 1 meter in diameter, tossed out of a primary crater 46 meters in diameter, to distances of 30 to 120 meters. The value is 1 to 2 $kg/cm^2$ or 1 to 2 tons/square ft.

A discussion of the terrae will be held brief because they are older than the maria and have experienced many more lunar impacts per unit area, making the study of their "original" surface very difficult. They are probably *not* analogous to terrestrial continents which appear to be chemically differentiated from the mantle. Instead, they appear to be remnants of the original lunar surface that have never melted and thus do not show the aftereffects of solidification observed on the maria: rilles, ridges, domes, volcanolike structures, and lineaments large and small. There are, however, some areas on the terrae that deserve meticulous investigation since they appear to come as close as can be found to the "original" premare lunar surface. Examples are found south and west of the craters Vlacq and Hommel and south of Mutus in the SE quadrant of the moon, where apparently no major impacts ever have occurred.

We return to the maria and briefly review the knowledge gained as a result of the successful mission of Ranger VII, July, 1964. Over 4300 TV records of the lunar surface were transmitted and these are being published in two editions: a limited hand-printed photographic one and a general library edition produced by the Government Printing Office. Three atlases cover this series: (*1*) The 199 frames obtained with the A camera, (*2*) the 200 frames obtained with the B camera, and (*3*) 170 composite sheets containing four frames each obtained with the four P cameras. All six cameras were equipped with RCA vidicons. For the A and B cameras the full detector surface was used, 11 × 11 mm, and with 1150 scan lines covering this dimension. The P frames measured approximately 3 × 3 mm, and were covered by 300 scan lines each. The six cameras had focal lengths of 25 and 76 mm, three of each kind. The short-focus cameras were $f/1$, the others $f/2$. The images are of remarkably good quality. With the large number of scan lines used, the frames have every appearance of good photographs. The picture-taking sequence started about 1000 seconds before impact on the moon. The exposure times were 2 and 5 milliseconds, i.e., so short that no visible trailing of the image occurred, except on the last two P frames. The transmission time was 2.5 seconds for the full scans (A and B cameras) and 0.2 second for the partial scans (P cameras). The last A frame was taken in time to be almost completely transmitted back to earth (2.5 seconds before impact) whereas the last B frame was taken 5 seconds before impact. Because of the short transmittal time of the smaller P frames, the highest resolutions were obtained on them. The last two, on P979, achieved resolutions of 0.25 and 0.4 meter, respectively.

The following results are based on a preliminary analysis of the Ranger VII data together with the published and unpublished earth-based observations. For a fuller account, the reader is referred to the writer's [1965] Ranger VII Report.

*a. The nature of crater rays and the origin of ray craters.* Major crater rays are found to be normally composed of several ray elements that are irregularly distributed along the direction of the ray. Each of these elements is composed of a small cluster of "secondary" bright-rimmed craters at the head from which flows, as a comet tail flows from the comet head, the ray element itself, a structureless streak of bright material. This streak is always pointed away from the central ray crater, be it Copernicus or Tycho. The inference is that the secondary craters were formed in conjunction with the primary ray crater and that the whitish material stems from ejecta by these secondary impacts, blown radially away from the primary crater by gases that originated at the primary impact site. This mechanism appears self-consistent if the impacting body is a *comet*, composed largely of frozen volatiles which after the impact explosion remain gaseous long enough to travel the 1000-km path (flight time $\simeq 1000$ seconds). It has been shown that only the new or parabolic comets are of sufficient frequency in the inner parts of the solar system to qualify as the impacting bodies.

Also their masses ($10^{17}$ to $10^{19}$ grams) and impact velocities ($\sim 54$ km/sec for parabolic comets moving radially with respect to the sun) are adequate to produce craters such as Copernicus and Tycho (energy required $\sim 10^{30}$ ergs).

The ray craters have two kinds of secondary craters: (1) the normal kind, heavily concentrated around the primary crater, thinning out very rapidly beyond 3 to 4 crater diameters, invisible at full moon, and presumably caused by lunar rock fragments that were ejected from the primary crater; all large post-mare craters, whether attended by rays or not, possess this class of secondaries; and (2), the white secondaries in the rays, already referred to, usually found in clusters at the head of the ray element; these white craters stand out very prominently at full moon. These bright secondaries are so different from the secondaries of class 1 that they are supposed to be of cometary origin, either due to icy fragments accompanying the comet nucleus that gave rise to the main crater, and thus in a sense "associated primaries"; or due to fragments of the nucleus tossed out by the central explosion. In either case the fragments must have arrived in frozen condition on the moon to produce craters so unlike those of class 1, and so much as the primaries themselves. Further analysis of the ray systems may define the radial mass distribution within comet nuclei, whereas the size distribution of ray craters should define the frequency distribution of masses among the parabolic comets. This analysis should also provide some information on the frequency increase or decrease of parabolic comets during the history of the solar system, the present rate being approximately known. This case illustrates the value of the moon as a fossil record.

*b.* The post-mare *non-ray craters and craterlike depressions* appear to belong to the following subclasses:

(i) *Primary craters*, due to impacts by bodies belonging to the inner fringe of the asteroid ring (small asteroids and meteorites). These craters are crisp in outline and form a well-defined class.

In a log-log plot, of number versus diameter, they follow a $-2.0$ slope down to $d = 100$ meters and a $-3.5$ slope to 10 meters. Below 10 meters the negative slope diminishes again, down to $d \simeq 1$ meter. Table 1 gives the total fraction of the mare surface occupied by these craters above the diameter limit stated. Almost identical curves were later found for Ranger VIII and IX, confirming the interpretation that on the whole these crisp craters are indeed primaries, uniformly distributed over the lunar surface. The two slopes are in accord with the inferred mass distributions of small asteroids and meteorites as recently derived by F. L. Whipple. The gradual drop-off in numbers well below 10 meters is attributed to erosion (gradual crater destruction) in the upper meter or so during post-mare history (the last 4.5 billion years), the initial depth of a 10-meter primary being about 2.5 meters.

## Table 1

FRACTION OF MARE SURFACE OCCUPIED
BY PRIMARY CRATERS ABOVE DESIGNATED DIAMETER

| Diameter (meters) | Per Cent |
|---|---|
| > 1000 | 1.0 |
| > 100 | 1.3 |
| > 10 | 3.8 |
| > 1 | 18.0 |

(ii) *Secondary craters*, due to impacts by fragments ejected by large primary craters. Many of these secondaries have one or more rock masses on their floors. A good example is shown later in Figure 13. These masses are of such dimensions as to confirm the secondary origin of the craters, provided the primaries were some 100 to 300 km away. Secondary craters are not numerous except near large primary craters.

(iii) *Depressions and dimple craters*. The last Ranger VII frames show the mare surface to be covered, for some 50 per cent of the area, with shallow, nearly circular depressions, most of them from 50 to 500 meters in diameter. Part of the last A camera frame is shown in Figure 11. In addition, there are numerous depressions that at their centers are not flat and roundish, but pointed, almost funnel-shaped; these have been called "dimple craters." Examples are shown in Figures 11 and 13. The first discussions held that the depressions were eroded impact craters; but it soon became evident that they are of internal origin and due to collapse [Kuiper, 1965; Whitaker, 1965]. The Ranger VIII and IX impact regions, representative areas of which are reproduced in Figures 12 to 14, showed much the same pattern as Ranger VII, with many of the collapse structures clearly related, particularly on the Ranger IX records, to other negative structures of internal origin (rilles, "crater" chains). The depressions are nearly absent on pre-mare crater walls (such as Alphonsus) and on mare ridges, for reasons that are understandable if they are indeed due to collapse. The absence is not compatible with the erosion hypothesis because the respective ages are nearly the same.

THE MOON AND THE PLANET MARS 45

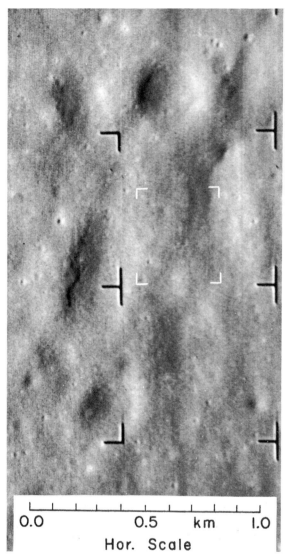

FIG. 11. Part of frame A199 of Ranger VII showing six large and several smaller depressions. Cf. Figure 24 for detailed view of area bracketed.

After the depressions had been found on the Ranger pictures, they were looked for on terrestrial flows having dimensions similar to those found on the moon. Because no mention of this phenomenon is made in standard geological textbooks, it came as a surprise that the depressions were indeed found on nine of eleven flows examined so far. The positive results were on extensive flows composed of pahoehoe. They are: the McCartys Basalt Flow, a 30-mile flow South of Grants, New Mexico; the older Laguna Flow in the same area; the 18-mile flow located 40 miles due South of Socorro, New Mexico; the 43-mile (70 km) basalt flow just West of Carrizozo, New Mexico; an

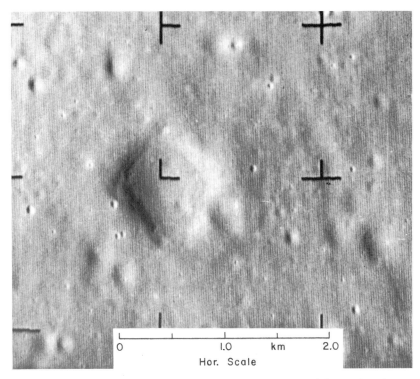

Fig. 12. Large square depression and several small round depressions in one of last frames of Ranger VIII.

older flow near the northwest tip of the Carrizozo flow; the 6-mile flow 18 miles West of La Mesa, New Mexico; an older flow just East of it; a flow in the Fort Rock area in Oregon, and a flow in Iceland. The negative results are on two short flows of aa, the Bonito Flow near Sunset Crater, and the flow of the SP Crater 10 miles North of Humphreys Peak, both in Arizona.

The only reference to the depressions found in the geological literature is contained in the study by R. I. Nichols [1946] who noted

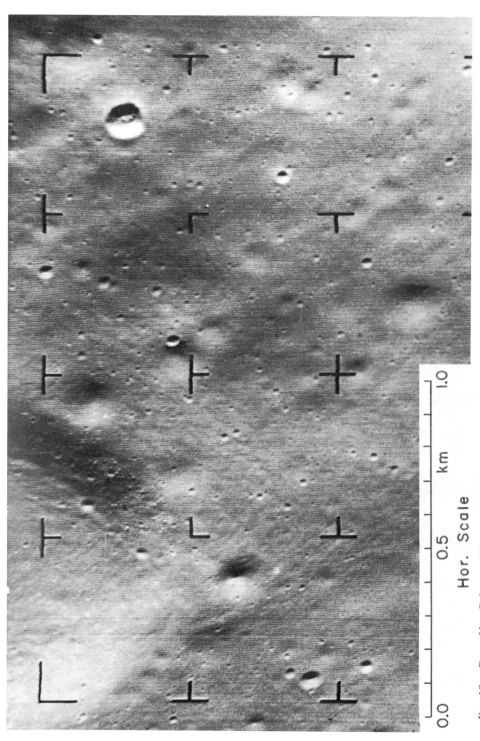

FIG. 13. Part of last B frame of Ranger VIII showing large circular depression, one prominent dimple crater, and several small circular depressions; and one secondary crater with internal rock mass.

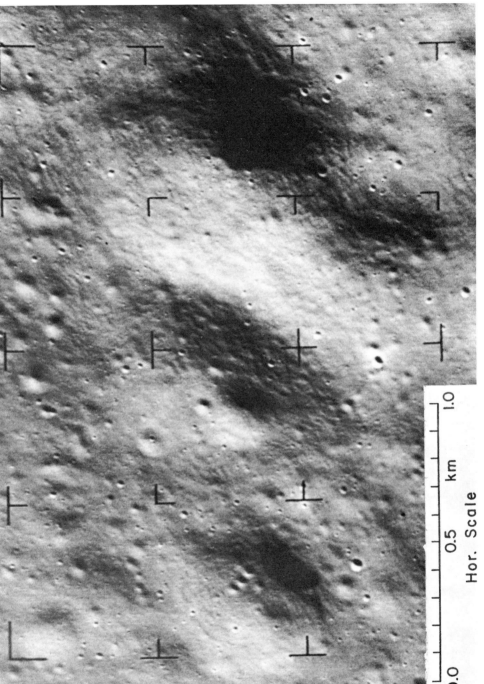

Fig. 14. Part of last B frame of Ranger IX showing three major and numerous minor depressions, "tree-bark" structure of slopes,

Fig. 15. Two adjacent views of northern strip of basaltic lava flow between Grants and McCartys, New Mexico, area described by R. I. Nichols [1946].

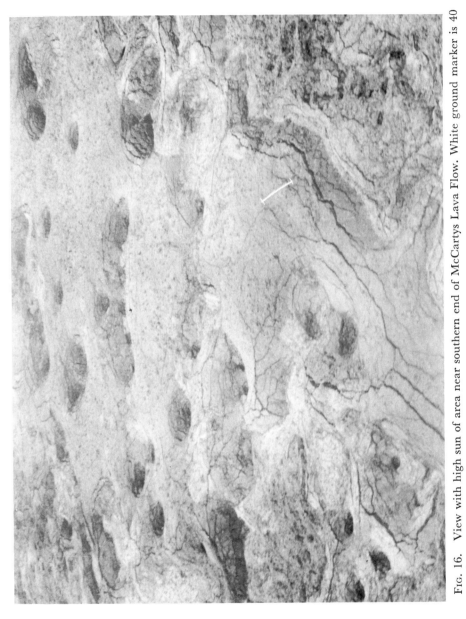

FIG. 16. View with high sun of area near southern end of McCartys Lava Flow. White ground marker is 40 meters long.

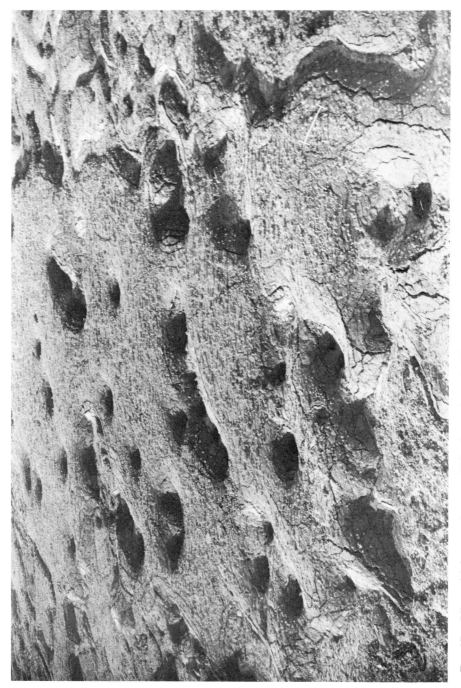

FIG. 17. View with low sun (approx. 12°) of area shown in Figure 16. White ground marker 40 meters length. Extreme areas on right and lower left are aa flows, resulting from breaks in terminal walls of main pahoehoe flow. Note topography of fracture systems.

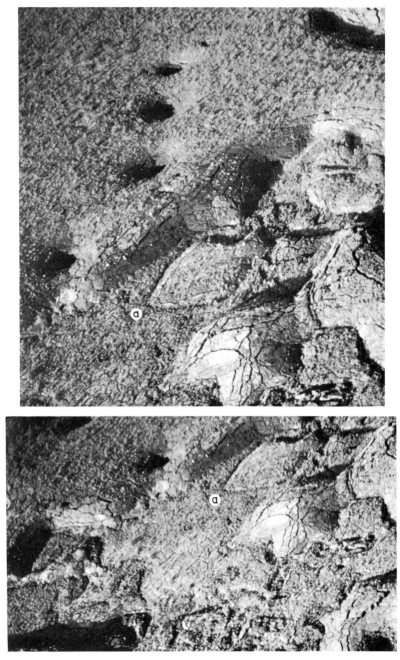

Fig. 18. Two adjacent views of round and square sinks near southern end of McCartys Flow. Note nearly square furrow (a) on floor of one square depression.

them in the upper strip of the McCartys Flow: "In an area approximately 2 miles long, near the terminus of the flow, there are about 100 collapse depressions." These occur adjacent to Highway 66 and shown in Figure 15. In Figures 16 to 18 views are presented of an area in the same flow some 25 miles to the South, where the lava bed is thicker and the depressions are less disturbed by subsequent events. Examples of the Socorro and La Mesa flows are shown in Figures 19 and 20. The patterns in all flows are similar. Most depressions measure about 40 to 100 meters at the top but some are larger. The sharper funnels at the bottom are around 10 to 30 meters. Prominent crustal fractures usually outline the contours. Sometimes the sinks are multiple, two or more occurring in a common trough. Figures 16 and 17 show five circular sinks nearly aligned in the trough that also contains a sixth member offset at right angles. Another sextuple set is seen in the La Mesa flow of Figure 20. Nearly square depressions occur also, sometimes outlined by narrow furrows, as seen in Figure 18.

All the aspects previously noted, including the square depressions and associated furrows, are found on the Ranger photographs. There are some differences also: The lunar sinks are on the whole somewhat larger and shallower and there is a greater spread in their dimensions.

The explanation of the nearly circular sinks on the terrestrial lava fields presents a fascinating problem. Ground studies have shown that withdrawal of subcrustal lava is the immediate cause. A freshly deposited flow appears to have resembled first a brittle bag containing a heavy liquid, with the skin formed by solidified lava of the pahoehoe type. As cooling proceeded, the pahoehoe skin thickened and cracked by cooling. Local drainage resulted at the lower levels of the flow front, causing areas of collapse in the lava cushion. The results of this mechanism can be seen in the foreground of Figure 17; the lower flow is aa. One observes that the frequency of circular sinks is often increased near the periphery of the lava cushion, consistent with their origin by drainage. Lava tunnels or caves are normally found beneath the sinks. The ceilings of these cavities are of uneven thickness, from a few inches to several feet, and this gives a clue to the shape and the pattern of the sinks. If one assumes finite random fluctuations in the temperature and water content in the magma chamber, there will be columns in it that are more fluid than others. Above the more fluid columns, the solid ceilings will thicken more slowly, if at all, and might temporarily even bulge upward as mounds. Later, when the lower walls crack, the very fluid columns will be the first to drain. The weak spots in the ceiling will then coincide with the centers of the cavities and nearly circular sinks may be expected.

FIG. 19. Area approximately one square mile in size of flow 40 miles south of Socorro, New Mexico, perhaps the most moonlike of lava flows recorded.

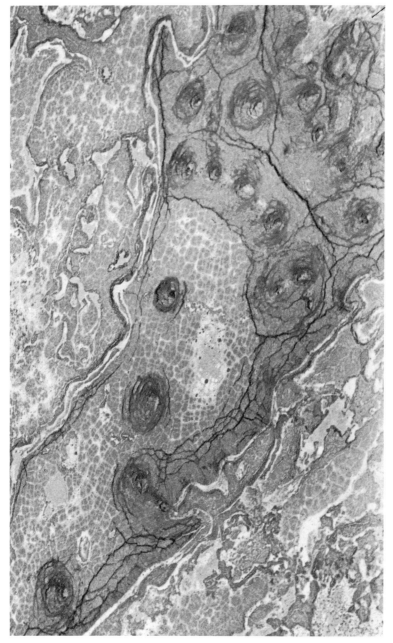

Fig. 20. Part of flow west of La Mesa, New Mexico. Note precise funnel-shaped depressions and roughly orthogonal fracture system in crust.

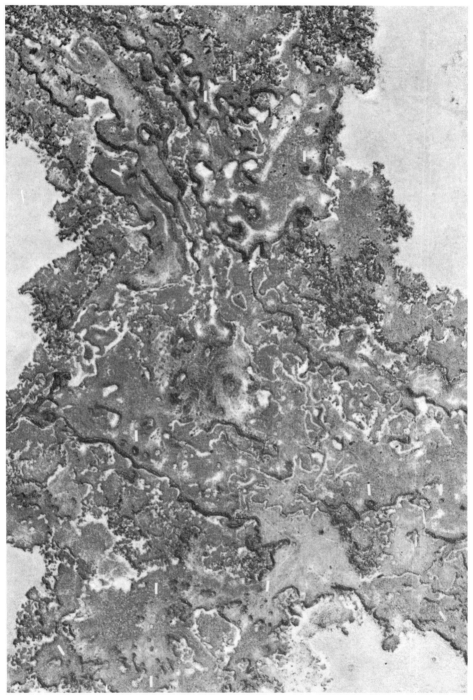

FIG. 21. Part of basaltic flow in Oregon (sheet CHZ 9-84) showing succession of flows and groups of near-circular depressions, some of which are indicated by short white bars.

The depths of the terrestrial sinks often approach the thickness of the flow itself. Sometimes a succession of five or six overlapping flows may be observed, such as in the Oregon Flow shown in Figure 21. Then each level may have its own sinks, all seemingly limited in depth

FIG. 22. View of southern tip of McCartys Flow, a single flow unit resembling gush of molten metal, with elevated areas showing depressions, and lower extensions presumably aa.

to the thickness of the component flow. Single flows usually have a "puffed-up" appearance such as shown in Figure 22, not unlike a gush of molten metal.

The Ranger pictures do not show clear evidence of flow units, with the possible exception of frame A 199 of Ranger VII. On this frame two units are weakly indicated, both running N-S, one on each side of the frame. If this appearance is real, the depth of the sinks is again comparable to the height of the flow. On Ranger VIII and IX, however, no such units are seen. One appears to be dealing here with extensive "lakes," not unlike those observed in large terrestrial calderas, and not with a succession of flows such as are observed on Mare Imbrium and Mare Serenitatis. The rock-froth layer expected to cover the lunar maria and the small lunar surface gravity (one-sixth of earth) both will have produced scale differences between the lunar sinks and those observed on the earth.

Parenthetically, the terrestrial collapse craters and their associated caves in the southwestern United States were known and used by the Indians, and some have become archaeological sites, according to advice received from Professor Emil Haury of the University of Arizona. One may speculate on future applications of the lunar collapse craters!

*c.* The fine structure of *mare ridges* has also been revealed by the Ranger records. By fortunate coincidence one impact crater was formed squarely on top of one of the ridges crossing the impact area; this may be seen on frame B196 and several P frames. The records show that the concept of a ridge being a strip of mare surface pushed up by a lunar dike is correct; and that mare dikes may have several branches, stemming from a single intrusion at greater depth and leading to a spread of several dikes upward, with a vertical cross section of the dike system somewhat resembling the branches of a river delta streaming upward. It is apparent that this spreading and branching of the dikes cause the width of the mare ridges and also their *en échelon* structure. In this manner the "braided" appearance of mare ridges (cf. Figures 5, 6, 7, and 9) is explained, and also the crossing-over of ridges, such as may be seen north of Plinius on Figure 7. A further widening of the ridges is apparently accomplished by the formation of laccoliths or sills, as may be seen from the detailed structure of the walls of the crater referred to on frame B196 of Ranger VII.

*d.* One of the most remarkable properties of the lunar surface is the presence of the two diagonal and a N-S *grid system*; and it is very significant that this system can be traced to the smallest dimensions covered by Ranger VII. Mr. Robert Strom has made an analysis of the approximately 10,000 lineaments he has charted on the visible face of the moon [Strom, 1964], and he has also analyzed the Ranger records and nearby comparison fields including some near the crater Copernicus. He has found that the Copernican "scars" are closely aligned with the direction from Copernicus itself, with a 20° total spread in the angles. In the Ranger VII impact area three lineament directions are present, as shown in Figures 23*a* to *e*. The directions to both Tycho and Copernicus differ enough from the N-S direction that it can be confidently stated that the small lineaments present are not due to Copernican or Tycho scars but are structurally controlled as

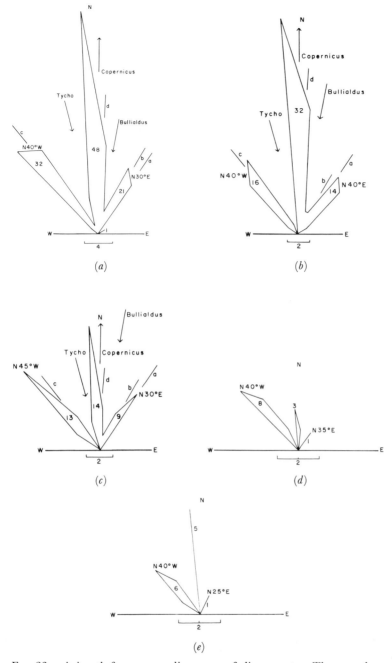

FIG. 23. Azimuth-frequency diagrams of lineaments. The number of lineaments represented by each peak, the directions to Tycho, Copernicus, and Bullialdus and the azimuths of certain lineaments, as indicated. (a) Area R1: Lineament length = 0.4 to 6 km. Average length = 1.1 km. Lineament density = 0.04 lineament/km². (b) Area R2: Lineament length = 50 to 380 m. Average length = 114 m. Lineament density = 4.7 lineaments/km². (c) Area R3: Lineament length = 12 to 53 m. Average length = 21.4 m. Lineament density = 252 lineaments/km². (d) Last $P_3$ photograph: Lineament length = 2.8 to 9.3 m. Average length = 5.4 m. Lineament density = 7094 lineaments/km². (e) Last $P_1$ photograph: Lineament length = 3 to 17.5 m. Average length = 4.5 m. Lineament density = 9876 lineaments/km².

are the diagonal patterns. Apparently, the triple global lineament pattern is very basic in the structure of the lunar crust and pervades the maria down to the smallest observable dimension, in the meter range. The cause of the global grid pattern has not been determined but it may be formally thought of as having been caused by a N-S compression. Perhaps the phenomena are a consequence of the tidal history of the lunar globe.

The presence of lineaments down to the meter range, not related

FIG. 24. Photograph of model of Ranger VII impact area made by sculptor Ralph Turner, based on last P frames and viewed with 23° sun angle as on Ranger photography. Cf. Figure 11.

to the ray systems of Tycho and Copernicus, indicates that the cover of the mare surface with lunar debris is less than about 2 meters thick. This result is compatible with the meteoritic impact rate derived by McCracken and Dubin [1963].

*e.* The last P frames of each of the Ranger missions possess the highest resolution for the three impact areas. The Ranger VII P frames have been used with great care in the production of a three-dimensional model, scale 1:500, of the impact area. The model is reproduced in Figure 24, seen under the same solar illumination, 23°, as existed during the mission. The area covered by Figure 24 has been outlined in Figure 11. The gain in resolution is seen to be at least fivefold.

Inspection of Figures 11 and 24 suggests a close parallel with the type of topography shown on Figures 16 and 17, lower right. In both cases the elevated "lava cushion" contains sinks and the wall of the terrace is marked by cracks or furrows running roughly along the contours.

*f.* Loose rocks have not been found on the moon except in two instances: (a) the last B frame of Ranger VIII shows some rock masses, 3 to 5 meters in size, resting on the slope of a large collapse crater (Figure 13); (b) the last P frames of Ranger IX show an area within 120 meters from a primary impact crater about 46 meters in diameter. A composite of these frames is shown in Figure 25. About 50 rocks have been marked by arrows. They are about 1 meter in diameter each, are nearly completely buried in the lunar crust, projecting only some 15 to 30 cm, as found from their shadows (sun angle 10°). Since their distances from the parent crater are known, a lower limit can be found for the impact velocity. If the mean bulk density of the rocks is assumed to be 2, the limiting bearing strength of the upper 2 feet of the lunar surface is found to be about 1 kg/cm$^2$ or 1 ton/square foot, somewhat like the strength of wet sand. This strength will support modest landing operations.

The absence of observable scattered rocks in general is attributed to this quite limited bearing strength of the lunar surface. Rocks of 1 meter in size and larger, tossed from distances well above 100 meters, will have been buried.

As Mr. W. Hartmann has pointed out to the writer, particle erosion on the lava surface might be conveniently classed into three categories.

*1. Cratering,* caused by the impacts of large masses. Cratering does not erode between the craters except by secondary material ejected during the impacts.

FIG. 25. Composite of last P frames of Ranger IX showing 46-meter diameter impact crater on top, accompanied by some 50 visible rocks, each about 1 meter in diameter and indicated by arrows in the figure.

2. *Sandblasting*, caused by impacts of masses of about 1 gram down to molecular size. Most of the sandblasting mass influx is concentrated in particles of $10^{-4}$ to 1 gram. A total mass of about 1 gram cm² has fallen on the maria during their entire history in the form of "sandblasting" particles [McCracken and Dubin, 1963]. Secondary particles of lunar material thrown out onto the lunar surface in the sandblasting processes form a layer of perhaps 20 grams/cm². When impacting particles of mass less than 1 gram strike an open, porous surface such as hypothesized for the moon, ejection of secondary particles will be severely impeded. Therefore, although impacting particles may dislodge up to 100 or 1000 times their own mass, secondary throwout will be much less than this. Each sandblasting particle may knock at least its own mass off the moon and the net effect of sandblasting erosion may be a slight mass loss by the moon.

3. *Sputtering*, caused by atoms and radiation on the lunar surface material. It has been estimated that a layer on the order of 10 cm deep has been lost in this way.

The net effect of "sandblasting" and "sputtering" will be to cause a softening of relief on a scale of 10 to 100 cm. The effect of secondary throwout by "cratering" in the region of the Ranger VII impact is to blanket the surface by a layer probably less than 1 meter in depth. This follows from the fact that not even the brightest rays are as much as 10 meters deep, as their relief cannot be detected from the earth, and the Ranger VII impact site is only weakly covered with ray material.

It is concluded that the total effect of the direct influx of cosmic material and of secondary throwout material in the region of the Ranger VII impact has been a softening of vertical relief and a mixing of surface layers on the order of 10 to 100 cm in depth.

While at present the shapes of the secondary craters remain to be interpreted, as indicated, the *impact origin* of lunar craters as a class is given strong support by both the *frequency distribution* and the *form* of the primary craters as revealed by the Ranger VII photographs. Since the frequency distribution of meteoritic and asteroidal masses is known, one may use the theory of cratering to predict the diameter distribution of primary lunar craters. This has been done and it is found that a straight line in a log frequency–log diameter plot results, with a slope that is in excellent agreement with observed crater counts, from the largest telescopic craters down to the meter size observed on the Ranger photographs, provided, however, that the clusters of secondary craters in the crater rays are avoided. If the lunar craters were volcanic, these conclusions would have no explanation, and one would still have to explain why no meteoritic craters were present.

In addition, the Ranger VII photographs give detailed views of craters of about one kilometer diameter, which we may compare to terrestrial cases. The lunar craters are shown to be bowl-shaped, just as are the nuclear explosion craters on the earth. *They do not have vertical walls*, as do volcanic collapse craters such as Halemaumau in Hawaii. Furthermore, there is no evidence of cinder-cone structures, which are common in regions of collapse craters on the earth. Yet in the continental regions, which are not covered by mare material, one can commonly see small craters (up to a few kilometers diameter) arranged in chains, graben, and other structures which attest to volcanic activity. Further, the maria themselves give every evidence of being lava flows, eroded by the same rain of material discussed earlier. Therefore, the moon's surface, considered *in toto*, has been shaped both by internal agencies, responsible for the flooding of the maria and small volcanic structures, and by external agencies, responsible for the most craters and the surface erosion.

The Air Force Aeronautical Chart and Information Center has undertaken the construction of a series of maps from the Ranger VII records. The scales of these maps are 1:1,000,000, 1:500,000, 1:100,000, 1:10,000, 1:1000, and 1:350. They are the most convenient summary available of the new surface data obtained. They are especially useful for locating the lineaments which are so important in the interpretation of the lunar surface.

The foregoing survey is concerned mainly with the larger structures of the lunar surface, structures such as on the earth would be dealt with by the geologist; and even with this restriction, several techniques are not considered (selenodesy, photogrammetry, height measurements from shadows, spectrophotometry and rock identification, discolorations by cosmic weathering). It has not been possible to include either a description and analysis of the radio and radar data of the moon, or of the important thermal measurements in and out of an eclipse, which together with the radio and radar data define the average physical properties of the upper layers of the moon in the cm-meter range. Nor has it been possible to review the data on photometry and polarization which define the nature of the uppermost surface in the micron range. Reference is made to recent texts [General References] on several of these subjects.

In conclusion it may be added that through the development of the Space Program the techniques now exist to obtain large quantities of vital data on the moon that were totally inaccessible only a decade ago. It should be borne in mind that Ranger VII was made to impact into as *unspectacular* an area as could be found on the moon, an area of the type where eventually man could safely land. Dramatic sights, such as a close-up of the Straight Wall, a crater-volcano such as Timocharis, a nest of volcanoes such as near Hortensius, or mountain peaks such as Pico, were thus knowingly avoided. The tools exist, however, to view these objects now at close range.

It may be hoped that the national interest in these primarily scientific problems will sustain a program that has already led to novel data and fresh insights concerning the nature and history of our satellite.

## 3. The Planet Mars

The planet Mars moves about the sun in a distinctly elliptical orbit ($e = 0.093$) which causes large distance variations among the different conjunctions with the earth. Thus at perihelion the Martian solar distance is $(1 - e)a = 1.382$ astronomical units, whereas at aphelion it is $(1 + e)a = 1.666$. The corresponding angular diameters as seen from the earth are about $25''.4$ and $13''.8$. When Mars is on the far side of the sun, its diameter is less than $4''$, so that useful observations of the planet are then nearly impossible. Good visual observations have an angular resolution of about $0''.2$, corresponding to about 1 per cent of the planet's diameter at favorable oppositions. Most existing photographic coverage has a resolution of 2 to 3 per cent, although occasionally 1 per cent has been attained. Since the closest views of the planet are made at full phase, when no shadows are shown, little is known on the vertical relief except from indirect arguments. These are based on the systematic patchy withdrawal of the polar snow caps which indicates the existence of plateaus about 1 km above the surrounding terrain. Conclusions about relief can, in principle, be based also on locally recurrent cloud phenomena and on spectroscopic measures of the bulk of the overlying atmosphere. The fractional resolution of the Martian image, about $10^{-2}$, is so much worse than, e.g., Tiros records of the earth, $10^{-3}$ to $10^{-4}$, that one might well question whether any real information about the surface of Mars can be obtained. The following brief summary lists some of the facts that are known or inferred on good grounds.

The diameter of the planet is 0.523 times the earth diameter; the mass is 0.1067 earth mass, the mean density 4.12, or, corrected for gravitational compression, about 4.0. The planetary composition therefore differs from stony meteorites (3.3), and a small metallic core is almost certainly present. The period of rotation is $24^h37^m22\overset{s}{.}67$, $41^m18^s$ longer than for the earth; the obliquity is $25\overset{\circ}{.}2$, $1\overset{\circ}{.}7$ larger than for the earth. The dynamical oblateness is 1/192 against 1/298 for the earth. The apparent oblateness of the disk is larger, approximately 1/90, apparently due to the fact that the eye sees the top of the atmospheric haze layer, not the planet itself, and that the haze is higher over the equator than over the poles.

Mars has no oceans and no lakes. In fact, liquid water cannot exist on this planet because of the low frost point, about $-75°C$. The polar caps are $H_2O$ ice (or snow), not $CO_2$; they do not melt in summer but *evaporate*. The total amount of precipitable water vapor in a vertical column of the atmosphere is about 0.01 mm. The principal gas observable spectroscopically is $CO_2$, and the amount is about 50 m atm NPT, as against 2.2 m atm for the earth. From the pressure broadening of the $CO_2$ bands a total atmospheric pressure of $0.017 \pm 0.003$ (probable error) atm is deduced. Almost certainly the bulk of the Martian atmosphere is nitrogen, as is true of the earth atmosphere; the inferred amount is 300 m atm, as compared to 6250 m atm on the earth. Gases whose presence has been tested but found absent are listed in Table 2, together with the upper limits set [Owen and Kuiper, 1964].

### Table 2

PRELIMINARY COMPOSITION MODEL OF MARTIAN ATMOSPHERE

| Gas | Abundance (cm, NPT) | Volume Fraction | Gas | Abundance (cm, NPT) |
|---|---|---|---|---|
| $N_2$ | 30,000 | 0.85 | COS | < 0.2 |
| $CO_2$ | 5000 | 0.14 | $CH_2O$ | < 0.3 |
| Ar | 400 | 0.01 | $N_2O$ | < 0.08 |
| $H_2O$ | 1. | | NO | < 20. |
| $O_3$ | < 0.05 | | $NO_2$ | < 0.0008 |
| $O_2$ | < 7. | | $H_2S$ | < 7.5 |
| $SO_2$ | < 0.003 | | $CH_4$ | < 0.4 |
| CO | < 1. | | $NH_3$ | < 0.1 |

The atmosphere also contains *particles*. When observations of the planet are made at increasingly short wavelengths, the atmosphere "closes in" and all surface features become invisible at about 4500 Å. This limit is not quite constant and is occasionally lowered by perhaps 300 Å ("blue clearing"). The photometric and polarimetric properties of the haze layer indicate that the haze consists of ice particles about

0.3 micron in diameter. Such a haze can be produced artificially by blowing very dry air past liquid nitrogen, a familiar experience to astronomers working at a high-altitude observatory with infrared instrumentation. The haze is blue, its shadow brown, somewhat like thin wood smoke. The "blue clearings" on Mars are probably caused by slight changes in the mean particle size of the haze, caused in turn by variations in the planetary atmospheric convection.

The Martian atmosphere occasionally contains also *dust* clouds that may be either local and confined, or spread and cause an almost planet-wide dusty condition of the atmosphere that may persist for one or several weeks. The color of the dust is yellowish and the particles are probably mostly 1 to 3 microns in size. Regular *clouds* may also be observed at times. They occur over the "forming" polar cap (i.e., fall and winter) and obscure the cap itself, and they occur sporadically elsewhere on the disk. Because of perspective they are seen more often on the Martian limb than at the center of the disk; sometimes they protrude somewhat beyond the terminator and are then striking in appearance. These clouds appear to consist of somewhat larger ice crystals, probably in the 1 to 3 micron range.

Because of the 25° obliquity, the *seasons* on Mars are quite marked and the alternation of snow deposits on the two polar areas makes for a fascinating spectacle. It needs to be borne in mind that the temperature on the edge of the snows is not 0°C but $-75°C$ approximately.

The observations of the Martian surface must be made through this variable atmosphere. Except during dust storms, however, the interference by the atmosphere is not severe in yellow, red, and near-infrared radiations. As was stated in Section 1, the surface displays light and darker areas (*deserts* and *maria*). But unlike the moon, where each lunation pattern is precisely the same, the Martian pattern shows changes, both with the seasons, roughly periodically, and progressively, over periods of decades. The basic mare pattern remains, however, and indicates definite terrain differences that require interpretation. The maria have sometimes been described as "green" or having colors changing with the seasons, much like terrestrial vegetation. This picture is grossly inaccurate. The green color of the maria, so often reported, is not observed in large telescopes when used under the best observing conditions. The colors appear to have been mostly contrast effects in the eyes of the observers, apparently due to the proximity of the orange or deep-ochre deserts.

It is not possible, in the limited space of this chapter, to portray and describe adequately the topography of the planet and its seasonal and progressive variations in tone. Reference is made for this purpose to a monograph by Slipher [1962] and particularly to a summary by Dollfus [1961a]. For excellent color reproductions of the planet reference is made to the results of Finsen [1961]; and to a fine pastel drawing by Antoniadi made in 1909, reproduced recently [Kuiper, 1964].

Some information on the nature of the Martian surface may be gained from spectrophotometric comparisons of Mars with various terrestrial rocks over the entire accessible wavelength range (0.6 micron where the atmosphere begins to be transparent to 3.5 micron where planetary heat radiation begins to exceed the reflected sunlight); and from *photometry* and *polarimetry* for the full range of accessible phase angles (0° to 48°). The data so far obtained [Binder and Cruikshank, 1964; Dollfus, 1961b] indicate that the Martian surface is probably a slightly weathered and powdery acidic rock, such as rhyolite. Further work on this subject is scheduled and is likely to be productive.

The available information is inadequate to form a clear picture of the Martian surface and in particular the nature of the maria. Except for small nearly circular patches, such as Fons Juventae, they do not seem to resemble the impact maria of the moon. Fons Juventae is about 170 km (100 miles) in diameter, intermediate in size between Mare Crisium and Plato on the moon. Presumably the meteoritic impacts are more frequent on Mars than on the moon because of the planet's position near the asteroid belt. However, the early phase of lunar bombardment, the first $10^8$ years or so responsible for nearly all the craters on the terrae, may well have been absent from Mars, because this phase was presumably not due to asteroidal bodies, but to a ring of small satellites originally surrounding the earth, through which the moon made its way outward after oceans had condensed on the earth and tidal friction became very large [Kuiper, 1954]. Nearly all of the Martian maria would therefore be expected to be due to flooding, not impact, though a few small impact maria or large craters might exist.

Mars, being intermediate between the moon and the earth and with a higher internal temperature than the moon, might have shown, just as the earth, plutonic activity substantially later than the critical $4.5 \times 10^9$ year epoch. In that case, dark-floored impact craters could have formed later in Martian history. Another possibility is that impact craters, formed all through Martian history, have dark floors because of eolian sediments.

The seasonal variations in the Martian maria could conceivably be due in part to low organisms able to exist in the rigors of the Martian climate. It is this possibility which has excited the imagination of all thoughtful persons and has given the impetus to far-reaching programs of exploration by means of spacecraft.

Earlier in this chapter it was implied that, given the solar history, mass appears dominant in determining the nature and the history of a planet. Taken literally, this would imply that the period of rotation, the mean density, the internal mass distribution, the density and composition of the surface rock, the density and the composition of the atmosphere, the water content, the development of life, all would, in principle, be predictable. It will be interesting, in the coming decade, to pursue this thought and to discover to what extent a planet has a personality of its own, determined by some accident during its development, not predictable on the basis of its mass alone. Clearly, in the case of the earth-moon system such an event has occurred, presumably caused by a peculiar distribution of angular momentum within the earth-moon protoplanet.

## References

Binder, A. B., and D. P. Cruikshank, Comparison of the infrared spectrum of Mars with the spectra of selected terrestrial rocks and minerals, *Comm. LPL*, 2, (37), 1964.

de Marcus, W. C., *Hdb. Physik*, 52, 419, 1959.

de Marcus, W. C., *Astron. J.*, 63, 2, 1958; Wildt, R., in *Planets and Satellites*, edited by G. P. Kuiper and B. M. Middlehurst, p. 208, University of Chicago Press, Chicago, 1961.

de Marcus, W. C., and R. T. Reynolds, *Mém. Soc. Roy. Sci. Liège*, 7, in-8°, Ser. 5, 51, 1963.

Dobar, W. L., O. L. Tiffany, and J. P. Gnaedinger, *Icarus*, 3, 323, 1964.

Dollfus, A., Chapter 15, in *Planets and Satellites*, edited by G. P. Kuiper and B. M. Middlehurst, University of Chicago Press, Chicago, 1961a.

Dollfus, A., Chapter 9, in *Planets and Satellites*, edited by G. P. Kuiper and B. M. Middlehurst, University of Chicago Press, Chicago, 1961b.

Finsen, W. S., Chapter 17, in *Planets and Satellites*, edited by G. P. Kuiper and B. M. Middlehurst, University of Chicago Press, Chicago, 1961.

Hartmann, W. K., and G. P. Kuiper, Concentric structures surrounding lunar basins, *Comm. LPL*, 1 (24), 51, 1962.

Hartmann, W. K., Radial structures surrounding lunar basins, I: The Imbrium system, *Comm. LPL*, 2 (24), 1, 1963; II: Orientale and other systems; Conclusions, *Comm. LPL*, 2 (36), 174, 1964.

Kuiper, G. P., *Proc. Nat. Acad. Sci. U.S.*, 40, 1110, 1954.

Kuiper, G. P., in *Vistas in Astronautics*, vol. 2, edited by M. Alperin and H. F. Gregory, Pergamon Press, London and New York, 286, 1958.

Kuiper, G. P., *Comm. LPL*, 2, No. 35, 1964, frontispiece colorplate.

Kuiper, G. P., Interpretation of Ranger VII records in *NASA Tech. Report No. 32-700*, Ranger VII, Part II, Experimenters' analyses and interpretations, Feb. 10, 1965, Chapter III, pp. 9–73. Also, Lunar results from Rangers 7 to 9, *Sky and Telescope, Special Supplement*, 293–308, 1965.

McCracken, C. W., and M. Dubin, Dust bombardment on the lunar surface, *NASA TN D-2100*, 1963; also in *Lunar Surface Layer; Materials and Characteristics*, edited by J. W. Salisbury and P. E. Glaser, Academic Press, New York, 179–214, 1964.

Nichols, R. I., McCartys Basalt Flow, New Mexico, *Bull. Geol. Soc. Am.*, 57, 1049–1086, 1946.

O'Keefe, J. A., and W. S. Cameron, *Icarus*, 1, 271, 1962.

Owen, T. C., and G. P. Kuiper, *Comm. LPL*, 2, 132, 1964.

Ringwood, A. E., *J. Geophys. Res.*, 67, 857, 4005, 4473, 1962.

Ross, C. S., R. L. Smith, and R. A. Bailey, Outline of the geology of the Jemez Mountains, New Mexico, *N. Mex. Geol. Soc. Twelfth Field Conference*, 1961, p. 139; also *Geol. Survey Research, Profess. Paper* 424-D, 1961.

Slipher, E. C., *A Photographic History of Mars, 1905–1961*, Lowell Observatory, Flagstaff, 1962.

Strom, R. G., Analysis of lunar lineaments, I: Tectonic maps of the Moon, *Comm. LPL*, 2 (39), 1964.

Whitaker, E. A., *NASA Tech. Rept. No. 32-700*, Chapter VI, 149–154, 1965.

## General References

Antoniadi, E. M., *La Planète Mars, 1659–1929*, Hermann, Paris, 1930.

Baldwin, R. W., *The Measure of the Moon*, University of Chicago Press, Chicago, 1963.

de Vaucouleurs, G., *Physics of the Planet Mars*, Faber and Faber Ltd., London, 1954.

Fielder, G., *Structure of the Moon's Surface*, Pergamon Press, New York, Oxford, London, Paris, 1961.

Fielder, G., *Lunar Geology*, Lutterworth Press, London, 1965.

Gehrels, T., *Astron. J.*, 69, 826, 1964.

Hapke, B., *J. Geophys. Res.*, 68, 4571, 1963.

Kopal, Z., and Z. K. Mikhailov, editors, *The Moon*, Academic Press, London, New York, 1962.

Kuiper, G. P., and B. M. Middlehurst, editors, *Planets and Satellites*, Chaps. 6–12, University of Chicago Press, Chicago, 1961.

Kuiper, G. P., and B. M. Middlehurst, *The Moon, Meteorites, and Comets*, Chaps. 1–5, University of Chicago Press, Chicago, 1963.

MacDonald, G. J. F., *Science*, 133, 1045, 1961.

Salisbury, J. W., and P. E. Glaser, editors, *The Lunar Surface Layer; Materials and Characteristics*, Academic Press, New York, 1964.

Shoemaker, E. M., R. J. Hackman, and R. E. Eggleton, in *Advances in the Sciences*, vol. 8, Plenum Press, New York, 1962.

van Diggelen, J., *Recherches Astron. Obs. Utrecht*, XIV, 2, 1958; also reprinted as *NASA TT F-188*, 1964.

# THE INTERPLANETARY MEDIUM AND SOLAR PLANETARY RELATIONS

*Ludwig F. B. Biermann*

Institut für Astrophysik, Max-Planck-Institut für Physik und
Astrophysik, Munich, Germany

In the context of the earth sciences, the interplanetary medium has long been looked upon as the environment through which the earth moves in its orbit around the sun. As has become clear in the last two or three decades, it is however not a static or even stationary environment; the interplanetary plasma, which is, from the point of view of solar-terrestrial relationships, its most important constituent, actually flows through interplanetary space with a speed very much in excess of that with which the earth moves. As we now believe, it does so continuously, though with varying density and velocity even at the lowest level of solar activity. The measurements to be made during the International Year of the Quiet Sun, on the occasion of the forthcoming sunspot minimum of 1965, should establish whether or not this present picture is correct.

There is of course the question of where the boundary between the earth's atmosphere—the lowest layers of which are the seat of the atmospheric motions to be discussed in this chapter—and interplanetary space should be placed. The measurements made by means of space vehicles since 1958 have shown that the ionized gaseous matter in a region extending out to at least 10 earth radii—this minimum value being reached at the subsolar point—is effectively coupled to the earth by its magnetic field. This region, which contains the so-called radiation belts, is now commonly called the magnetosphere [Gold, 1959]. As we shall see, there is good reason to regard its outer boundary as the frontier, beyond which we reach interplanetary space, in spite of the fact that this frontier is not quite fixed in position and that there is a transition region enveloping the magnetosphere in which the temperature and the state of motion are greatly affected by the nearness of the earth.

The atmospheric layers above the stratosphere, and the magnetosphere itself, the outer layers of which, in regard to their density, resemble interplanetary space, are the seat of many important geophysical phenomena, such as magnetic storms. The latter are believed to be caused by the sun's corpuscular activity. Their detailed interpretation as well as that of the radiation belts is, however, still seriously incomplete.

In this chapter I shall attempt to give a consistent picture of the interplanetary medium and of some aspects of the solar-terrestrial relations, which can be related to the sun's corpuscular activity.

The two main constituents of the interplanetary medium are the plasma, which is mainly of solar origin, and the zodiacal dust cloud, which originates from the comets (Figure 1). In addition to the plasma,

Fig. 1. Zodiacal light.

# INTERPLANETARY MEDIUM

Fig. 2. Comet Mrkos 1957d with plasma tail (straight, filamentary) and dust tail (curved, diffuse).

there is probably some neutral gas though its amount must ordinarily be rather small—otherwise it would produce visible absorption lines in the solar spectrum. There are, furthermore, energetic charged particles, that is, cosmic rays of solar or galactic origin, spiraling around the lines of force of the interplanetary magnetic fields, with radii of gyration that for most of them are small on the scale of the extent of interplanetary space.

In solid form, we have in addition to the dust layer, which is most clearly seen as the zodiacal light, the meteoric material observed as meteors and meteorites, and, as transient visitors from the outer parts of the solar system, the comets, the long-lived nuclei of which are now usually believed to be conglomerates of dust and meteoric matter and of compounds of C, N, O, and H in frozen form, according to a picture proposed by F. Whipple some ten years ago [Whipple, 1963].

If a comet approaches the sun, it releases, from its surface to the interplanetary medium, gas in molecular form, which soon becomes ionized, as well as dust. The comets with visible plasma tails, such as the recent comets Arend-Roland 1956h and Mrkos 1957d (Figure 2), are particularly valuable, however, as natural probes for the solar plasma, the presence of which they indicate [Biermann, 1951, 1953,

1957]. Well over a hundred comets are known to have possessed plasma tails. Our present picture of the nature and the over-all properties of the interplanetary plasma rests to a considerable extent on the interpretation of the observations of such comets, as does also the interpretation of the direct measurements from space vehicles, which so far have been carried out only in the vicinity of the ecliptic plane.

In this summary discussion of the constituents, I have already mentioned some possibilities of obtaining information on the interplanetary medium. Though I wish mainly to discuss results of such observations, a few remarks on the methods by which information is obtained may be in order. Direct visual or photographic observation of the zodiacal light gives, as was mentioned already, information on

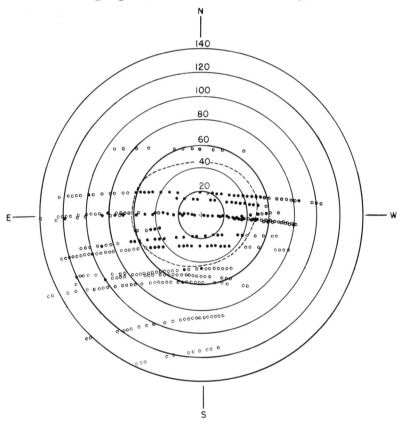

FIG. 3. Diagram showing scattering suffered by 13 radio sources in the vicinity of the sun. Distance scale defined by circles with radii increasing in increments of 20 solar radii. The closed contour, shown dashed, marks the average extent of the scattering corona [Slee, 1961].

the small dust particles which form the zodiacal cloud; however, all attempts to determine the electron density in interplanetary space from the observed polarization of the zodiacal light have so far failed. The very careful work of Blackwell and coworkers has provided essentially upper limits, which are somewhat higher than the values derived from the direct measurements from space vehicles.

Observations of radio point sources in the vicinity of the sun, e.g., of the Crab Nebula source, have given evidence for scattering of the radio waves, which in all probability is due to multiple scattering in the local density fluctuations of interplanetary electrons (Figure 3) [Hewish and Wyndham, 1963]. Though no absolute values for the electron density can be derived, it can be shown that for the vicinity of the ecliptic plane the value derived from the direct measurements are entirely consistent with the radio observations. Furthermore, it can be concluded that the electron density distribution in interplanetary space is probably somewhat flattened, but only moderately so.

The detailed properties of the scattered light indicate furthermore that the density distribution of the scattering electrons must at least at times be filamentary. This evidence from the observations of radio sources pertains to distances from the sun out to 60 or perhaps 100 solar radii, or approximately $\frac{1}{4}$ to almost $\frac{1}{2}$ astronomical units, that is, to a part of interplanetary space for which so far all other methods fail.

Extended direct measurements of the plasma, after the pioneering work done with Lunik II and III, and Explorer X (Figure 4), have so far been made and reported from the space probe Mariner II on its way to Venus in the fall of 1962, and from the first Interplanetary Monitoring Platform launched in November 1963 [Proc. Symp. Plasma Space Sci., 1963; Proc. Conf. Solar Wind, 1964; I. G. Bulletin, 1964]. The Mariner II measurements pertained essentially to ions coming from the solar direction within a rather small cone (of a few degrees of arc); for the reduction it was assumed that the particles came exactly from that direction, taking account, however, of the motion of the space probe itself, which causes an aberration effect. The energy, in the range of a fraction of 1 kev to almost 10 kev, was measured in steps such that the velocity dispersion and thus a kind of temperature (which should include small-scale turbulence) could be determined.

Most of these vehicles also carried several magnetometers, with which the magnetic fields in the magnetosphere and in interplanetary space could be measured, and furthermore instruments for measuring energetic particles of $\sim 10$ Mev and more for positive ions, and 30 kev and more for electrons.

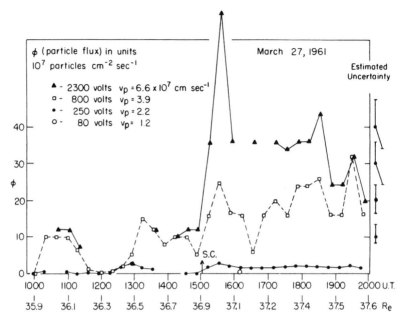

Fig. 4. Plasma measurements made with the eccentric orbit satellite Explorer X (S.C. indicates the "sudden commencement" of a magnetic storm).

The fine dust particles with a radius of a few 1000 Å, which are believed to produce most of the zodiacal light by scattering the solar light, could not be detected so far by any instrument in outer space. Only rather bigger particles, the impact of which on the vehicles is more easily felt but which of course occur much more rarely, have been encountered occasionally at a large distance from the earth.

The most important indirect ways for obtaining information on the interplanetary plasma are observations of geomagnetic phenomena and of plasma tails of comets. The geomagnetic data which have been regularly recorded for about a hundred years, give specifically information about individual storms. According to Chapman's and Ferraro's work of 1931, these storms give evidence for the arrival of clouds of solar plasma emitted during big solar flares with a velocity of the order of 1000 to 2000 km/sec and reaching the earth after about a day; recurrent magnetic disturbances, which tend to repeat themselves after 27 days, the apparent period of rotation of the sun, indicate the presence on the sun of long-lived active regions from which solar plasma is emitted with smaller velocities (600 to 800 km/sec), but for long periods of time [Chapman and Bartels, 1951].

The visible plasma tails of comets, which point radially away from the sun and the length of which is often of the order of 10 to 100 million km, are now believed to derive the momentum of their motion with a velocity of the order of one to several hundred km/sec from the solar plasma. They may hence be looked upon as natural probes, which reveal the presence of solar plasma streaming approximately radially through interplanetary space with a velocity of several 100 km/sec or more. The aberration effect due to the motion of the cometary nucleus around the sun is also seen; its amount is usually several degrees of arc and is in agreement with what one would expect from the ratio of the comets' orbital velocity and the normal velocity of the solar plasma of 300 to 500 km/sec [Biermann, 1951, 1953, 1957]. The observed properties of these tails indicated those properties of the solar corpuscular emission, that is, its relative steadiness and approximate isotropy, which in conjunction with considerations on the dynamics of the solar corona gave rise to the now familiar terminology "solar wind," proposed by E. Parker in 1958.

To illustrate these points, Figure 5 shows the regions in inter-

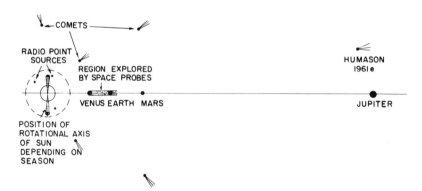

Fig. 5. Regions in interplanetary space explored by direct measurements from space probe and for which indirect evidence is available from the observation of radio point sources and of comets with plasma tails.

planetary space for which we have information from the radio data, from direct measurements, and from the comets. If we also take into consideration the time over which measurements or observations are available and the fact that cometary observations do not indicate any difference between the vicinity of the ecliptic plane, for which we have direct measurements, and the regions more distant from that plane, it is easily seen that the comets are likely to retain their role as main

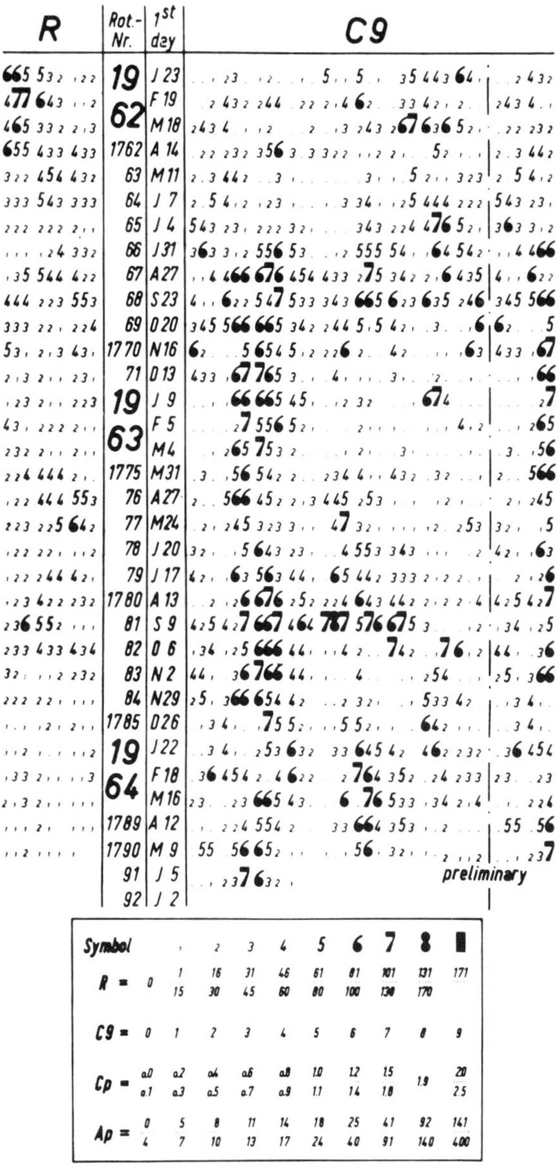

FIG. 6. Daily geomagnetic character figures C9 and sunspot numbers R.

probes for the larger part of interplanetary space still for some time to come, although it is to be hoped that out-of-the-ecliptic studies will become possible before too long.

We discuss now the characteristic properties of the solar plasma wind outside the magnetosphere as given by the direct measurements for the vicinity of the ecliptic plane for the fall of 1962 (Mariner II) and for December 1963 and the first months of 1964 (IMP I); at those times the sunspot numbers were between 10 and 80 respectively up to 40 (IMP I); the magnetic indices C9 were between 1 and 7 during both periods of time, though the high values of C9 had become less frequent by the end of 1963 and after (Figure 6).

The flux of solar ions was during both periods $\gtrsim 10^8$ cm$^{-2}$ sec$^{-1}$ under generally quiet conditions, the average energy per ion being around 1 kev, which for protons would correspond to a velocity of about 450 km/sec. In these properties there was apparently little change between the two periods of time in question, in spite of the decrease of the general level of solar activity, which of course showed up in the frequency of disturbed periods. The ion density derived from these figures, which is obviously less accurate, would be a few ions per cubic centimeter up to about 10 or 20 ions per cubic centimeter (at disturbed periods up to $10^2$ cubic centimeters) (Figures 7, 8).*

The average direction of the ion flux coincided, within the errors of the measurements from IMP I, usually with the apparent direction from the sun, in agreement with the prediction based on the observations of the plasma tails of comets. The fluctuations of direction seem however, according to new data obtained by Professors Rossi, Bridge, and coworkers with the IMP I, to be larger than was apparent from the cometary data, which necessarily refer to averages over at least a few hours.

In regard to the chemical composition, some of the data obtained from Mariner II indicate the presence of $\alpha$-particles which, when moving with the same mass velocity as the protons as expected from plasma physics, would have twice the kinetic energy per unit charge as compared to the protons; this appears to show up in the energy spectrum obtained from the Mariner II data (Figure 9). Unfortunately, the ratio of protons to $\alpha$-particles derived in this manner is rather uncertain, because the time required for taking one energy spectrum

---

* The author wishes to thank Professors B. Rossi and H. S. Bridge and their coworkers at Massachusetts Institute of Technology as well as Dr. N. F. Ness and coworkers at Goddard Space Flight Center for having made available to him important results gained with the IMP I, prior to publication.

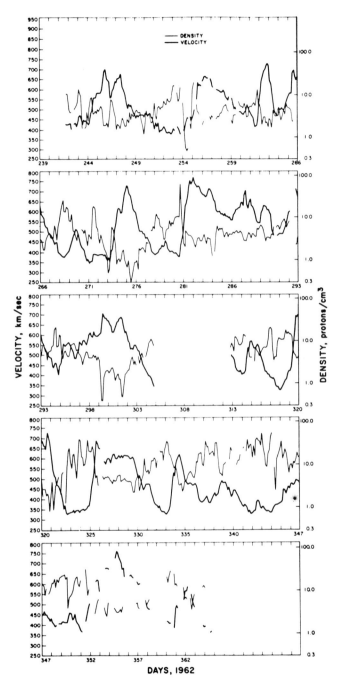

FIG. 7. Measurements of velocity and density of the interplanetary plasma, made with the space probe Mariner II, November/December 1962.

*INTERPLANETARY MEDIUM* 81

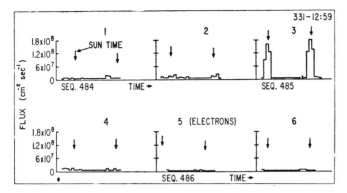

FIG. 8. Plasma flux measured with the spin-stabilized eccentric orbit satellite IMP I showing proton flux in one energy band (260 to 600 ev) from the solar direction only.

is fairly long ($\gtrsim 5$ min); with some caution it may be said that somewhat less than 10 per cent of the ions, by numbers, appear to be α-particles, but whether there is really a difference between the He:H ratio in interplanetary space and in the sun's atmosphere has still to be seen.

The dispersion in the energy as measured during the time interval in question is determined mainly by the thermal motion and by small-scale turbulence. During quiet periods it may be mainly due to the former, and the temperatures derived from the Mariner II data were $\lesssim 10^5$ degrees. This, at the same time, indicates a sound velocity of

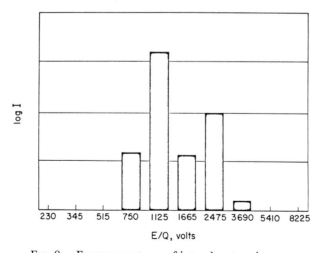

FIG. 9. Energy spectrum of interplanetary ions.

$\gtrsim 50$ km/sec, which would yield a Mach number of $\approx 10$. For computing a Mach number, one should, however, take into account the energy density of the interplanetary magnetic fields; if that is done, on the basis of the data still to be discussed, smaller values are obtained for the Mach number, typically of $\approx 7$ to 8.

The variability of the flux is apparent from the data obtained by Neugebauer and Snyder from the Mariner II measurements as well as from those obtained from the measurements gained with IMP I. Figure 4 showed data already obtained during the sudden commencement of a magnetic storm; typical for such events is the sudden

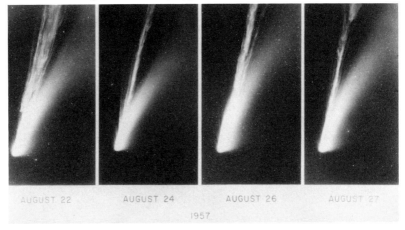

Fig. 10. Tail of comet Mrkos 1957d on August 22, 24, 26, and 27, showing changes between successive nights. Photographed with the 48-inch Schmidt telescope.

appearance of particles of much higher energy, $\approx 2$ kev and $\approx 3$ kev, which are practically absent during quiet periods. As yet the details of these observations are not understood.

The variability of the plasma flow in interplanetary space also shows up in the plasma tails of comets; Figure 10 shows a sequence of pictures of comet Mrkos 1957 obtained while the comet was at heliographic latitudes of $\sim 45°$. Comet Morehouse of 1908, the plasma tail of which was particularly prominent, showed similar phenomena at heliographic latitudes between 40 and 50°.

In general, one has also to take into account the intrinsic "activity" of the comet itself, that is, the tendency to produce spontaneous outbursts of gas, which might simulate effects due to solar activity; as regards the latter, that is, phenomena in the tail due to the solar plasma, a number of cases of solar-cometary-terrestrial relations, analogous to the well-known solar-terrestrial relations, have been established, e.g., by Mme. Rhea Lüst [1961]. A few cometary phenomena, such as the one shown in Figure 11 (shock front in comet Mrkos), can very probably be related to the action of a shock front

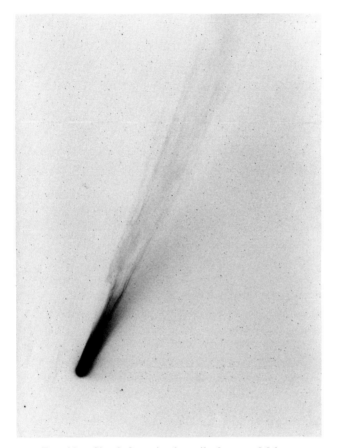

FIG. 11. Shock front in the tail of comet Mrkos.

moving through interplanetary space. The small thickness of these fronts, as well as the sharpness of other features visible in the plasma tails of comets, for instance, the narrow filaments characteristic for many comets, can be understood only by the action of magnetic fields, which cause the ions to gyrate with a radius of only a few 1000 km or less.

We turn hence to the results of the magnetometer measurements.

There is a measurable interplanetary magnetic field, which is during quiet periods around 4 to $7\gamma$ ($1\gamma = 10^{-5}$ oersted). This corresponds to an energy density of $\gtrsim 10^{-10}$ erg/cm³, as compared to a kinetic energy per cubic centimeter of the solar wind of $\gtrsim 10^{-8}$ erg/cm³. From this an "Alfvén" speed, that of magnetohydrodynamic waves, of $\approx 70$ km/sec and the Mach number of 7 to 8 mentioned already may be derived. This, at the same time, means that the magnetic

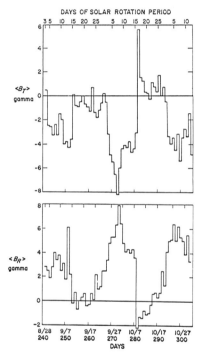

FIG. 12. Azimuthal and radial component of the interplanetary magnetic field showing spiral pattern during the first part of each solar rotation.

fields, which are so to speak "frozen" into the plasma, are moved about by the latter, without reacting too much on plasma motions. Also the magnetic field is often rather variable, as we have seen already. Averages taken over longer periods of time or during quiet periods often indicate, however, a large-scale spiral structure (Figure 12), which on theoretical grounds had been predicted by E. Parker. This Archimedean spiral corresponds to the one obtained by the superposition of the sun's rotation with the velocity of the radial flow outward of $\approx 500$ km/sec. It can be seen without great difficulty that there exist stationary solutions of the equations of magnetofluid dynamics, which have precisely this character. The sign of magnetic field in these solutions is of course determined by that of the field on the sun's surface, from which the field line emerges.

About 40 per cent or so of the data obtained with Mariner II showed this kind of structure, while the data for the remaining periods of time were difficult to interpret. Also no clear correlation with magnetic fields observed simultaneously on the sun's surface could be established. The more recent data obtained with the IMP I, which referred to a somewhat quieter period, gave, however, a clearer picture.

According to recent work done by N. F. Ness and J. M. Wilcox [1964], the data obtained with IMP I for the generally quiet 75-day period December 1963 to February 17, 1964, showed a 27-day autocorrelation with an amplitude of $0.85 \pm 0.10$. For this period the correlations with the polarity of the magnetic fields on the sun's surface determined by R. Howard at Mount Wilson, as far as data were available, were also discussed. If plotted as a function of the time lag from the central meridian passage of the area to the time at which the field was measured in interplanetary space, a definite correlation was found for a time lag of $4\frac{1}{2}$ days, which corresponds to a radial flow velocity of almost 400 km/sec. The fact that both signs occur alternatively with similar, though not equal probability, indicates a filamentary structure in interplanetary space, the scale of which is, however, still uncertain.

For the periods of time during which the spiral structure is not seen, other possibilities have to be considered, such as the presence of magnetized clouds with magnetic lines of force detached from the sun, a picture which was recently rediscussed by R. Lüst and the author [Biermann and Lüst, 1964].

Both from the measurements with Mariner II and with IMP I it was found that the field component perpendicular to the ecliptic plane was more often directed southward than northward. Though the average value, taking regard of the sign, was only of the order of $1\gamma$, the result appears to be significant; it is not yet understood.

The interaction of the solar wind with the earth's magnetosphere has been the subject of theoretical work by quite a number of authors, for instance, Petschek and coworkers here in Cambridge [Levy et al., 1963], and of experiments made with space vehicles. The models which have emerged and which are in quite good agreement with the observations assume approximate magnetohydrostatic equilibrium on the surface of the magnetosphere, such that on the inside there is only the pressure of the geomagnetic field deformed by the shielding currents in the surface, balanced on the outside by the pressure of the solar wind. Since the latter at large distances from the earth is supersonic, whereas the velocity at least near the stagnation point must be small, there should be a shock front, analogous to the bow show of the supersonic flow around an obstacle which should surround the magnetosphere at some distance [Axford, 1962].

There is of course the question whether under such conditions, with a kinetic mean free path of the order of a hundred million km (1 A.U.), a shock front can exist. Theory of a plasma with low but finite resistivity has shown that, in the presence of magnetic fields with a component at right angle to the plasma velocity, as indeed observed in the interplanetary plasma, sharp changes of the flow characteristics resembling shocks, but without genuine generation of entropy in this apparent shock, can appear over distances of the order of the radii of gyration of the charged particles. For Mach number $>2$, the radii of gyration of the ions appear to be the relevant ones, which under these conditions would be of the order of several hundred or a thousand km only; the expected thickness of this apparent shock would hence be a few hundred km or so.

The theory of magnetohydrodynamic shock fronts in the absence of collisions is a difficult subject; in laboratory experiments it is hardly possible to create sufficiently pure conditions, so that theoretical predictions could be checked. For this reason the behavior of the

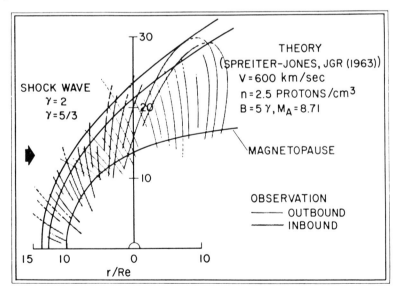

FIG. 13. Outer boundary of the magnetosphere and position of shock front according to theory and observation.

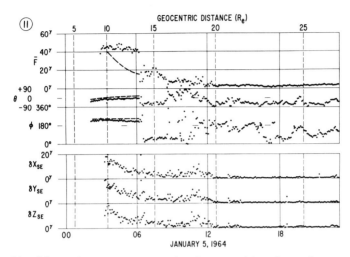

Fig. 14. Magnetic measurements showing transition from the magnetosphere through the transition layer into free interplanetary space.

interplanetary plasma in the vicinity of the geomagnetic field boundary is of particular interest.

The observation from space vehicles, especially those obtained from the IMP I, have given clear indications of the existence of the shock front demanded by theory [I. G. Bulletin, 1964]. The most detailed computations of the position of the shock front and of the boundary of the magnetosphere are probably those of Spreiter and Jones [1963], the results of which are seen in Figure 13 together with the results of the plasma measurements. From the magnetic measurements (Figure 14) [Ness, 1964] it is seen that, approaching this boundary from the inside, the magnetic field intensity rises above the value derived from the theory of the undisturbed geomagnetic field. Near the boundary approximately twice the undisturbed value is reached; the shielding currents, which enclose the geomagnetic field by compensating its exterior part, necessarily strengthen it on the inside. Outside the boundary, there is a transition region with greatly varying fields, in which also the plasma flux and its direction are found to be very unsteady. Only after passing the outer boundary of the transition region, that is, the shock front, the comparatively quiet interplanetary space is reached.

The main features of the situation including the narrowness of the front agree reasonably well with the theoretical prediction, but some details are still unexplained. Hence the existence of a collision-free shock outside the magnetosphere may be regarded as established, and with it the applicability of magnetofluid dynamics to extremely low densities and very weak magnetic fields.

The first indication that very sharp fronts are carried through interplanetary space with the solar plasma came from the observation of the so-called sudden commencements of magnetic storms, an example of which was shown in Figure 4. These, however, refer to individual events, not to a stationary situation.

Especially interesting are the observations of energetic particles in the vicinity of the shock front, by which the acceleration of some electrons to $\approx 30$ kev has been established [Fan *et al.*, 1964]. This is the first *in situ* observation of such an acceleration process. The number of these electrons is much smaller than of those found in the radiation belts inside the magnetosphere, the origin of which is still not really clear; it seems likely that most of them somehow result from the interaction of the plasma flow around the magnetosphere with the latter, but by which mechanisms is still an open question.

Another very important observation, which tends to confirm the applicability of the concepts of magnetofluid dynamics, is that of what appears to be the wake of the moon in the solar wind [Ness, 1964]. Figure 15 shows the relative position of the earth and the moon

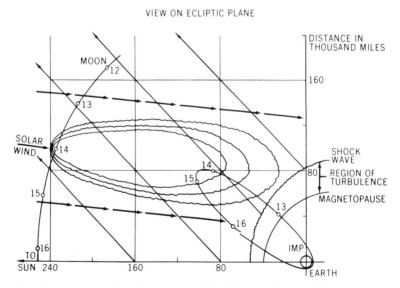

FIG. 15. Position of earth, moon, and IMP I around the middle of December 1963.

and the direction toward the sun around the 14th and 15th of December 1963. Very violent fluctuations of the magnetic field were observed during 20 hours between these dates, during an otherwise generally quiet period, as judged from geomagnetic evidence and from the measurements on IMP I during the days before and after the event. After careful consideration of possible alternatives, N. F. Ness and coworkers concluded that the existence of the moon's wake had been established. It seems likely that around new moon, if the moon is near the ecliptic, this wake could reach the vicinity of the magnetosphere; hence it is conceivable that hitherto undetected geophysical effects may result.

According to a hypothesis proposed by T. Gold at the Pasadena Conference on the solar wind in April 1964 [Proc. Conf. Solar Wind, 1964], the existence of this wake may be related to the finite electric conductivity of the moon, which should cause the lines of force of the interplanetary magnetic fields to pile up in front of the moon and the solar wind to flow at some distance around it. The lunar wake may of course in a general sense be compared with the plasma tails of comets, or rather with the larger region around the visible tail, in which the plasma velocity is seriously affected; the relative extent of both seems similar. Also we would expect the earth to have a similar wake, which should be many times longer than that of the moon, from the cross sections involved. However, no observations at large distance from the earth in the antisolar direction seem to be available as yet. Such observations should also decide whether the so-called Gegenschein, which looks like an extension of the zodiacal light to distances beyond the earth, is an interplanetary or a geophysical phenomenon.

From the standpoint of solar-terrestrial relations it may be asked which phenomenon in interplanetary space precedes individual magnetic storms or the recurrent magnetic disturbances believed to be caused by solar $M$-regions. Magnetic storms are usually connected with a gradual rise in the velocity of the solar plasma and its dispersion, in such a way that the kinetic pressure is considerably increased; whether and how part of this increased energy flux becomes available for maintaining a magnetic storm during its main phase, is not yet understood; quite recently, Akasofu has proposed that the main part of the energy of a magnetic storm comes from a neutral component of the solar wind [Akasofu, 1964]. The magnetic character figures $Kp$ could, for the plasma measurements obtained with Mariner II, be related with the plasma velocity. A persistent solar $M$-region appears to emit plasma with above average velocity, 600 to 800 km/sec, which must push slow plasma in front of it through interplanetary space.

All the measurements made hitherto refer to the point in space occupied by the vehicle at the time of the measurement. It may be asked whether it is possible to follow the path of a plasma volume element in interplanetary space in the same way as we get information on the flow pattern of wind in the earth's atmosphere, for instance, by observing smoke. For this we would have to seed the interplanetary plasma, which is invisible, with ions which absorb and reemit solar light in the ordinary spectral range, such as the ions of Ba or of similar elements [Biermann *et al.*, 1961; Föppl *et al.*, 1965]. This experiment can be looked upon as creating an artificial analog to the plasma tail of a comet, the visible ions of which are predominantly $CO^+$ or $N_2^+$. It has been found that quantities of the order of a few kilograms, say of $Ba^+$, should be observable from ground at a distance of the order of 30 earth radii, half the distance of the moon. This experiment is actually being prepared at the Munich Max-Planck-Institut; preliminary trials at ionospheric altitudes give us hope that similar experiments in interplanetary space would give valuable information on the behavior of the interplanetary plasma and how ions of nonsolar origin are coupled to the solar wind [Proc. Conf. Solar Wind, 1964].

Observations of such artificial plasma clouds would also help us to make still better use of the comets with plasma tails for obtaining information on the interplanetary plasma in regions which cannot yet be reached by space probes. As an example for this situation, I would like to mention the recent comet Humason, which showed a plasma tail out to the distance of Jupiter, that is 5 A.U. Though a detailed study has not yet been made, present indications are that the behavior of the tail of comet Humason does not suggest any drastic change in the properties of the solar wind out to that distance.

In closing I would like to emphasize two more general aspects of the physics of interplanetary space. The first has to do with the current laboratory research on plasma. Here we work with plasmas, usually confined by magnetic fields, with densities for largely ionized plasmas down to approximately $10^8$ or $10^9$ particles per cubic centimeter. Certain phenomena, such as hydrodynamic shocks in the absence of collisions, the existence of which has been inferred theoretically, are difficult to observe under laboratory conditions. In interplanetary space, however, where the densities generally are in the range 1 to 100 ions per cubic centimeter, we have almost ideal conditions for the observation of such shocks, the result of which we have seen. Measurements or experiments in this cosmic laboratory may therefore be looked upon as an important extension of laboratory plasma physics. The general applicability of current theoretical concepts, particularly of the model called supersonic magnetofluid dynamics, could be investigated for these densities in the presence of magnetic fields in range 1 to a few $10\gamma$; though many detailed questions are still open, it seems fair to say that the application of the theory has been successful.

This leads me to my second and last point. The interplanetary plasma is the only plasma of cosmic dimensions with densities and magnetic fields similar to those in interstellar space in which we can make measurements *in situ* and can even carry out planned experiments. For this reason such studies are not only basic ones for the earth sciences but also for the physics of the space between the stars, that is, of interstellar space, and beyond. Cosmic plasma physics appear thus to be a link which connects in a rather singular way large parts of astrophysics, of laboratory physics, and of the earth sciences.

## References

Akasofu, S., *Planetary Space Sci.*, *12*, no. 9, 801, 1964.
Axford, W. I., *J. Geophys. Res.*, *67*, 3791, 1962.
Biermann, L., *Z. Astrophys.*, *29*, 274, 1951.
Biermann, L., *Mem. Soc. Roy. Sci. Liège*, *XIII*, Parts I–II, 291, 1953.
Biermann, L., *Observatory*, *77*, 109, 1957.
Biermann, L., R. Lüst, Rhea Lüst, and H. U. Schmidt, *Z. Astrophys.*, *53*, 226, 1961.
Biermann, L., and R. Lüst, The problem of the plasma flux and the magnetic fields in interplanetary space, contribution to a volume dedicated to Prof. S. Rosseland on the occasion of his 70th birthday, *Astrophys. Norveg.*, *IX*, 61 (1964).
Chapman, S., and J. Bartels, *Geomagnetism*, Oxford University Press, 1951.
Davis, L., and J. E. Midgley, *J. Geophys. Res.*, *67*, 499, 1962.
Fan, C. Y., G. Gloeckler, and J. A. Simpson, *Phys. Rev. Letters*, *13*, no. 5, 149, 1964.
Föppl, H., *et al.*, Preliminary experiments for the study of the interplanetary medium by the release of metal vapor in the upper atmosphere, *Planetary Space Sci.*, *13*, 95 (1965).
Gold, T., *J. Geophys. Res.*, *64*, 1219, 1959.
Hewish, A., and J. D. Wyndham, *Monthly Notices Roy. Astron. Soc.*, *126*, 469, 1963.
*I. G. Bulletin*, Natl. Academy of Sciences, no. 84, June 1964.
Kellog, P. J., *J. Geophys. Res.*, *67*, 3805, 1962.
Levy, R. H., H. E. Petschek, and G. L. Siscoe, Aerodynamic aspects of the magnetospheric flow, *Res. Rep. of the Office of Naval Research*, Washington, December 1963.
Lüst, Rhea, *Z. Astrophys.*, *51*, 163, 1961 (see also L. Biermann, 1951, 1953, 1957).

Ness, N. F., C. S. Scearce, and J. B. Seek, Initial results of the IMP I magnetic field experiment, preprint, Goddard Space Flight Center, April 1964.

Ness, N. F., and J. M. Wilcox, On the solar origin of the interplanetary magnetic field, preprint, Goddard Space Flight Center, August 1964.

Ness, N. F., Observation of the magnetohydrodynamic wake of the moon, preprint, Goddard Space Flight Center, September 1964.

Papers given at the first symposium of COPERS, Paris, June 1962, *Space Sci. Rev.*, *1*, 485.

Parker, E. N., *Interplanetary Dynamic Processes*, Interscience Publishers, Inc., New York, 1963.

Proc. of the Conference on the Solar Wind of the Jet Propulsion Laboratory, Pasadena, April 1964, in press.

Proc. of the Study Week on the Problem of Cosmic Radiation in Interplanetary Space, Rome, 1962.

Proc. of the Symposium on Plasma Space Science of the Catholic University of America, Washington, June 1963, in press.

Slee, O. B., *Monthly Notices Roy. Astron. Soc.*, *123*, 223-231, 1961.

Spreiter, J. R., and W. P. Jones, *J. Geophys. Res.*, *68*, 3555, 1963.

Whipple, F., The moon, meteorites, and comets, in *The Solar System*, vol. IV, edited by B. M. Middlehurst and G. P. Kuiper, Chap. 19, p. 639, University of Chicago Press, 1963.

# PART II

# ATMOSPHERIC MOTIONS

# LARGE-SCALE MOTIONS OF THE ATMOSPHERE: CIRCULATION

*Edward N. Lorenz*

*Massachusetts Institute of Technology, Cambridge, Massachusetts*

The atmosphere which surrounds our earth is in a state of continual motion. Energy received from the sun ultimately maintains this motion against the dissipative effects of friction. The atmosphere thereby offers a challenging collection of problems to the theoretician. Among these problems, a central one is the following: Given an atmosphere which surrounds a planet and is driven by heat received from a sun, to deduce the resulting global circulation within the atmosphere of the planet, on the basis of the governing physical laws. Other closely related problems are the deduction of specific features of the circulation from the governing laws, and the deduction of certain features of the circulation after taking for granted the existence or nature of certain other features.

On the surface these are straightforward problems in fluid dynamics. In problems of this sort the governing laws are generally expressed as a system of partial differential equations. The motion is inevitably influenced by the fields of pressure and density, and it is logical to regard these fields along with the motion as constituting the circulation.

Our central problem begins to appear more formidable when we express it in more detail. For our own atmosphere, we might state it somewhat like this. "Let there be given an atmosphere, which consists of a specified total mass of a gas of specified chemical composition, together with smaller variable amounts of certain impurities. Let the atmosphere surround an earth of specified mass, radius, and rate of rotation, which serves as both a source and a sink for some of the atmospheric impurities, and whose surface possesses mechanical and thermal properties with specified geographical distributions. Let the earth and its atmosphere receive energy from a sun whose total energy output and energy spectrum are specified. From the physical laws which govern the mechanical and thermodynamic state of the atmosphere and its environment, and the acquisition and removal and internal changes of phase of the impurities, we are to deduce the nature of resulting circulation."

Let us examine this problem for a while. Is it really possible to deduce the properties of the circulation from the laws which govern it? In fact, do the laws even determine the circulation?

There are many facets to these questions. I shall begin by enumerating some of the prominent features of the circulation which have been revealed by observations, and which an acceptable solution to the problem would have to contain.

Possibly the most noteworthy feature of the circulation is hydrostatic equilibrium—an approximate balance between gravity and the vertical pressure-gradient force. This balance demands that the density be proportional to the vertical derivative of pressure. As a result, the pressure and density fields are for practical purposes separate manifestations of the same field—the field of mass.

Nearly as prominent in middle and high latitudes is geostrophic equilibrium—an approximate balance between the Coriolis force and the horizontal pressure-gradient force. This balance, which is the chief feature distinguishing the motion of many rotating fluids from that of nonrotating fluids, demands that the flow be parallel to the isobars on horizontal surfaces, so that, in the northern hemisphere, the winds blow clockwise around high-pressure centers and counterclockwise around low-pressure centers. Further properties of geostrophic motion have been described in detail by Phillips [1963].

The hydrostatic and geostrophic equations may be combined to yield the familiar thermal wind equation, which states that the vertical shear of the wind is directed parallel to the lines of constant density, or, for practical purposes, to the isotherms, on horizontal surfaces.

One feature of the circulation which is very much in evidence is the presence of motions of vastly different scales. Organized circulation systems range all the way from intense high-level vortices covering a major fraction of a hemisphere, through the familiar upper-level waves and migratory cyclones and anticyclones, thunderstorms, and individual cumulus cloud circulations, to the smallest turbulent eddy occurring in the lee of a rock or some other obstacle.

If one takes the pains to analyze the total field of motion and the accompanying field of mass into components of various scales, perhaps by some scheme of harmonic analysis, one finds that the different scales possess different properties. Only the larger-scale motions are approximately geostrophic in their behavior, and the very smallest scales, taken by themselves, do not even satisfy the hydrostatic equation. The total field of motion is nearly geostrophic simply because the major fraction of the kinetic energy of the atmosphere is in the larger scales.

Although the smaller scales of motion sometimes appear rather chaotic, there are certain regularities to be found in the larger scales. One of the most familiar is the occurrence of the trade winds—the easterly winds which are found nearly all the time over the oceans in low latitudes. The famous paper of Hadley [1735] concerning the trade winds marks one of the earliest attempts to account for a major feature of the general circulation. Between the trade-wind belts of the two hemispheres is the intertropical convergence zone, a relatively narrow band in which much of the equatorial storminess is concentrated. Poleward from the trade-wind zones are regions where westerly winds are favored.

At higher levels in middle latitudes are the strong westerly winds, identifiable through the thermal wind equation with the decrease in temperature from low to high latitudes. Some twenty years ago, upper-level observations became plentiful enough to reveal that the westerly winds often culminate in the jet stream, a fairly narrow meandering belt in which the very strongest westerlies are concentrated. The general decrease of temperature from the ground upward, and the existence of a tropopause, above which there is a stratosphere where the temperature no longer decreases with elevation, must certainly be included among the outstanding regularities.

Although some of these large-scale features are virtually always present, it is common experience in most regions of the world that the weather changes from day to day, if not from hour to hour, so that we are not dealing with a steady-state atmosphere. Many of the variations are associated with changes in the positions and intensities of readily identifiable circulation systems, ranging from intense tropical hurricanes and extratropical storms down to minor small-scale disturbances.

If the atmosphere is not a steady-state system, it is just as certainly not a periodically varying system. To be sure, there are periodic components—notably the annual and diurnal periods—just as there is a long-term average state. But although investigators claim to have detected innumerable other periodic components in the observations, I am not aware of any serious claims today that the circulation consists entirely of a superposition of periodic oscillations.

A further significant quantity is the intensity of the circulation. The total atmospheric kinetic energy is roughly equal to the amount of solar energy intercepted by the earth in one hour.

If the statement of the problem and the complex combination of features which must be present in the solution fail to convince us that the problem may be rather difficult, we need only note that the governing system of equations is nonlinear. To many applied mathematicians, no further evidence is needed. The dominating nonlinear terms are those which represent advection—the displacement of features of the field of motion or mass by the motion itself. Since the field of motion is ordinarily not uniform, adjacent portions of any feature will receive unequal displacements, and the feature as a whole will be distorted as well as displaced.

Some of the difficulties may be alleviated by abandoning the exact form of the central problem, and replacing the system of equations governing the true atmosphere by an idealized system of equations. Let us examine some of the ways in which the equations have been modified in various studies of the general circulation.

A common simplification replaces the atmosphere by an ideal gas of uniform composition. Water in its gaseous, liquid, and solid phases is completely disregarded. A parallel simplification replaces the underlying earth by a level surface with uniform mechanical and thermal properties. Oceans and continents with their contrasting heat capacities, and mountains, hills, and smaller irregularities which act as mechanical obstacles, are omitted. Still another idealization regards the incoming solar radiation as a function of latitude only. Differences between summer and winter, and day and night, are eliminated.

There is no question but what the inhomogeneities and asymmetries of the atmosphere and its environment profoundly affect the circulation. Yet the uniform atmosphere moving over a uniform earth has probably received more theoretical attention. Not only is it less complicated, but to many theoreticians it is aesthetically far more pleasing, and, particularly for those whose principal interest is general fluid dynamics, more meaningful.

At this point I wish to digress and mention quite a different type of study. I refer to laboratory models of the atmosphere. The most familiar to the meteorologist are those of Fultz *et al.* [1959] and Hide

[1958]. Here a cylindrical or an annular vessel containing water is rotated about its axis of symmetry, which is vertical, and is heated near its outer radius and cooled near its center or inner radius. The resulting circulation in the liquid is then studied. Although the complete apparatus for the experiment has always been rather elaborate, the actual container in some of the earlier experiments was an ordinary dishpan, and the experiments are probably still best known as the "dishpan experiments."

The circulation in the dishpan may be regarded as another idealization of the atmosphere. Within the limits of experimental control it is the circulation of a uniform fluid over a uniform surface, driven by a heat source which varies only with the distance from the axis. Theoreticians who favor aesthetically pleasing problems but still prefer to study real physical systems have found that the dishpan answers their needs. Outstanding studies are those of Davies [1953, 1956].

The idealizations which we have so far discussed, including the dishpan, all replace the real atmosphere by some physically conceivable system. To be sure, we do not expect to encounter in nature a rotating planet without days and nights, but, in principle at least, we can picture such astronomical monstrosities as a ring-shaped sun with a cold rotating planet at the center. By contrast, many additional common idealizations lead to systems of equations which do not describe realizable physical systems.

One of the most familiar of these is the substitution of the exact hydrostatic equation for the vertical equation of motion. Mathematically this is a major simplification. The new system describes an atmosphere which is incapable of propagating sound waves. Although it is unrealistic because real fluids do propagate sound, it is very practical because sound waves do not seem to interact appreciably with motions of meteorological interest.

A similar idealization is the substitution of a form of the geostrophic equation for the equation expressing the time derivative of horizontal velocity divergence. The new system is incapable of propagating gravity waves [Charney, 1948]. These waves appear not to interact appreciably with motions of meteorological interest, but here the lack of interaction is less certain.

One further idealization is one of the most familiar. The spherical surface of the earth is replaced by a plane, sometimes of limited horizontal extent. The Coriolis parameter is assumed constant, except in those terms where its derivative with respect to latitude appears; this derivative is assigned another constant value, ordinarily denoted by $\beta$. The resulting surface is the familiar beta plane, first introduced by Rossby [1939]. Physically one would be hard put to construct a beta plane, yet the beta plane has enjoyed such constant popularity that it seems to have acquired all the status of a physical reality in the meteorological world.

Using some or all of these idealizations, let us see how we might now attack our problem. One straightforward procedure involves expanding each dependent variable in a power series in a parameter $\epsilon$, which characterizes the intensity of the thermal forcing. Since there will be no motion in a dissipative system without forcing, the constant terms in the series for the motion will vanish, and the remaining terms may be found in turn by solving linear equations. It can be shown that the series converge for small values of $\epsilon$, although not necessarily for those values of greatest interest.

It turns out, however, that each term in the series, and hence the solution itself, is invariant with time if the heating is also invariant with time. Small-scale horizontal structure tends to be absent if it is also absent in the underlying surface, although the boundary layer will possess rapid variations in the vertical direction. What we shall have deduced, then, is a steady-state atmosphere devoid of small-scale circulations.

Why does this solution so completely fail to conform with the observations? Have our idealizations created an atmosphere which is incapable of oscillating with time or maintaining a small-scale structure, except perhaps temporarily? The occurrence of nonperiodic motion in the dishpan, which is a similarly idealized atmosphere, indicates that this is not the case. Apparently the solution would be realistic for very weak heating, but for values of $\epsilon$ representative of the atmosphere it is unstable with respect to further modes of oscillation, which possess small-scale structures and oscillate nonperiodically. If the steady circulation could be established, and then disturbed slightly, the disturbances would grow until they became dominating features. In the process of growing, the new modes would acquire their energy from the already established modes, and the transfer of energy from one mode to another would be expressed by the nonlinear terms in the governing equations.

We find, then, that the idealizations which we have so far introduced do not overcome the basic difficulties which are associated with the small-scale features and the nonperiodicity of the variations, which in turn result from nonlinearity. Let us try to assess the full effect of nonlinearity upon the problem of deducing the general circulation. If the circulation were steady, we could hope to find this steady circulation analytically. Even if it were periodic, we could hope to solve for a complete cycle. But nonlinearity has rendered the circulation nonperiodic. The equations therefore possess an infinite number of distinct time-dependent solutions, only one of which can duplicate the observed history of the atmosphere, and the probability of selecting this particular solution by chance is zero. If we are to deduce the circulation from the governing laws alone, without using the observed circulation at some special time as a basis for choosing a particular solution, the most that we can hope to do is to duplicate the set of characteristic features of the circulation, which we may call the global climate.

Even this task may not be feasible; there are some systems of equations with the property of intransitivity. For a transitive system, almost all solutions possess the same climate, or set of statistics, but, for an intransitive system, there are two or more climates which a randomly chosen solution has a positive probability of possessing. Whichever climate becomes established will persist forever. There are also real physical systems which are intransitive; under certain conditions the dishpan is such a system.

There is no general rule for determining whether a given system of equations is intransitive, and we must remain in the dark concerning the atmospheric equations. Observations give us no further enlightenment; the atmosphere is a one-shot experiment, and we cannot turn it off and then start it again to see whether another climate develops. Should the atmosphere be intransitive, the most that we can hope to do is to deduce the various different global climates, one of which should conform with observations.

Having decided to find only the statistical properties of the circulation, we might attempt to derive a new set of equations whose dependent variables are the statistics. But the same nonlinearity which has led us to seek only the climate also renders it impossible to obtain a closed system of equations with the statistics as variables. Every effort to complete the system by deriving an additional equation inevitably introduces a new statistic. The only alternative procedure seems to be to return to time-dependent solutions, even though we may not care about them for their own sake, and then compile statistics from them. These statistics will be derived from finite samples, and perhaps not be representative, but this is true also of statistics derived from observations.

We have not finished complaining about nonlinearity. Because the solutions of interest are nonperiodic, we cannot express them in terms of familiar analytic functions. Our one remaining method of solving the equations is by numerical means. To solve partial differential equations numerically, we must first replace the instantaneous field of each dependent variable by a finite set of numbers. Ordinarily these numbers are the values of the variables at a chosen grid of points, although they may be the coefficients occurring in the expansion of the field in a series of orthogonal functions. In essence we have introduced a further idealization.

But now the effect of small-scale features, another product of nonlinearity, comes into play. It is clearly not economically feasible to choose enough grid points or orthogonal functions to give even a remotely adequate picture of the small-scale features. How many cumulus clouds, for example, are in the sky at any one time? We must therefore omit any description of the small-scale features, and deduce the characteristics of only the large-scale motions of the atmosphere.

However, it behooves us not to omit the effects of the small-scale features upon the large-scale motions which we are studying. We may incorporate these effects through the introduction of coefficients of eddy viscosity and eddy conductivity.

It is common experience that we do not possess appropriate numerical values for these coefficients, even to within a factor of 2. The most certain thing we know about them is that they are not constant, so that the use of constant coefficients is a further idealization. Our inadequate knowledge stems from the lack of an adequate theory of the smaller scales of motion; this in turn is partly due to nonlinearity.

Following these modifications which nonlinearity has demanded, further idealizations may be in order. With small-scale features eliminated, the remaining large-scale circulation generally possesses a far less detailed structure in the vertical direction than in the horizontal. It has therefore proven feasible to replace the entire three-dimensional field of motion by two two-dimensional fields—one in each of two layers. If the geostrophic approximation is also used, a single two-dimensional temperature field may be identified with the difference between the two fields of motion.

Many of these idealizations were originally devised for the purpose of dynamical weather forecasting. Phillips [1956] was the first to apply the methods of dynamical forecasting to the problem of deducing the general circulation. His system of equations contains all of the idealizations which we have enumerated.

Still more recent is the common use of low-order models. Here the number of degrees of freedom, that is, the number of numerical values which must be specified to describe an instantaneous state of the system, is reduced to the point where only the very largest scales of motion remain. These models are so simple that one may actually inspect the few columns of numbers representing a time-dependent solution and get some feeling for what is taking place. They are especially useful for determining just what physical features must be retained in order that specific climatic features may be reproduced [Lorenz, 1962, 1963b].

Our latest idealizations all leave us with systems of equations which are readily handled by digital computing machines. In retrospect, it appears that numerical methods afford the only method by which the central problem can be solved, and, even then, it can be solved only when the system of equations is idealized. It is always dangerous to claim that a task is impossible, unless, like the trisection of the angle with the compass and straightedge, it can be proven to be impossible, but the writer's [1964] experience with the very simplest of nonlinear equations, a first-order quadratic difference equation in one variable, strongly indicates that the statistics of nonperiodic solutions cannot in general be found except by numerical means.

Let us recall, then, that dynamic meteorology was a well-developed discipline long before digital computers were dreamed of by most meteorologists. When Richardson [1922] foresaw the power of numerical methods nearly half a century ago, he visualized a team of 64,000 men working together, rather than a computing machine. Even today many of the most capable meteorologists have little access to machines. What, then, were the problems which dynamic meteorologists of earlier generations could hope to attack successfully?

We have already alluded to some of these. First there are problems of deducing specific features of the general circulation. It might appear illogical that we could deduce part of the circulation while neglecting the remainder, when all parts are interrelated. Probably we cannot deduce exact numerical values of any feature without implicitly considering all features, but we can certainly deduce upper and lower bounds.

Take, for example, the problem of explaining why the equatorial regions are warmer than the polar regions, as opposed to the determination of exact temperatures in these regions. Perhaps everybody knows the answer; is it not simply that there is more solar heating in the equatorial regions? The problem may seem too simple to merit further consideration, until we recall that at certain elevations, notably the lower stratosphere, the equatorial regions are ordinarily colder than the polar regions. Perhaps the problem is not so trivial after all.

The problem can be rigorously treated within the framework of a two-level model, such as the one used by Phillips [1956]. Here the variance of temperature serves as available potential energy, the only direct source for kinetic energy. Kinetic energy is continually being destroyed by friction, and must be continually replaced at the expense of the temperature variance. The outgoing radiation, if not regarded as a constant, is taken as an increasing function of temperature, and thereby further serves to decrease the temperature variance. Incoming radiation must then maintain the temperature variance in the face of these other processes, and it can do this only by heating the warmer regions more than the cooler regions. But incoming radiation in this model depends upon latitude alone; it follows that by and large the warmer regions are the low latitudes, and the cooler regions are the high latitudes.

In the real atmosphere the poleward temperature decrease probably occurs for the same basic reasons, but the arguments as we have presented them are no longer rigorous. For example, the incoming radiation, excluding that part reflected back to space, is no longer a function of latitude alone, so that a correlation between temperature and heating no longer demands a correlation between temperature and latitude. To make our arguments rigorous, we probably must incorporate the fact that the effect of the asymmetries does not exceed some critical value. I believe that the argument can be made rigorous for the real atmosphere. I have not seen this done, possibly because the problem has seemed too trivial at first glance.

Similar methods, involving careful consideration of the asymmetries, can perhaps be used to establish the necessity for the existence of other qualitative features of the circulation—possibly the trade winds or the jet stream. I do not know of any such studies where completely rigorous proofs have been offered. Here is a wide open field of research.

There are also problems of deducing certain features when other features are taken for granted. Problems of this sort have perhaps occupied the major efforts of those theoreticians who prefer analytic to numerical procedures. Here alone among the problems we have mentioned is the opportunity to work with linear equations, whose mathematical theory has been so highly developed.

The most familiar examples are problems involving the stability of a steady field of flow. The equations may be idealized to the point of neglecting viscous dissipation and thermal forcing, since it is no longer necessary to explain how the basic flow originated.

Numerous investigators [e.g., Kuo, 1952] have obtained the result that patterns containing six, seven, or eight waves around the circumference of the globe will grow more rapidly than any other modes of oscillation, when superposed upon basic flow patterns resembling those found in the atmosphere. The additional hypothesis [Eady, 1949] that those wave patterns which grow most rapidly when small are those which are favored to remain after reaching finite amplitude affords an explanation for the observed prevalence of waves of this scale.

Somewhat different is a recent study of Saltzman [1963], who assumed statistical properties of the waves and deduced with fair accuracy the observed distribution of the zonally averaged component of the circulation. Perhaps it may be possible to combine this sort of study with the studies of stability, and solve simultaneously for the basic flow and the superposed waves.

Let us now return to the central problem, and observe what success has been attained in deducing the climate compatible with various idealizations. We begin with the simplest model of a forced dissipative system which has been studied, and, so far as I know, the only one whose general nontransient solution has been found analytically [Lorenz, 1962]. It is a low-order model, with 8 degrees of freedom. From our previous remarks, we can already infer than the general solution must be periodic.

Even in this simple model there are upper-level westerly winds, and also cyclones and anticyclones and upper-level waves when the flow is unsteady. Moreover, as in the dishpan experiments, the rate of rotation and the intensity of heating determine whether a steady or an unsteady flow will occur. The model by its very construction is incapable of maintaining trade winds and surface westerlies. Two somewhat less restricted models, one with 14 and one with 28 degrees of freedom, oscillate nonperiodically under suitable conditions [Lorenz, 1963$b$, 1965].

The prototype of general-circulation models, that of Phillips [1956], has 480 degrees of freedom. The features which develop in the low-order models occur here with more realistic structures, and, in addition, trade winds and middle-latitude westerlies occur.

The most recent and least highly idealized general-circulation models are those of Leith [1965], Mintz [1965], and Smagorinsky [1963, 1965]. Each of these models has several thousand degrees of freedom. None of them uses the geostrophic approximation, and all of them have oceans and continents. Leith's atmosphere contains water vapor, while Mintz's earth has mountains. Together, the models show most of the principal features of the large-scale motions, except the prevalence of tropical storms. Smagorinsky's model, with the most detailed vertical structure, exhibits a fairly realistic tropopause.

We have therefore nearly solved the problem of deducing the large-scale motions of the atmosphere from the governing physical principles, within the framework of the idealizations which we have introduced. Some of the principal remaining deficiencies, such as the absence of tropical hurricanes and their influence upon the larger features, can possibly be attributed to faulty parameterization of the small-scale processes, notably cumulus convection. If, a few years from now, we succeed in properly reproducing all of the principal features, we shall, in the opinion of some meteorologists, have attained the Ultima Thule in the study of the general circulation.

There are others who would feel that by merely duplicating the circulation as we have observed it we have done little more than verify the validity of the equations. To appreciate this point of view more fully, let us note that, in a certain sense, the dishpan may be regarded as an analog computing machine, solving its own idealization of the equations governing the atmosphere. To someone who seriously asks, "Why does the atmosphere have trade winds?", the answer, "because the dishpan has trade winds," would have all the earmarks of an elephant joke.

But is the answer, "because the computing machine says so," or even the more informative answer, "because trade winds occur in all the known solutions of systems of equations which closely resemble the atmospheric equations," any more satisfactory? Undoubtedly it is the desire to discover *why* some phenomenon occurs, rather than merely showing that it must occur, which has led many theoreticians to avoid purely numerical problems. As we have noted, there is no shortage of worthy problems of an analytic character available to them.

A little reflection will reveal that there really is much to be learned about the atmosphere from numerical computations, regardless of whether they answer the questions which we consider most fundamental. A particularly welcome feature is the opportunity to perform controlled experiments upon the atmosphere in simulated form, with no risk of rendering the earth unfit for further habitation. We might even learn whether the atmosphere is transitive. I should like to conclude by describing a series of experiments of this sort which is just now bearing fruit.

These experiments are concerned with the extent to which the future state of the atmosphere, or the future weather, can be predicted. Some of the factors involved have been discussed by Thompson [1957], and more recently by the writer [1963a]. A mathematical proof that a stated task, such as the trisection of an angle, is impossible constitutes an acceptable solution to a problem, and the possibility of proving that the weather cannot be predicted has appealed to some theoretical meteorologists whose own attempts to forecast the weather have met with less than complete success. We shall not be so much concerned with the philosophical question of determinism, or whether the atmosphere has decided what to do, as with whether it has signaled its intentions to us.

We begin with the now familiar observation that the atmosphere is nonperiodic, except for the annual and diurnal periods and possibly a few minor periodic components. It can be proven that the nonperiodic component of a system varying with periodic and nonperiodic components is completely unpredictable in the far distant future, unless its present state or some past state is known exactly.

In brief, a nonperiodic system is essentially one which never does the same thing twice; any approximate repetition of its previous behavior must be of finite duration. It therefore cannot be predicted to do the same thing twice by any acceptable forecasting scheme. But if there are uncertainties in observing the system, a state which is indistinguishable from some previous state will eventually occur, and no forecasting scheme will have any rational basis for predicting that the system will not do the same thing twice.

There are always appreciable uncertainties in observing the state of the atmosphere, due not so much to faulty observation as to the impossibility of observing the weather everywhere. Even in the United States and Europe, where the observing weather stations are most closely spaced, a system as large as a thunderstorm occurring between stations can go completely unnoticed. Over most of the ocean, except in the principal shipping lanes, there are often gaps in the observations large enough to conceal a fully developed hurricane. It follows that there will be a finite limit to how far into the future we can predict, as long as these conditions remain.

Of course, our observations do not eliminate the possibility that the atmosphere is periodic, with a period longer than the duration of recorded history. However, the established nonperiodicity of some of the simpler idealized systems suggests that this is not the case.

The theory which so neatly asserts the ultimate unpredictability of the atmosphere does not tell us how far into the future we can predict. The rate at which initial errors will amplify may be estimated through numerical experiments.

The first experiment [Lorenz, 1965] was based on the low-order model with 28 degrees of freedom. Here the growth rate of errors varied as greatly as the weather situation itself, but, on the average, errors doubled in about four days.

Slightly more encouraging news has just been communicated by Mintz, following experiments with his system of equations with many thousand degrees of freedom. He finds that on the average about five days are required for the size of the errors to double. The doubling period for the real atmosphere, whatever its length may be, is yet another characteristic feature of the general circulation.

If it really requires as long as five days for typical errors to double, moderately good forecasts as much as two weeks in advance may some day become a reality. If it requires no longer than five days for errors to double, accurate detailed forecasts for a particular day a month or more in advance belong in the realm of science fiction.

## References

Charney, J. G., On the scale of atmospheric motions, *Geofys. Publikasjoner Norske Videnskaps-Acad. Oslo*, *17*, 2, 1948.

Davies, T. V., The forced flow of a rotating viscous liquid which is heated from below, *Phil. Trans. Roy. Soc. London*, A, *246*, 81–112, 1953.

Davies, T. V., The forced flow due to heating of a rotating liquid, *Phil. Trans. Roy. Soc. London*, A, *249*, 27–64, 1956.

Eady, E. T., Long waves and cyclone waves, *Tellus*, *1*, 35–52, 1949.

Fultz, D., R. R. Long, G. V. Owens, W. Bohan, R. Kaylor, and J. Weil, *Studies of thermal convection in a rotating cylinder with some implications for large-scale atmospheric motions*, Meteorol. Monographs, 1959.

Hadley, G., Concerning the cause of the general trade winds, *Phil. Trans. Roy. Soc.*, *39*, 58, 1735.

Hide, R., An experimental study of thermal convection in a rotating liquid, *Phil. Trans. Roy. Soc. London*, A, *250*, 441–478, 1958.

Kuo, H. L., Three-dimensional disturbances in a baroclinic zonal current, *J. Meteorol.*, *9*, 260–278, 1952.

Leith, C. E., Lagrangian advection in an atmospheric model, *Proc. Sympos. Long-Range Forecasting*, World Meteorol. Org., Tech. Note 66, 168–176, 1965.

Lorenz, E. N., Simplified dynamic equations applied to the rotating-basin experiments, *J. Atmospheric Sci.*, *19*, 39–51, 1962.

Lorenz, E. N., The predictability of hydrodynamic flow, *Trans. N. Y. Acad. Sci.*, *25*, 409–432, 1963a.

Lorenz, E. N., The mechanics of vacillation, *J. Atmospheric Sci.*, *20*, 448–464, 1963b.

Lorenz, E. N., The problem of deducing the climate from the governing dynamic equations, *Tellus*, *20*, 1–11, 1964.

Lorenz, E. N., A study of the predictability of a 28-variable atmospheric model, *Tellus*, *21*, in press, 1965.

Mintz, Y., Very long-term global integration of the primitive equations of atmospheric motion, *Proc. Sympos. Long-Range Forecasting*, World Meteorol. Org., Tech. Note 66, 141–167, 1965.

Phillips, N. A., The general circulation of the atmosphere: a numerical experiment, *Quart. J. Roy. Meteorol. Soc.*, *82*, 123–164, 1956.

Phillips, N. A., Geostrophic motion, *Rev. Geophys.*, *1*, 123–176, 1963.

Richardson, L. F., *Weather prediction by numerical process*, Cambridge University Press, 1922.

Rossby, C. G., and collaborators, Relation between variations in the intensity of the zonal circulation of the atmosphere and the displacements of the semi-permanent centers of action, *J. Marine Res.*, *2*, 38–55, 1939.

Saltzman, B., A generalized solution for the large-scale, time-average perturbations in the atmosphere, *J. Atmospheric Sci.*, *20*, 226–235, 1963.

Smagorinsky, J., General circulation experiments with the primitive equation. I. The basic experiment, *Monthly Weather Rev.*, *91*, 99–164, 1963.

Smagorinsky, J., Implications of dynamical modeling of the general circulation on long-range forecasting, *Proc. Sympos. Long-Range Forecasting*, World Meteorol. Org., Tech. Note 66, 131–137, 1965.

Thompson, P. D., Uncertainty of initial state as a factor in the predictability of large scale atmospheric flow patterns, *Tellus*, *9*, 275–295, 1957.

# MOTIONS OF INTERMEDIATE SCALE: FRONTS AND CYCLONES

*Arnt Eliassen*

University of Oslo, Blindern, Norway

We know that the changing wind and weather typical of middle and high latitudes of our globe are connected with a restless motion of tropospheric fronts and cyclones. To be able to understand and predict the behavior of these features has for a long time been a challenge to meteorologists, and still is. We certainly cannot yet claim to have a complete theory of fronts and cyclones, but considerable progress has been made in recent years, and I would like to add that the scientists of Massachusetts Institute of Technology have contributed brilliantly to this progress.

In this chapter I would like to discuss some recent advances in the theory of fronts and cyclones. The subject is tremendous, and I shall not even try to cover everything, but rather give a subjective account of what I personally feel is particularly important.

As a Norwegian, I may perhaps be excused for taking as my starting point the Norwegian cyclone theory, which came into being at the end of the First World War in Bergen. Vilhelm Bjerknes [1933] has given an interesting account of his earlier views upon the cyclone problem, before the Bergen school. Influenced by Helmholtz, he arrived at the conclusion that cyclones in their early stage of development must be tractable by means of linear perturbation theory. He therefore began a systematic study of the solutions of the linear perturbation equations, hoping to find a growing cyclonic vortex; but instead he found that such solutions would invariably describe some sort of wave motion. From this he concluded that cyclones must originate as growing waves of some kind. He says further that he had no idea of how the transition from this as yet unobserved wave into a vortex should be visualized.

Then came Jacob Bjerknes' [1919] frontal cyclone model (Figure 1)

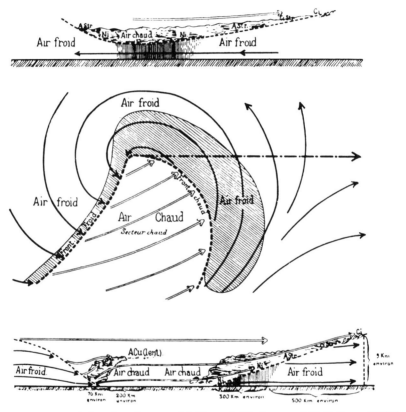

FIG. 1. J. Bjerknes' cyclone model. From J. Bjerknes and H. Solberg [1922].

and the discovery by J. Bjerknes and Solberg [1922] that new cyclones may form in the crest of growing waves on a preexisting polar front [Figure 2]. The cyclone wave grows in amplitude, occludes, and ends as an almost symmetric vortex. Vilhelm Bjerknes could conclude that his missing mysterious wave motion had been found empirically.

It was an essential feature of the Norwegian cyclone theory that the fronts were considered as the primary, and the cyclones as secondary

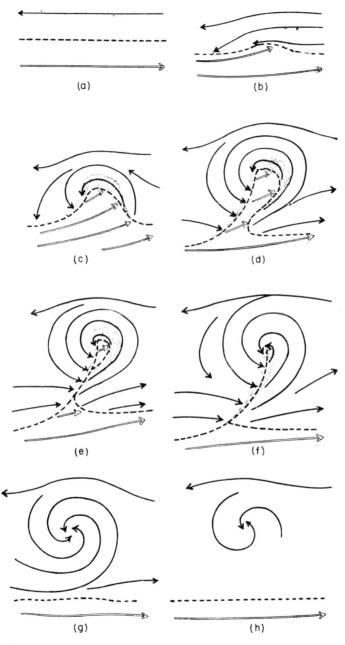

Fig. 2. Formation and life cyclus of a frontal cyclone. From J. Bjerknes and H. Solberg [1922].

formations. The front was the primus motor, which gave birth to the cyclones and was, in addition, the carrier of the important cloud and precipitation systems.

The next step would be to supply the theory of the growing frontal waves. This work was taken up by Solberg [1928, 1933]; he studied the solutions of the linear perturbation equations for the particular basic state of two statically stable air masses of different density, in relative sliding motion, separated by an idealized, sloping frontal surface on a flat, rotating earth. Both air masses were taken to be barotropic so that the baroclinicity was wholly concentrated in the frontal surface.

The problem turned out to be mathematically difficult, especially because of the awkward geometry of the sloping interface intersecting

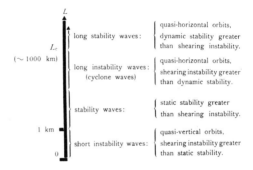

Fig. 3. Wavelength regions of stability and instability. From J. Bjerknes and C. L. Godske [1936].

the ground. Solberg bypassed this difficulty by considering instead a two-layer system bounded by two rigid planes parallel to the undisturbed front. For this system he found, besides short amplifying waves of the Helmholtz type, several types of long amplifying waves with wavelength of the order of 1000 km. Their existence seemed to depend upon the presence of some sort of internal stability (gravitational or rotational, or both) in each of the two fluid layers. One of these wave types seemed to have several kinematic features in common with a young cyclone wave.

The interpretation of Solberg's pioneering work is not easy. J. Bjerknes and Godske [1936] proposed to interpret the results as a combined effect of Helmholtz instability, static stability of the interface, and rotational stability (Figure 3). For very short waves, Helmholtz instability dominates; for somewhat longer wavelengths, the Helmholtz instability is suppressed by the static stability of the interface, and the waves are neutral. For still longer waves, however, the particle orbits become tilted to an almost horizontal orientation as a result of the earth's rotation; consequently the effect of the static stability is reduced, and the Helmholtz instability again dominates, resulting in Solberg's long amplifying waves. The very longest waves are again stable because of the rotational stability.

This interpretation seems questionable, however, since Solberg [1933, Section 109] demonstrated that long amplifying waves would occur even without rotation, provided that the layers had internal static stability. This result has been verified and discussed by Höiland [1948].

The problem of the frontal waves was also attacked by Kotchin [1932] in the case when both fluid layers are homogeneous and incompressible. Using the quasi-static approximation, Kotchin was able to satisfy the boundary conditions for a horizontal lower boundary as well as an upper horizontal lid (Figure 4). He found the system to be

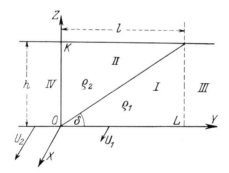

FIG. 4. Kotchin's [1932] two-layer system with a sloping front.

unstable and that for a sufficiently high density contrast, amplification would take place for long waves as well as for short waves, whereas waves in an intermediate band of wavelengths, with Rossby number of the order of unity, would not grow.

When the amount of aerological data increased so that analyses could be made at higher levels, it became apparent that the idealized notion of a sharp front, separating almost barotrophic air masses on both sides, was too restrictive. Instead one got the picture of a westerly current increasing with height with the baroclinicity distributed in a broad band. Small frontal cyclones are hardly detectable in the upper troposphere, whereas larger frontal cyclones or cyclone families manifest themselves as waves in the upper current. The shorter cyclone waves seem to be damped with height, while the longer waves, with wavelengths of the order of 5000 km, are accentuated in the upper troposphere. The study of these long waves in the upper westerlies was taken up in particular by J. Bjerknes and Rossby. Rossby [1939] even ignored the baroclinicity altogether and thus arrived at the simple two-dimensional nondivergent, or barotropic, model of atmospheric motion. By applying this model to waves on a uniform zonal current, Rossby derived his famous wave formula; however, these waves are always stable. Also Blinova [1943] developed a theory of the atmospheric disturbances on this basis.

The basis of our present-day ideas on the cyclone problem are three important theoretical advances made shortly after the Second World War. These are the introduction of the quasi-geostrophic method and the discovery of the two different mechanisms by which disturbances of cyclone scale can grow, namely, barotropic and baroclinic instability.

The quasi-geostrophic equations were formulated, first of all, by Charney [1947, 1948] and Obukhov [1949]. In their most compact formulation, these equations reduce to a single prognostic equation, expressing the conservation of potential vorticity; from this quantity, all other variables may be derived, including the vertical motion and the divergent part of the horizontal velocity field. The quasi-geostrophic equations are valid only for motions of large and intermediate scales. They describe the low-frequency motions in which we are primarily interested while eliminating irrelevant wave motions of high frequency, and have played an important role in the recent development of dynamic meteorology.

The *barotropic* instability was derived by Kuo [1949] and Fjørtoft [1950] by applying the barotropic model to a zonal basic current with horizontal shear. This is really an adaptation to atmospheric conditions of the classical problem of the stability of a linear current in an ideal fluid, dealt with by Rayleigh and generalized by Fjørtoft [1950] and Höiland [1953]. In the linear current, a necessary condition of instability is that the shear of the current must somewhere have a maximum, and a corresponding condition holds for a zonal current in a barotrophic atmosphere. Such an absolute vorticity maximum may be present on the poleward side of a strong west-wind belt. The wave growth is connected with a redistribution of the preexisting vorticity of the basic current so that vortices form. The energy of the growing waves is taken from the available *kinetic* energy of the basic current; therefore, the eddy flux of angular momentum executed by such waves

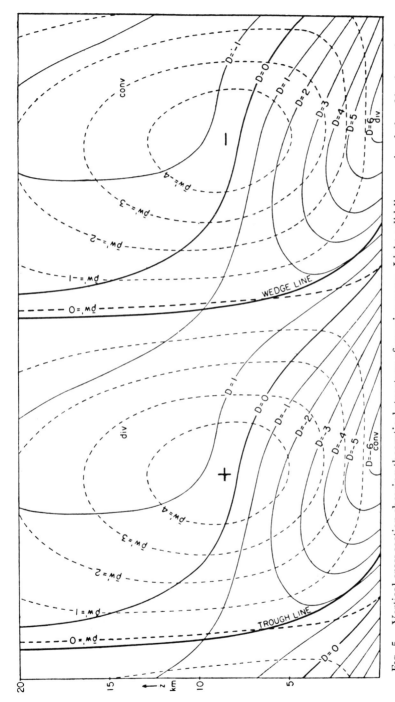

FIG. 5. Vertical cross section showing theoretical structure of growing wave. Light solid lines are isopleths of horizontal mass divergence (in $10^{-8}$ gm cm$^{-3}$ sec$^{-1}$); and dashed lines, isopleths of vertical mass flux density (in $10^{-3}$ gm cm$^{-2}$ sec$^{-1}$). From Charney [1947].

is directed away from the maximum west wind, tending to smooth out the angular velocity of the basic current.

The discovery of the *baroclinic* instability is due to Charney [1947] and Eady [1949]. They considered a stably stratified basic current with a horizontal temperature gradient, but without horizontal shear. They found amplifying waves with wavelengths of the same magnitude as cyclone waves. These waves take their energy from the available *potential* energy of the basic flow; but the way in which this is accomplished is rather subtle.

The qualitative agreement between the theoretical structure of these wave solutions (Figure 5) and the observed structure of large cyclone waves is impressive. The trough- and wedge-lines tilt westward with height. Ahead of the wave trough, a warm current moves North while it ascends, and in its rear cold air moves South and sinks. These vertical motions are so slow that the trajectories slope more gently than the isentropic surfaces; consequently, the warm air moving North remains warm in spite of its ascent, and the cold air moving South remains cold even though it sinks. The motion may be described, as Eady does, as quasi-horizontal overturnings in planes tilted slightly upward toward the North; in these overturnings, cold air sinks and warm air ascends, thus releasing potential energy. This instability mechanism gives maximum amplification for waves in a range of wavelengths of the order 4 to 5000 km, which is the size of the long waves in the westerlies, or large cyclonic systems.

There is a disagreement between the theories of Charney and Eady with respect to the behavior of relatively short waves. In contrast to Charney, Eady found that waves shorter than a certain critical wavelength would not grow because of the effect of the static stability. To my knowledge, this contradiction has not been clarified, but it may turn out to be just an apparent disagreement between "no growth" and "very slow growth."

Charney, on the other hand, using Rossby's $\beta$-plane approximation, obtained a maximum wavelength for growing waves, longer waves being stabilized due to the monotonic increase of the Coriolis parameter with latitude. Green [1960] and Burger [1962] have shown this to be incorrect and that also longer waves will amplify. Their results are mathematically interesting, but may not be relevant to the motions in the atmosphere, since the use of the $\beta$-plane approximation to such long waves, larger than the earth's radius, is not justified, as Phillips [1963] has pointed out.

The theories of barotropic and baroclinic instability previously referred to are still too restrictive for direct application to the atmos-

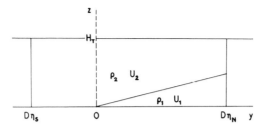

FIG. 6. Eliasen's [1960] two-layer system.

phere, since the one ignores baroclinicity and the other the horizontal wind shear. The atmospheric westerlies possess plenty of both, so that there is both available potential and available kinetic energy present which can be exchanged with perturbation energy. What is needed is therefore a combination of both theories.

There is not much hope that such a combined perturbation theory can be worked out analytically in the general case, because the equation for the amplitude in the meridional plane is nonseparable, as has been pointed out by Pedlosky [1964]. However, one special case of this kind has been solved and discussed in considerable detail by Erik Eliasen [1960], namely, the classical problem of wave disturbances on an idealized frontal surface. Eliasen takes a basic current consisting of two homogeneous and incompressible layers of different density and different zonal velocity, separated by a sloping zonal interface. This system possesses horizontal shear as well as horizontal density contrast, but these are both concentrated in the front. The system is bounded by two horizontal planes, and there is a northern boundary wall placed where the two layers are of equal thickness; the southern boundary

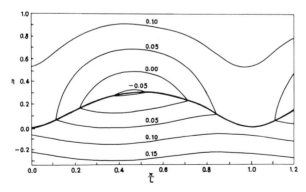

FIG. 7. Theoretical isobars and front at a level near the lower boundary in a growing frontal wave, according to E. Eliasen [1960].

was in the calculations placed at infinite distance (Figure 6). The equations were the same as used by Kotchin. The Coriolis parameter was considered constant.

Under these circumstances, waves longer than a certain minimum wavelength were found to grow, and there is a certain wavelength of the order 2 to 3000 km which gives a maximum growth rate corresponding to a doubling time of the order of one day. The calculated structure of the growing wave is in excellent agreement with the young frontal cyclone (Figure 7).

Eliasen also calculated the energy transformations for the growing waves, and found that the waves receive energy from the potential as

FIG. 8. Rotating tank experiment showing frontal waves in a system consisting of two fluid layers of slightly different density. From Fultz [1952].

well as the kinetic energy of the zonal flow. His numerical values for a particular case show that two-thirds of the wave energy is taken from the potential and one-third from the kinetic energy of the mean flow. Thus the system may be characterized as barotropically and baroclinically unstable.

In retrospect, it is likely that the instability found by Kotchin in 1932 is of this same kind. Solberg's instability is different, however; as pointed out by Eliasen, Solberg's growing waves take their energy entirely from the kinetic energy of the mean flow.

Fultz [1952] has produced beautiful frontal waves in a rotating tank, using two liquids of different density (Figure 8).

Recently Charney and Stern [1962] have discussed the stability of a baroclinic current with horizontal shear. Under the condition that the horizontal ground is an isentropic surface, they were able to prove that the current is stable with respect to quasi-geostrophic motions if the potential vorticity changes monotonically with latitude along isentropic surfaces. Thus a necessary criterion would be that the isentropic gradient of potential vorticity does not have the same sign everywhere. This criterion has been generalized by Pedlosky [1964] to the case where the isentropes intersect the ground. Because of its generality, Pedlosky's result forms a common viewpoint from which all the various stability theories can be judged, and I believe that it will prove very valuable as a guide in future work.

In 1951, Charney made a study of perturbations on a basic current with a jetlike structure, resembling the atmospheric westerlies. From a study of instantaneous energy variations he came to the conclusion that the wave disturbances receive energy from the potential energy of the mean flow, but surrender part of this energy to the kinetic energy of the mean flow.

As noted by Kuo [1951], this conclusion was definitely confirmed by the calculations from actual data of the meridional eddy flux of angular momentum, made by Starr and his coworkers [1954] and by J. Bjerknes and Mintz [1955]. These calculations showed that the disturbances transport angular momentum from latitudes of weak westerlies into latitudes of strong westerlies, thus increasing the kinetic energy of the mean zonal current.

It should be borne in mind that this result applies to the exchange of energy between the mean zonal current, obtained by averaging over all longitudes, and the observed fully developed system of disturbances. It does not follow, in my opinion, that individual cyclonic disturbances always grow as a result of baroclinic instability and not as a result of barotropic instability. The mean zonal current is a mathematical fiction; it does not really exist and therefore no small waves can grow on it.

If the instability theory is to be applied to the growth of individual disturbances, the basic current should be defined locally as the current prevailing over a limited area. The stability properties of such locally defined currents may differ considerably from case to case, and looked upon in this way, it is quite possible also that barotropic instability in some cases may contribute significantly to the formation of disturbances. One might suspect that short and shallow cyclone waves grow in accordance with Erik Eliasen's and Kotchin's theory of frontal instability, while the large and deep cyclonic systems grow primarily by baroclinic instability.

The linear theory seems to be able to explain why cyclones exist at all and has also given us important insight into the physical processes

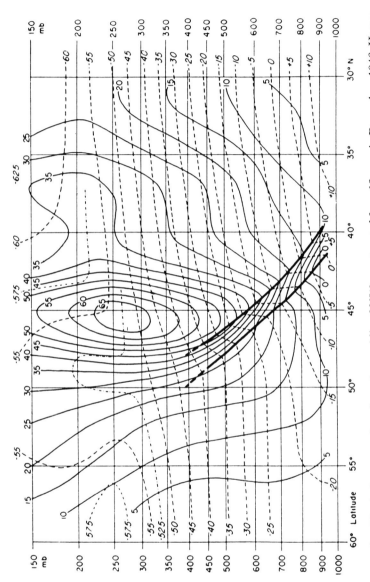

Fig. 9. Vertical cross section along the meridian 80°W, constructed from 12 cases in December, 1946. Heavy lines: frontal boundaries. Thin solid lines: isopleths of westerly wind component (in m/sec upright numbers). Thin dashed lines: isotherms (°C, slanting numbers). From Palmén and Newton [1948].

operating in cyclones, but it is highly inadequate for dealing with the fully developed system of disturbances in the atmosphere. These are truly nonlinear and can be treated only by numerical methods.

The concept of fronts has changed considerably since its introduction into meteorology by the Bergen school. They were for a long time considered merely as abrupt layers of transition between air masses of different density. The frontal cloud and precipitation systems were assumed to be due to an upgliding motion of the warm air, but there was no convincing theory which could explain such vertical motions.

After the Second World War, a much more lively picture of the front emerged from the discovery by the Chicago school that the major

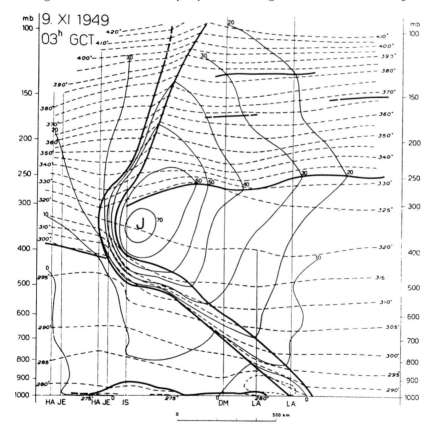

FIG. 10. Vertical cross section through a warm front at 0300 GMT 9 November, 1949. Heavy lines: frontal boundaries, inversions, tropopause. Thin solid lines: isopleths of observed wind speed (m/sec). Dashed lines: potential isotherms (°K). From Berggren [1952].

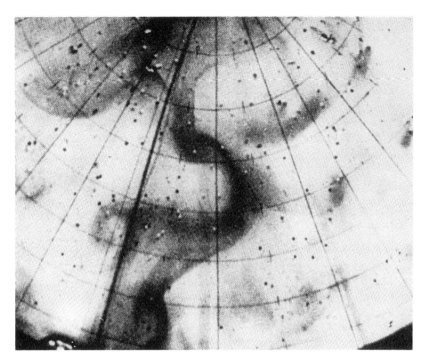

FIG. 11. Rotating tank experiment showing the formation of fronts in water. Fronts made visible by dye. From Faller [1956].

frontal zones are connected in the upper troposphere by strong jets of a peculiar structure (Figures 9 and 10). It is also significant that fronts were seen to form in the rotating tank experiments (Figure 11), as shown by Faller [1956]. This demonstrates that condensation heat is not a necessary factor for the formation of fronts.

Moreover, the quasi-geostrophic theory of atmospheric motions opened new possibilities for theoretical study of the vertical circulation in frontal zones, possibilities which have not yet been fully exhausted.

In order that the geostrophic balance, or thermal wind relation, between horizontal temperature gradient and the vertical wind shear shall persist, a certain vertical circulation is necessary. Roughly speaking, the quasi-geostrophic theory requires ascending motion wherever the vorticity advection by the thermal wind is positive, and descending motion where this advection is negative. From this simple principle, the nature of the vertical circulation systems may often be deduced qualitatively. In a more rigorous formulation, the principle may of course also be used in a quantitative sense.

Since the vertical motion depends upon the existence of a thermal wind and a vorticity gradient, it follows immediately that the strongest large-scale vertical velocities are to be expected in frontal zones, where both thermal wind and vorticity gradients are particularly strong.

I shall now illustrate the application of the quasi-geostrophic theory by considering certain typical patterns of geostrophic flow and temperature field in frontal zones and show how the main character of the vertical circulation can be deduced qualitatively. Bergeron's [1928] classical theory of frontogenesis explains the concentration of isotherms as a kinematic result of temperature advection in a horizontal deformation field (Figure 12). This advective process accounts for the formation

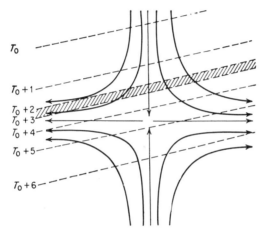

FIG. 12. Bergeron's theory of frontogenesis by differential advection in a horizontal deformation field. Solid lines: streamlines. Dashed lines: isotherms. From Petterssen [1956].

of the frontal temperature contrast, but it does not directly explain the corresponding increase of vertical wind shear which is necessary to maintain geostrophic equilibrium. It will be seen from Figure 12, however, that the thermal vorticity advection is negative on the cold side and positive on the warm side, showing that a thermally direct circulation in vertical planes normal to the isotherms is required from quasi-geostrophic theory when a band of isotherms is being stretched. The horizontal branches of this circulation are nongeostrophic, horizontally divergent winds which by Coriolis forces produce the required increase of vertical wind shear.

This process has been dealt with theoretically by Sawyer [1956] who has also demonstrated the existence of such vertical circulations by dew-point analyses in frontal cross sections, based on accurate measurements [1955]. One of his cases showing a remarkable difference in dew point indicating ascending and descending currents is reproduced in Figure 13.

It is noteworthy that a stretching of the isotherms generally takes place in developing cyclone waves as a result of amplification of the wave-shaped isotherms.

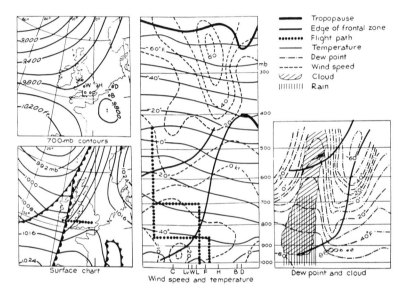

FIG. 13. Synoptic situation and vertical cross section through a warm front at 1500 GMT, 19 March, 1952. From Sawyer [1955].

There is also another type of horizontal advective process, which may produce and maintain a strong horizontal temperature gradient, namely differential advection along a shear line. This type of advection will typically operate along cold fronts; the cold air behind the front is maintained at a low temperature by advection of cold air from the North, and the warm air is kept warm by warm advection from the South. As seen from Figure 14, the thermal wind will in such cases have a component across the front from the cold to the warm air. With the maximum vorticity at the front itself, the vorticity advection by the thermal wind is seen to be negative on the cold side, and positive on the warm side. Therefore the quasi-geostrophic theory requires again a thermally direct vertical circulation as indicated by the dashed lines in Figure 15, which represents a schematic cross section. This

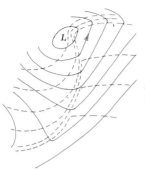

FIG. 14. Schematic diagram showing isobars (solid) and isotherms (dashed) along a cold front.

vertical circulation will counteract the effect of the increase with height of geostrophic wind normal to the front, which would otherwise cause the front to tip over. The streamlines obtained when the vertical circulation is added to the geostrophic flow normal to the front are shown as the fully drawn lines of Figure 15. This picture agrees with well-known models of cold front cloud systems. It also shows that the cold front at low levels moves into the warm air, being pushed by geostrophic winds at high levels.

A calculation from quasi-geostrophic theory of the vertical circulation in a cold front, based on real data, has recently been made by my coworker Mr. Todsen [1964],* using a method described by myself [Eliassen, 1962]. The surface map Figure 16 shows the cold front and

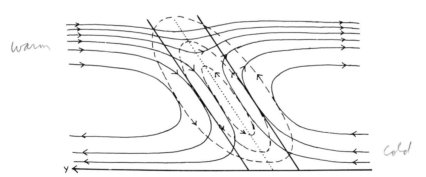

FIG. 15. Schematic vertical cross section normal to a cold front showing frontal boundaries (heavy lines), streamlines of vertical circulation (dashed lines), and of total geostrophic plus nongeostrophic flow (solid lines).

* The work was sponsored by the Air Force Cambridge Research Laboratories, OAR under Contract No. AF61(052)–525 with the European Office of Aerospace Research, United States Air Force.

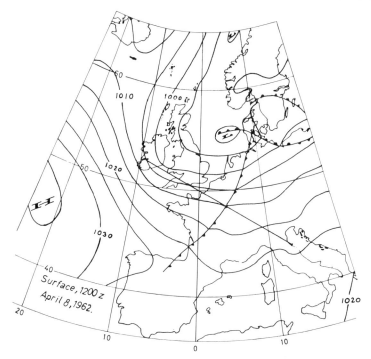

Fig. 16. Surface map of 1200 GMT, 8 April, 1962. Straight line shows position of analyzed cross section. From Todsen [1964].

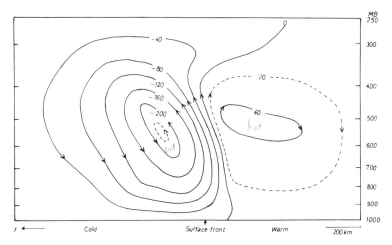

Fig. 17. Vertical cross section normal to the cold front (Figure 16), showing isopleths of the stream function for the vertical circulation (in $10^6$ gm sec$^{-3}$). From Todsen [1964].

the location of the vertical cross section used in the calculation. The streamline pattern of the vertical circulation is shown in Figure 17. It is a circulation in the direct sense, partly due to stretching of the isotherms, and partly to thermal wind across the front. The vertical motion is relatively weak, of the order 10 cm/sec, and checks fairly well with rainfall data. The horizontal nongeostrophic winds associated with the vertical circulation are of the order 10 m/sec. The strong horizontal convergence of these winds near the surface front will

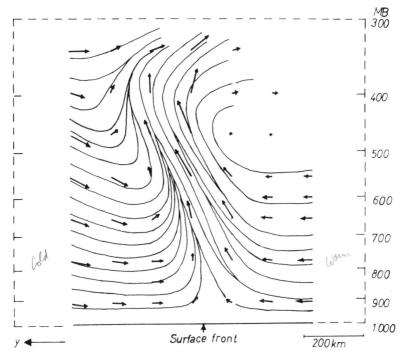

FIG. 18. Same vertical cross section as in Figure 17. Streamlines of total flow (geostrophic plus vertical circulation). From Todsen [1964].

strengthen the cyclonic wind shear and the horizontal temperature gradient, thus maintaining a sharp front near the surface. This frontogenetic effect is seen more clearly from Figure 18 which in the same vertical cross section shows the streamlines of the total velocity field (geostrophic wind plus vertical circulation) in a frame of references moving with the front. This two-dimensional velocity field is convergent, since the geostrophic wind parallel to the front is divergent. The stream-lines near the surface are seen to converge toward the surface front from both sides and ascend at a slope of about

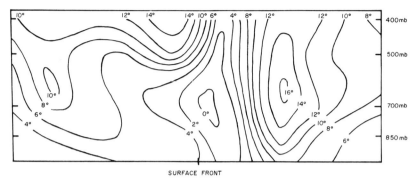

FIG. 19. Same cross section as in Figure 17. Dew-point depression (spacing 2°C). From Todsen [1964].

1:50, which corresponds roughly with the direction of the absolute vorticity. The motion pattern agrees well with the distribution of dew-point depression shown in Figure 19.

We shall now consider a different type of frontogenesis. Figure 20 shows schematically a band of isotherms through a cyclone. Since the vorticity is larger in the cyclone than in the anticyclone to the East, the vorticity advection by the thermal wind will be positive ahead of the cyclone. The quasi-geostrophic theory therefore requires ascending motion within the baroclinic zone ahead of the cyclone and a corresponding horizontally convergent motion in the lowest layers. So far, everything is in agreement with linear theory of cyclone behavior. But if we consider the development of the system to finite amplitude, certain second-order effects will take place, such as a concentration of

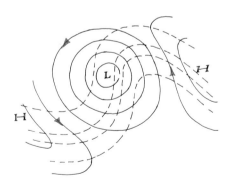

FIG. 20. Schematical distribution of isobars (solid) and isotherms (dashed) in an extratropical cyclone.

the band of isotherms to the East of the cyclone center at low levels, as a result of temperature advection by the convergent wind field. This isotherm concentration will in turn lead to a concentration of the field of ascending motion, since it is confined to the baroclinic zone of decreasing width. Thus we may visualize how the cyclone forms its

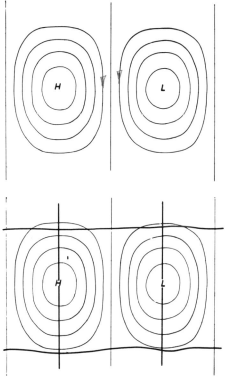

FIG. 21. From a computation reported by Edelmann [1963]. Initial situation. Upper part: surface isobars (heavy solid lines, spacing 5 mb) and 900 mb isotherms (weak solid lines, spacing 8°C). Lower part: surface isobars, and vertical motion (individual pressure change) at 800 mb (weak solid lines, spacing 1 mb/hr).

own low-level warm front. The converging motion will also produce the cyclonic shearing vorticity along the front.

A dramatic illustration of such a process can be seen from a remarkable calculation of the development of a cyclone wave reported by Edelmann [1963]. The calculation was based on a 5-level model of the atmosphere confined in a channel between two zonal walls. The initial situation was taken as a baroclinic zonal current, which vanishes at the ground and reaches a maximum of 105 km/hr at 300 mb in the middle of the channel, and superimposed upon it a harmonic disturbance. Figure 21 taken from Edelmann's paper, shows the innocent-looking initial disturbance. The upper part shows surface isobars and 900-mb isotherms; the lower part, surface isobars and vertical motion at 800 mb. Figure 22 shows the situation after 2 days. The wave has

FIG. 22. Same as Figure 21, situation after 2 days.

intensified, and the isotherms have assumed a wave shape. A slight tendency toward warm-front formation can be seen. In Figure 23, after 4 days, an extremely sharp warm front has formed, in spite of the fact that the numerical model includes horizontal as well as vertical diffusion of momentum and heat. The frontogenetic process is very active at the eastern end of the sharp warm front, where the vorticity gradient along the isotherms is particularly strong. Therefore, the eastward extent of the sharp front increases; the front seems to grow along the isotherms, splitting the anticyclone in two parts, as shown in

Figure 24, which represents the situation after 6 days. These maps also show the formation of a weaker cold front and other interesting features which I shall not go into.

Although the fronts in Edelmann's calculations appear to be strongly exaggerated, the result agrees qualitatively with an observation familiar to synopticians, that fronts tend to form and intensify in

FIG. 23. Same as Figure 21. Situation after 4 days. (Spacing of vertical motion 2 mb/hr.)

developing cyclones. This may seem to contradict the Bergen school idea that cyclones form at fronts, but these statements are really not mutually exclusive and may both be true. Which came first, the front or the cyclone, when the atmosphere was first started up is of little concern to us.

However, a natural conclusion from the various theories is that the larger-scale cyclonic systems, associated with the transient waves in the upper westerlies are the primary perturbation systems, since their existence can be accounted for by baroclinic instability even in the absence of fronts. These large cyclonic systems, in turn, produce the fronts, partly by setting up the deformation fields required in Bergeron's theory, and partly by the nonlinear process demonstrated by Edelmann. Finally, the fronts will give birth to the smaller, frontal wave cyclones.

Undoubtedly, condensation heat contributes to the formation of

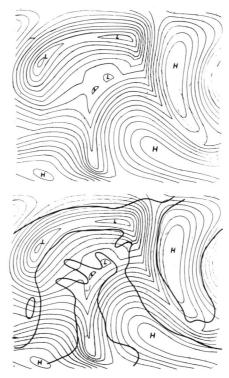

FIG. 24. Same as Figure 21, situation after 6 days. (Spacing of vertical motion 2 mb/hr.)

fronts, and to their remarkable persistence, although the existing theories and rotating tank experiments indicate that fronts would probably exist even in a cloudless atmosphere. Some years ago, I made an attempt to show that fronts near the surface would tend to sharpen as a combined effect of Ekman layer friction and condensation heat [1959]. The reasoning was this: Given a diffuse frontal zone with a large, but finite cyclonic shearing vorticity, the Ekman layer flow will then converge toward the frontal zone where a rising current is produced, as shown by the fully drawn lines of Figure 25. The condensation heat in this rising current will in turn produce an additional rising current with horizontal convergence below it. This convergence will sharpen the front by advecting momentum and entropy toward the front.

This is an interesting theory, but it is not quite correct. The Ekman convergence produces above the Ekman layer a horizontally divergent motion which I had neglected to take into account, and which will offset the convergent motion produced by the condensation heat unless the temperature stratification along the absolute vortex lines is

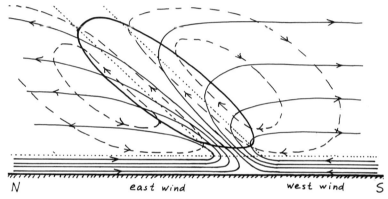

FIG. 25. Schematic vertical cross section normal to a front, showing streamlines of the vertical circulation produced by the Ekman convergence. From A. Eliassen [1961].

conditionally unstable. Therefore, the conclusions drawn concerning the frontogenetic effect of the Ekman layer–condensation heat combination were much exaggerated.

In the first part of this chapter, I have dealt only with the most common types of extratropical cyclones. In tropical cyclones, which I shall not discuss, the physical processes are quite different. I would like to conclude by drawing attention to the fact that, even in high latitudes, certain kinds of cyclones occur which seem to bear resemblance to tropical cyclones. Figure 26, obtained from my colleague, Mr. P. Dannevig, shows a surface map of 1500 GMT, 2 December, 1961. The cyclone in question is located off the Norwegian west coast, imbedded in deep cold arctic air flowing south over the warm Norwegian Sea. These storms are not uncommon in the Norwegian Sea in winter; they give very heavy snowfall and often carry winds of hurricane force in a small region near the center.

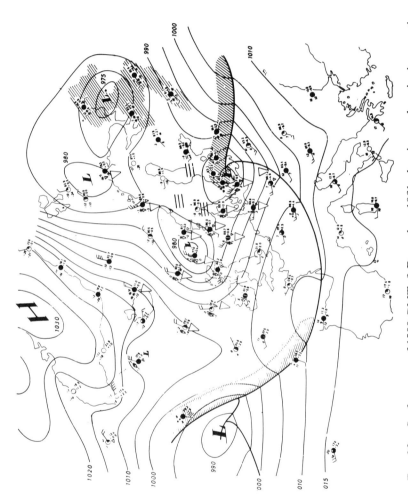

Fig. 26. Surface map of 1500 GMT, 2 December, 1961, showing cyclone in deep arctic air off the Norwegian coast. Courtesy of P. Dannevig.

Clearly there is still much to be done before we can claim to have a fully satisfactory theory of fronts and cyclones. I wish the scientists at Massachusetts Institute of Technology all success in their work toward this goal.

## References

Bergeron, T., Über die dreidimensional verknüpfende Wetteranalyse, *Geofys. Publikasjoner, Norske Videnskaps-Akad. Oslo*, 5 (6), 1928.

Berggren, R., The distribution of temperature and wind etc., *Tellus*, 4, 43, 1952.

Bjerknes, J., On the structure of moving cyclones, *Geofys. Publikasjoner, Norske Videnskaps-Akad. Oslo*, 1 (2), 1919.

Bjerknes, J., and C. L. Godske, On the theory of cyclone formation at extratropical fronts, *Astrophys. Norw.*, 1 (6), 1936.

Bjerknes, J., and Y. Mintz, Investigations of the general circulation of the atmosphere, *Final Report, Contract AF 19(122)–48*, 1955.

Bjerknes, J., and H. Solberg, Life cycle of cyclones and the polar front theory of atmospheric circulation, *Geofys. Publikasjoner, Norske Videnskaps-Akad. Oslo*, 3 (1), 1922.

Bjerknes, V., J. Bjerknes, H. Solberg, and T. Bergeron, *Physikalische Hydrodynamik*, Springer, Berlin, 1933.

Blinova, E. N., 1943, reviewed in I. A. Kibel: *An introduction to the hydro-dynamical methods of short weather forecasting*, pp. 103–109, translated from Russian, Pergamon Press, 1963.

Burger, A. P., On the non-existence of critical wavelengths in a continuous baroclinic stability problem, *J. Atmospheric Sci.*, 19, 30, 1962.

Charney, J. G., The dynamics of long waves in a baroclinic westerly current, *J. Meteorol.*, 4, 135, 1947.

Charney, J. G., On the scale of atmospheric motions, *Geofys. Publikasjoner, Norske Videnskaps-Akad. Oslo*, 17 (2), 1948.

Charney, J. G., On baroclinic instability and the maintenance of the kinetic energy of the westerlies, *Intern. Union Geodesy Geophysics*, Brussels, Assoc. de Météor, 47, 1951.

Charney, J. G., and M. E. Stern, On the stability of internal baroclinic jets in a rotating atmosphere, *J. Atmospheric Sci.*, 19, 159, 1962.

Eady, E. T., Long waves and cyclone waves, *Tellus*, 1, 33, 1949.

Edelmann, W., On the behavior of disturbances in a baroclinic channel, *Tech. Note no. 7, Contract AF61(052)–373*, 1963.

Eliasen, E., On the initial development of frontal waves, *Publ. Danske Meteorol. Inst.*, Meddelser No. 13, 1960.

Eliassen, A., On the formation of fronts in the atmosphere, in *The atmosphere and the sea in motion*, (*Rossby Mem. Vol.*), edited by Bert Bolin, New York, 277, 1959.

Eliassen, A., Current concepts of fronts, *Tenki (Japan)*, 8, 1, 1961 (in Japanese).

Eliassen, A., On the vertical circulation in frontal zones, V. Bjerknes Cent. Vol., *Geofys. Publikasjoner, Norske Videnskaps-Akad. Oslo*, 24, 148, 1962.

Faller, A. J., A demonstration of fronts and frontal waves in atmospheric models, *J. Meteorol.*, 13, 1, 1956.

Fjørtoft, R., Application of integral theorems in deriving criteria of stability for laminar flows and for the baroclinic circular vortex, *Geofys. Publikasjoner, Norske Videnskaps-Akad. Oslo, 17* (6), 1950.

Fultz, D., On the possibility of experimental models of the polar-front wave, *J. Meteorol., 9,* 379, 1952.

Green, J. S. A., A problem in baroclinic stability, *Quart. J. Roy. Meteorol. Soc., 86,* 237, 1960.

Höiland, E., Stability and instability waves in sliding layers with internal static stability, *Arch. Math. Naturvidensk., 50* (3), 1948.

Höiland, E., On two-dimensional perturbation of linear flow, *Geofys. Publikasjoner, Norske Videnskaps-Akad. Oslo, 18* (9), 1953.

Kotchin, N., Über die Stabilität von Marguleschen Diskontinuitätsflächen, *Beitr. Physik Atmosphare, 18,* 129, 1932.

Kuo, H. L., Dynamic instability of two-dimensional flow in a barotropic atmosphere, *J. Meteorol., 6,* 105, 1949.

Kuo, H. L., A note on the kinetic energy balance of the zonal wind systems, *Tellus, 3,* 205, 1951.

Obukhov, A. M., On the problem of the geostrophic wind, *Izv. Akad. Nauk SSSR, Ser. Geograf. i Geol.* (4), 281, 1949.

Palmén, E., and C. W. Newton, A study of the mean wind and temperature distribution in the vicinity of the polar front in winter, *J. Meteorol., 5,* 220, 1948.

Pedlosky, Joseph, The stability of currents in the atmosphere and the ocean: Part I, *J. Atmospheric Sci., 21,* 201, 1964.

Petterssen, Sverre, *Weather analysis and forecasting*, McGraw-Hill Book Company, Inc., New York, 1956.

Phillips, N. A., Geostrophic motion, *Rev. Geophysics, 1,* 123, 1963.

Rossby, C. G., *et al.*, Relation between variations in the intensity of the zonal circulation and the displacement of the semi-permanent centers of action, *J. Marine Res., 2,* 38, 1939.

Sawyer, J. S., The free atmosphere in the vicinity of fronts, *Geophys. Mem.*, Met. Office, no. 96, 1955.

Sawyer, J. S., The vertical circulation at meteorological fronts and its relation to frontogenesis, *Proc. Roy. Soc. London, A, 234,* 346, 1956.

Solberg, H., Integrationen der atmosphärischen Störungsgleichungen, *Geofys. Publikasjoner, Norske Videnskaps-Akad. Oslo, 5* (9), 1928.

Solberg, H., in *Physikalische Hydrodynamik*, by V. Bjerknes, J. Bjerknes, H. Solberg, and T. Bergeron, Springer, Berlin, 1933.

Starr, V. P., Studies of the atmospheric general circulation, *Final Report, Contract No. AF19-122-153,* 1954.

Todsen, M., A study of the vertical circulations in a cold front, *Part IV of Final Report, Contract No. AF61(052)-525,* 1964.

# ATMOSPHERIC TURBULENCE

*A. M. Obukhov*

Institute of Atmospheric Physics, Moscow, U.S.S.R.

At the outset I wish to present my thanks to the conveners of the Conference for the honor of delivering my lecture within the walls of the Massachusetts Institute of Technology.

About a hundred years ago Osborne Reynolds in his classical works determined two types of movement of liquid: laminar and turbulent. The movement of media under natural conditions, in particular the air of the earth's atmosphere, is, as a rule, turbulent. This can easily be seen in the spreading of the flow of smoke from a chimney or in the random swinging of the weather vane.

The long period of the study of atmospheric turbulence has been associated mainly with the application to atmospheric conditions of properties of the turbulent motion of a homogeneous liquid, studied by fluid dynamicists in laboratories. A well-known example is the application of a semiempirical theory of turbulence to the surface layer of the atmosphere and, in particular, of the logarithmic law of wind distribution supported by a long series of observations for conditions close to neutral stratification. The efforts during the last 10 to 15 years showed that the wind profile can deviate considerably from the logarithmic depending on thermal conditions in the surface layer of the atmosphere. Experimental researches in the surface layer are now well developed in a number of countries: Australia, United States, Soviet Union, Japan, England, etc.

The application of the theory of similarity in this instance proved quite useful for the generalization of empiric data and prediction of certoin extreme cases. The friction tension and heat transfer in the vertical should be used as external parameters in the study of wind and temperature distribution in the surface layer 20 to 30 meters thick. At present we are aware of the existence of a "geophysical variant" of the semiempirical theory of turbulence, which takes account of the thermal stratification and of the heat transfer connected with it. The summary of results obtained in this field is given in a recently published book by Priestley [1959]. The practical importance of the study of the surface layer lies in the construction of the physical foundation for the calculation of heat, water, and friction transfer over different types of the earth's surface. Special problems arise in the study of processes in the boundary layer of the atmosphere above the sea, where questions of atmospheric turbulence are closely interlaced with the problem of wind waves. This trend in research attracts much attention of geophysicists who specialize in the study of the physics of atmosphere and ocean.

I do not think we should dwell in detail on the purely phenomenological description of atmospheric turbulence by means of different coefficients of turbulence in the tasks of dynamic meteorology. Such a description is often helpful in the solution of actual problems, for instance, of the classical one about the turn of wind with altitude, but it shows little about the nature of the turbulence itself. The solution of different problems for the same conditions in the atmosphere is often carried out with application of considerably varying coefficients. This circumstance has been well defined by L. Richardson in 1926, when he gave an outline of a rational explanation of the seemingly surprising situation. Figure 1 shows a "collection" of coefficients of turbulent diffusion assembled by Richardson. The coefficient of diffusion can be formally defined by the formula

$$K = \frac{1}{2} \frac{d}{dt} \langle (x - \bar{x})^2 \rangle$$

where $\bar{x}$ is the coordinates of the center of the cloud and $\langle \rangle$ is the symbol of statistic averaging. With big clouds the diffusion is meant in the horizontal direction.

It is practical, as suggested by Richardson, to average according to the sampling of clouds of a certain size. The data on the plot show that the exchange (diffusion) coefficient depends considerably on the linear scale $\ell$. Richardson was the first who gave a qualitative explanation of this fact, i.e., the dependence of empirically determined characteristics of turbulence on the scale of the phenomenon $\ell$, by indicating the multiscale character of turbulence and an extremely wide range of corresponding scales in conditions of atmosphere. The largest vortices are formed as the result of instability of the main stream, and their sizes are comparable with the whole system. These first-order disturbances are also unstable because the Reynolds number is too large for them, and in disintegrating they produce second-order disturbances which in their turn cause the appearance of smaller "vortices," etc. This process of splitting of primary "macroscale vortices" goes on until viscosity interferes and destroys the smallest vortices in this chain. The Richardson scheme qualitatively not only explains the existence

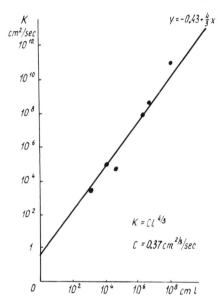

FIG. 1. Dependence of turbulent diffusion coefficient on the size of clouds. According to Richardson [1926].

of a wide spectrum of different scale disturbances, but also indicates a certain mechanism of dissipation of energy in media with considerably low viscosity.

In the problem of turbulent diffusion the cloud itself plays the role of "spectral instrument." The vortices, belonging to the scale considerably larger than the diameter $\ell$, only transport the cloud as a whole without changing its size. With the growth of the cloud the increasing range of the spectrum of turbulence affects the relative motion of particles, i.e., diffusion, which causes the effective coefficient of diffusion to grow progressively.

The works of A. N. Kolmogorov [1941] and the publications of Heisenberg and Weizsacker, which appeared a little later, added to the scheme just described a well-contructed hypothesis of physical character, thus allowing the scheme to be described in a mathematical form and formulating certain regularities of quantitative character.

Let me remind you of one of the major conclusions in the theory of Kolmogorov, which found a wide application in geophysics and which was supported by an extensive series of experiments in the atmosphere, ocean, and in laboratory wind tunnels. I refer to the "law of two-thirds" for the velocity field in the well-developed turbulent flow. Let us assume that $M_1$ and $M_2$ are two points of study and that the distance between them is small compared with a certain "external scale" $\ell_0$, for instance, with the distance to the boundary of the flow; and at the same time it is big compared with the "inner scale" of turbulence $\ell_1$, which depends on the viscosity of the medium. Finding support in rather natural hypotheses (local isotropy, speculations of the theory of similarity) A. N. Kolmogorov in 1941 indicated that the difference between velocities in points $M_1$, $M_2$ satisfies the statistical law

$$\langle (\mathbf{v}_2 - \mathbf{v}_1)^2 \rangle = c(\epsilon \zeta)^{2/3}$$

where $\zeta$ is the distance between points of observations, $c$ is the number constant (according to experiments $c \approx 6.9$), and $\epsilon$ is the average dissipation of the mechanical energy of flow per mass unit of the medium.

Figure 2 presents results of the first measurements of the structure of velocity field in the surface layer of the atmosphere, carried out in

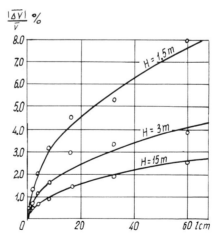

FIG. 2. Structural function of the velocity field in the surface layer of the atmosphere at different heights $H$. According to the data of measurements in 1951 [Obukhov, 1951].

the Soviet Union in 1951. A hot-wire anemometer was used as the measuring instrument. The dissipation values were controlled by measuring the mean wind profile. The observations were carried out above a smooth part of the earth's surface. The verification of Kolmogorov's law in the atmosphere has also been proved by McCready [1953] and by a number of other authors.

The study of dependence of the mean-square difference of a certain physical characteristic of the turbulent stream from the distance between observation points $\zeta$ is one of convenient ways of description of the statistic structure of the flow. Another method, widely applied in the theoretical and experimental researches of turbulence, is the spectral method of description. The field is represented in the form of superposition of disturbances of different scale providing additive contribution to the energy of the field (dispersion of the observed quantity). Under certain conditions of statistical homogeneity the well-known Fourier expansion can be used. The quantitative description of the statistical structure of the field is given by the distribution of the energy over spectrum $E(k)$, where $k$ is the wave number (frequency in space). The integral $\int_{k_1}^{k_2} E(k)\, dk$ represents energetic contribution of disturbances with wave numbers in the interval $(k_1, k_2)$. The value $\ell = (1/k)$ can be treated as the linear scale of corresponding "elementary disturbances."

The Taylor-Khinchin-Wiener theorem about the connection of correlation functions with spectral distributions, and its generalization for the case of scalar and vector fields serves as a dictionary for the interpretation of properties of statistical structure of turbulence from the correlation language to spectral one and back. In particular, the structural law of two-thirds corresponds to the spectral distribution of energy according to the law of "minus five-thirds":

$$E(k) = c' \epsilon^{2/3} k^{-5/3}$$

(According to the available experiments $c' \approx 1.4$.)

As an example of measurements of the spectrum of turbulent pulsations of velocity in the atmosphere, Figure 3 presents the data collected by S. L. Zubkovsky on spectral distribution for the vertical component of the velocity $w$ according to the observations from an aircraft (on a logarithmic scale). The measurements were carried out with a special acoustic anemometer and a special analyzer. The wave number was determined taking into consideration the speed of the aircraft. As we see, the application range of the law of "minus five-thirds" is sufficiently large. Spectral measurements provide possibilities

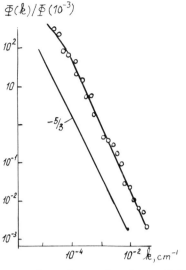

Fig. 3. Spectral distribution of the vertical component of velocity [Zubkovsky, 1963].

for convenient determinations of the $\epsilon$ characteristic in actual geophysical conditions.

The value $\epsilon$ is the mean dissipation energy related to the mass unit of the medium; it is an important characteristic of the disturbances "cascade" of various scales, which as a whole form what we call turbulence in statistical description. This very quantity determines on the average the value of transformation of energy of vortices of a given scale into energy of smaller vortices. At large Reynolds numbers in

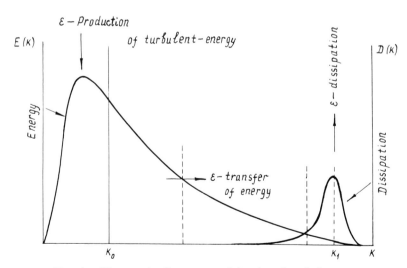

Fig. 4. The graph of spectrum of developed turbulence.

the atmosphere the income of energy into the spectrum of turbulence and its dissipation due to viscosity take place in essentially different parts of the spectrum. The income is in the low-frequency region $k < k_0$, where $k_0$ is the inverse value of the scale $\ell_0$, which characterizes large-scale disturbances. The loss of energy (viscous dissipation) takes place in the region of a rather high-frequency part of the spectrum with maximum for the frequency $k_1 = (1/\ell_1)$. In the intermediate region $k_0 < k < k_1$ which is called the inertia interval, only the transfer of energy along the spectrum takes place, which is brought about by the nonlinearity of the equations of hydromechanics. Figure 4 shows this situation by a graph.

According to the deductions of the theory of similarity, A. N. Kolmogorov [1941] received a definite expression for the scale $\ell_1$, the smallest vortices, responsible for the viscous dissipation [$\ell_1 = (1/k_1)$]. Assuming that this scale is determined by dissipation $\epsilon$ and viscosity $\nu$, and considering the size of these values, we can easily get

$$\ell_1 = \sqrt[4]{\frac{\nu^3}{\epsilon}}$$

(the number multiplier of the order of unity is omitted).

The inner scale $\ell_1$ can be roughly estimated on the average for the whole atmosphere of the earth. Let us assume that $\epsilon = \eta \cdot \epsilon_0$, where $\eta$ is the "efficiency coefficient" of the atmospheric engine and $\epsilon_0$ is the power intensity of arriving radiation per unit mass of the atmosphere. It is evident that

$$\epsilon_0 = \frac{w_0 g}{4 p_0}$$

where $w_0$ is the solar constant in mechanical units, $p_0$ is the mean pressure at the earth's surface, and $g$ is the acceleration of gravity. With $w = 1.4 \times 10[6 \text{ erg}/(\text{cm}^2 \cdot \text{sec})]$, $p = 10^6$ cgs, $g = 10^3$ cgs the calculation results in

$$\epsilon_0 = 350 \text{ cm}^2/\text{sec}^3$$

The choice of $\eta$ does not significantly influence the result because the coefficient $\eta$ is introduced with the sign of the root in fourth power. Let us assume together with Brunt that the efficiency of the heat engine of the atmosphere is $\eta = 0.02$, i.e., approximately the same as the first steam engine built by Stephenson in 1829. In this case $\epsilon = 7$ cm$^2$/sec$^3$, and Kolmogorov's formula gives $\ell_1 = 0.2$ cm.* This rather

* In the surface layer the character value of dissipation $\epsilon$ is one order of value greater. However, it is not felt in the estimation of $\ell_1$ due to the effect of the fourth power root.

rough estimation agrees, however, with recently received results of excellent experiments accomplished by Pond, Stewart, and Burling [1963] of Vancouver (Canada) (Figure 5). These experiments studied the fine structure of the field of velocities in the above-water layer of the atmosphere with a hot-wire anemometer and high-frequency equipment. As far as I know, the researches of Professor Stewart's group in Canada are the first successful attempt to approach the upper limit of the spectrum of atmospheric turbulence.

Thus the inner scale of turbulence is measured by some millimeters. Such is the mean value of "laminar grains" which comprise the turbulent atmosphere. This scale is sufficiently "macroscopic" in the sense that it exceeds by many times the mean free path of molecules and, of course, the distance between separate molecules. Therefore, the description of actual instantaneous states of turbulent atmosphere with the Navier-Stokes equations appears as completely justified, though some authors tend to doubt it. The inner scale determines the distance between the observation points, beginning with which the error of the linear interpolation of the real field becomes considerably smaller than the mean amplitude of changes of the field at the same distance. For distances much more than $\ell_1$, i.e., for the inertia interval, the interpolation error is already comparable with the natural variance of the field. Thus for the measurement of the actual pressure and velocity gradients, introduced into the equation of hydrodynamics, it should be necessary to have instruments installed at distances less than $\ell_1$, i.e., at distances less than 1 cm.

From the geophysical point of view the inner scale of turbulence is undoubtedly a very small value. Any practically possible network of observations will be too "rough" in the sense that the distance between observation points shall be $\ell \gg \ell_1$. This means, at the same time, that in the work with a rough network (macroscopic in the geophysical sense) the actual gradients are individually undefinable. The gradients of the average fields can be determined with the scale of averaging not less than $\ell_1$. Therefore, in order to calculate the evolution of the averaged fields (with the scale of averaging $\ell$), we should apply the equations of hydrodynamics which should also be correspondingly averaged and with the same scale of averaging. As the equations of hydrodynamics are nonlinear, this process of averaging results in the appearance of additional effects of transportation of the impulse, heat, and passive substances (humidity, dust, etc.). According to the set terminology, these effects are also called "turbulent": turbulent friction, turbulent diffusion. We should bear in mind that the characteristic value of the additional effects just mentioned,

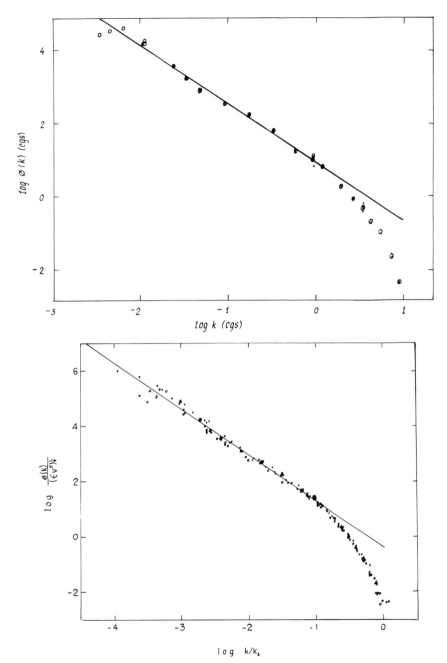

Fig. 5. Spectral measurements in the region of the high-frequency part of the spectrum. Pond's data [1963].

stipulated by averaging, greatly depends on the scale of averaging. Only in the case of considerably large averaging with the scale $\ell \gg \ell_0$ (corresponding to the time averaging on a sufficiently large period) can we approach to a certain "state of saturation." Thus, for instance, in the case of the processes of turbulent transfer along the vertical, the experiment shows that we need averaging over about 10 minutes in order to approximate the usual characteristics of turbulence (mean profiles of wind, temperature, transfer of impulse, and heat) to certain stable values. And we can speak about "saturation" rather provisionally and only for very large scales of averaging of the order of thousands of kilometers, if we mean the irregular changes of characteristics of the atmosphere in the horizontal direction and the effects of horizontal exchange connected with them.* As a rule, when solving certain problems which need the consideration of atmospheric turbulence, we and our instruments are somewhere "within" a very wide spectrum of disturbances present in nature.

The idea of "scale relativity" in the theory of turbulence, suggested by Richardson some forty years ago and developed by A. N. Kolmogorov and his followers, is at present a useful orientation in various problems, and especially in the studies of the geophysical and astrophysical character. In particular, for turbulent diffusion it results in the law

$$K(\ell) = c\epsilon^{1/3} \ell^{4/3}$$

where $\ell^{4/3}$ is the effective coefficient of diffusion caused by vortices of the scale less than $\ell$. The coincidence of the result received theoretically in 1941 with the Richardson empiric data [1926], mentioned in the beginning of the chapter, was in that period an encouraging circumstance.

During recent years the radiophysicists and astronomers have shown certain interest in the studies of the microstructure of the atmosphere. The pulsations of the refractive index for the electromagnetic waves are connected with the turbulent state of the atmosphere, which cause the observed effects of scattering of radio waves in the troposphere. The phenomenon of tropospheric scattering allows reception of ultrashort radio waves beyond direct visibility. Astronomers know well the phenomena of scintillation and quivering of stars, which are also caused by turbulent fluctuations of the refractive index of the atmospheric air in the optical range. In the case of light, the pulsations of the refractive index of the atmosphere are associated mainly with

* In another limit case of "dense" network with distance between observation points of the order of $\ell_1$ (mm), additional turbulent effects will be very small and can be compared with molecular effects.

the pulsations of temperature; for the case of radio waves the humidity pulsations are an important contribution.

The theory of microstructure of the thermal field and concentration of the passive contaminants in the turbulent flow was developed in 1949–1951 [Obukhov, Corrsin], which led to the statistical laws similar to the structure laws of the field of velocities. For the mean square of difference in temperature (humidity) in close points we receive the law of two-thirds, and for the corresponding spectral density the law of minus five-thirds. The available observations in the atmosphere indicate that the range of spectrum in which the law of minus five-thirds is true (in application to the temperature field) is sufficiently wide.

Figure 6 presents experimental data about the spectrum of temperature pulsations in the atmosphere at the altitude of 70 meters according to the simultaneous observations from an aircraft (white spots) and on a tower (black spots) [Tsvang, 1963]. Figure 7 shows some results

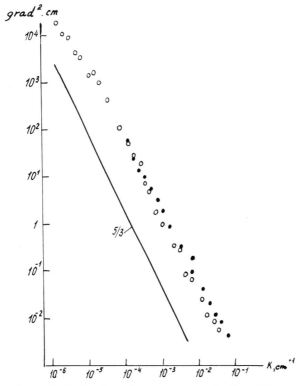

FIG. 6. Spectral distribution of temperature pulsations [Tsvang, 1963].

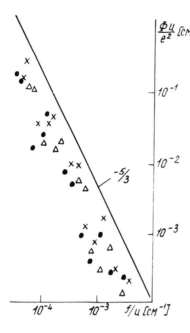

Fig. 7. Spectral distribution of humidity pulsations [Yelagina, 1962].

relating to the spectrum of humidity pulsations obtained by L. G. Yelagina [1962] of the Institute of Atmospheric Physics, Academy of Sciences of the Soviet Union, with a special optical hygrometer.

The problem of "turbulence and propagation of waves" is investigated in many theoretical and experimental researches. These problems are set forth in a book by V. I. Tatarski [Moscow, 1959] which was also published in English by McGraw-Hill [1961].

In this chapter I have mainly dwelt on the "philosophy" of atmospheric turbulence, restricted to the demonstration of only a few results of actual measurements. In reality we already possess a number of direct measurements of characteristics of atmospheric turbulence, in particular, of pulsation spectra of wind components, temperature, cross-correlation of these values (heat flux), and other characteristics at the surface of the earth and in the free atmosphere. These data provide a certain conception about the dependence of properties of turbulence on external conditions, for instance, on conditions of stratification stability and wind shear. More detailed information can be found in the book by Lumley and Panofsky [1964], and in the Proceedings of the Symposium on Turbulence, which took place in Marseilles in 1961.

And still we know too little about atmospheric turbulence. The studies of many important problems are only being started. As an example, the theory of anisotropic turbulence needed for the description of the large-scale processes could be mentioned here. However, this branch of atmospheric physics is successfully developing. Researchers in a number of countries are intensively active in the wide program of study of different aspects of this complicated problem.

I very much believe that some time hence a more competent speaker in a similar conference will give a more complete answer to the question: What, after all, is atmospheric turbulence in reality?

### References

Kolmogorov, A. N., Local structure of turbulence in an incompressible viscous fluid at very high Reynolds' numbers, *Dokl. Acad. Nauk SSSR*, *30*, 1941. German translation of this article together with other Soviet works is contained in the book: *Statistische Theorie der Turbulens*, Akademie-Verlag, Berlin, 1958.

Lumley, I. L., and H. A. Panofsky, *The Structure of Atmospheric Turbulence*, Interscience Publishers, Inc., New York, 1964.

MacCready, P. B., Atmospheric turbulence measurement and analysis, *J. Meteorol.*, *10*, 1953.

Obukhov, A. M., The investigation of wind microstructure in the atmospheric surface layer, *Proc. Acad. Sci., USSR, Geophys. Ser.* no. 3, 1951.

Pond, S., R. W. Stewart, and R. W. Burling, *J. Atmospheric Sci.*, *20*, no. 4, 1963.

Priestley, C. H. B., *Turbulent Transfer in Lower Atmosphere*, The University of Chicago Press, Chicago, 1959.

Richardson, L. F., Atmospheric diffusion shown on a distance-neighbour graph, *Proc. Roy. Soc. (London)*, *A110*, No. 756, 1936.

Tatarski, V. I., *A Theory of Fluctuational Phenomena in the Presence of Waves Propagating through a Turbulent Atmosphere*, Moscow, 1959. English translation of this book: Tatarski, V. I., *Wave Propagation in a Turbulent Medium*, McGraw-Hill Book Company, Inc., New York, 1961.

Tsvang, L. R., Certain characteristics of the spectrum of thermal pulsations in the boundary layer of the atmosphere, *Proc. Acad. Sci., USSR, Geophys. Ser.*, no. 10, 1963.

Turbulence in Geophysics, *Proceedings of Symposium at Marseilles (1961)*, edited by P. N. Frenkiel, Washington, D.C., 1962.

Yelagina, L. G., An optical instrument for the measurement of turbulent pulsations of humidity, *Proc. Acad. Sci., USSR, Geophys. Ser.*, no. 8, 1963.

Zubkovsky, S. L., Experimental investigation of the spectrum of pulsations in the vertical component of wind velocity in free atmosphere, *Proc. Acad. Sci., USSR, Geophys. Ser.*, no. 8, 1963.

# PART III

# DYNAMICS OF THE OCEANS

Thalasso chlora

# SEDIMENTARY RECORD OF LONG-PERIOD PHENOMENA*

*Gustaf Arrhenius*

Scripps Institution of Oceanography, University of California, La Jolla, California

An apocryphal Greek legend tells us that Poseidon, the god of the ocean, became jealous of the tribute paid by the human beings to his terrestrial colleagues on Olympus. He decided for this reason to create a new center for worship on another mountain, towering over the plain of Troy, much as the giant building inaugurated by this conference rises over the Commonwealth of Massachusetts. The guardianship over the shrine was embodied in the nymph Ida, whose name also came to be used for the mountain itself. At the spot where the priestess sacrificed during an earthquake, a gushing spring is said to have come forth, known because of its remarkable aquamarine color as Θαλασσιοχλωρά, or Sea-Green. Hence the close association ever since between this name, that of Ida, and the earth sciences.

It was by drinking from this spring that the Titaness Mnemosyne became miraculously pregnant and, when her time had come, gave birth to the nine Muses, the spirits of arts and sciences. There is a symbolic meaning attached to this Hellenic concept of Mnemosyne or Memory as the mother of man's intellectual and artistic endeavors. Particularly in the earth sciences the earth's own memory of its past history, displayed to us in the form of the geological record, is continually stimulating our inquiry into the processes that governed our

* Contribution from Scripps Institution of Oceanography, New Series.

planet in the past and is urging us to clarify the nature of these processes as they are active today.

On this basis it is understandable why the arrangers of the International Earth Sciences Conference have wished the profound contributions of Stommel and Munk to be preceded by some memory data, and I feel honored to pay this tribute to Mnemosyne.

The memory of the earth is its blanket of layered rocks, which represent a more or less continuous time series, with the geophysical and geochemical conditions at the various time levels reflected by the properties of the corresponding sedimentary strata. The part of the blanket that covers the ocean floor is particularly interesting since it shows much less wear and tear, holes, and missing or distorted layers than the ragged parts which still remain draped over the continents. Furthermore, the pelagic sediments in protected basins far away from land have received only relatively small amounts of diluting debris from the continents, and as a result the subtle details of long-period phenomena in the ocean, the atmosphere, and outer space are more faithfully recorded than in any other environment.

Among the most revealing ocean-atmosphere controlled sedimentation phenomena are those in the subpolar and equatorial organic productivity zones. The latter, in particular, have been extensively discussed during the last decade. Since not much fundamental information has been added lately, I will dwell on this specific aspect only to the extent that it is of importance for interpreting the processes to be mentioned later.

Diagrams like Figure 1, representing the vertical equatorial circulation in the ocean, have appeared in press in various stages of development [Arrhenius, 1952, 1959, 1963], and I should like to add only a few comments at this time. It is interesting to note that the most significant features of the equatorial circulation mechanism were originally suggested by their manifestations imprinted in the sediments below. A number of profiles through surface layers of the deep sea floor in the East Pacific were investigated in the period 1948 to 1952, after collection by the Swedish Deep Sea Expedition led by Hans Pettersson in 1947–1948. It became obvious that there was a regular and pronounced layering of the deposits, recognizable in detail along the Equator over about a quadrant of the earth's circumference. In an oversimplified way the sediment in this zone can be said to consist of two components—one inorganic, the other produced by living organisms. The inorganic component is in some areas deposited with remarkable uniformity, in others with observable variations in the rate. The rate of production of the biotic component, which consists mainly

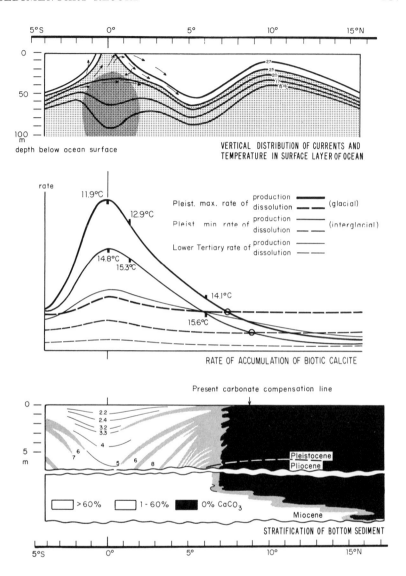

Fig. 1. The upper graph shows the thermal stratification and the vertical and horizontal circulation in the surface layer of the East Equatorial Pacific. The equatorial undercurrent is indicated by dark shading. The lower graph indicates the stratification of the bottom deposits below the Equatorial Zone and the middle diagram shows an interpretation in terms of rate of biological production represented by the rate of accumulation of calcium carbonate. From Arrhenius [1963].

of microscopic shells and other skeletal structures, is controlled by lifting of nutrient-rich deep water into the sunlit zone, where primary production of plant life is possible.

The individual sediment layers, representing discrete time periods, were observed to become increasingly rich in this biotic component, and consequently thicker toward the Equator. It was, therefore, concluded that the productivity reaches a maximum at the Equator and drops monotonically to the North and the South. Existing dynamic theory could qualitatively explain the equatorial maximum, but required a second maximum at about 8°N, which was definitely not indicated by the sedimentary record. Some important detail appeared to be missing in our knowledge of the Equatorial Current system. This feature was discovered by Cromwell, Montgomery, and Stroup [1954] in the form of a mighty undercurrent, locked on the Equator by the rotational forces of the earth and running downhill from the West to the East with a core speed exceeding 1 m/sec. Our

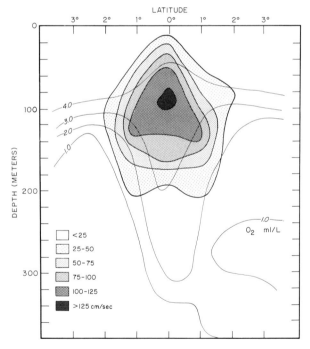

FIG. 2. Horizontal velocity distribution in the surface layer of the East Equatorial Pacific, showing velocity contours of the Equatorial Undercurrent. The figure also shows the distribution of dissolved oxygen, indirectly indicating the vertical circulation rate. From Knauss [1960].

knowledge of the detailed characteristics of this current (Figure 2) is largely due to John Knauss [1960].

What in Figure 2 appears as a cylindrical jet due to vertical exaggeration, is actually a thin ribbon, 300 km wide and 200 meters thick, that shoots through the sea with the core at a depth of about 100 meters. With a transport rate of the same order of magnitude as that of the Gulf Stream, this Equatorial Undercurrent is one of the major dynamic structures of the world ocean. It maintains a steady state by returning to the East much of the water driven over to the western side of the ocean by the trade winds.

It is probably instabilities on the undercurrent that cause efficient vertical mixing through the otherwise exhausted and sterile warm surface layer so that the Equator in this part of the ocean is marked by a narrow, cold, nutrient-rich, and turbulent strip of water with an organic productivity greatly increased over that of the blue ocean desert to the North and South.

The sediment layers below the equatorial zone of vertical mixing further reveal that this trade-wind-driven mechanism has been turning over at greatly varying speed during the last million years. The sequence of layering and the age relationships of the strata indicate that the alternating ice ages and warm interglacials which are characteristic of the last million years in the history of the earth were associated with pronounced increases (during the ice ages) and decreases (during the interglacials) of the trade-wind intensity, and probably of the meridional circulation in general. The oceanic record of these global events provides probably the most detailed information now available of any parameter reflecting the climatic change during the Quaternary.

Figure 3 shows the productivity zones in the ocean as indicated by measurements of standing crop and turnover in the surface sea water. From this it is evident that, in addition to the equatorial zone, intense vertical mixing or upwelling also occurs in the subarctic and subantarctic convergences. Again, this is reflected in the composition of the sediment below by increased concentration of organic remnants such as opaline silica (Figure 4) mainly from diatoms. The subantarctic region particularly provides a potential memory record of climatic evolution in the high latitudes. This record has not yet been extensively studied. However, it is clear from our present information that the subantarctic convergence during the last ice age was displaced a considerable distance toward the Equator in comparison with its present and interglacial location. It is also evident from the stratigraphic record in the South Pacific and Indian Ocean and in the North

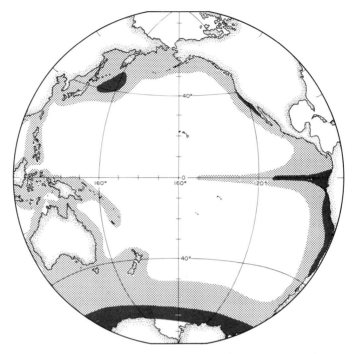

Fig. 3. Productivity of Pacific Ocean surface waters, estimated from observations of living planktonic organisms. From Sverdrup [1954].

Pacific, both investigated by Jousé [1960, 1961], that such invasions of the subpolar regimes into lower latitudes has occurred at regular intervals in the last geological period; these expansions appear to be, at least approximately, contemporaneous with the ice ages in the high northern latitudes. To put the geophysical interpretation of these sequences on a quantitative basis, measurements are needed both of the time and the amplitude factors by analysis of the proper chemical and structural parameters of the sediment, which are capable of providing the memory information. Such studies should provide, when combined with the already existing equatorial observations, a solution to the outstanding problem concerning the simultaneous or out-of-phase occurrence of glaciations in the Northern and Southern Hemispheres. They should also prove or disprove the Milankovic theory of climatic oscillations, give further details on the circulation of the ocean and the atmosphere during ice ages and interglacials, and define the location of the Equator and the poles in the past. Concerning this latter problem, we can state with certainty at the present time that the Equator has remained within one or two degrees of its present location in the East and Central Pacific throughout the last half-million years. This is as far back in time as we have been able to secure reliable records; the technology for probing at least thirtyfold deeper into time has been, however, available for several years.

As we have seen, the atmospheric circulation is responsible in an indirect way for the mixing-production mechanisms discussed. It would be interesting to obtain some direct evidence of the change in the paths and intensities of winds in the past. A suitable memory input for this purpose is furnished by the amount and distribution over the ocean of atmospheric dust from the continents.

Such dust consists of the major low-density continental minerals in size ranges depending on the time of flight from the source. Among these minerals the members of the mica group are particularly useful

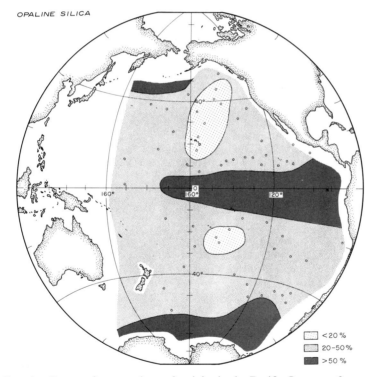

FIG. 4. Present-day organic productivity in the Pacific Ocean surface waters as indicated by the concentration of opaline silica (mainly from diatoms and to some extent from radiolaria) in the surface layer of the bottom deposit. From Bonatti and Arrhenius [in preparation].

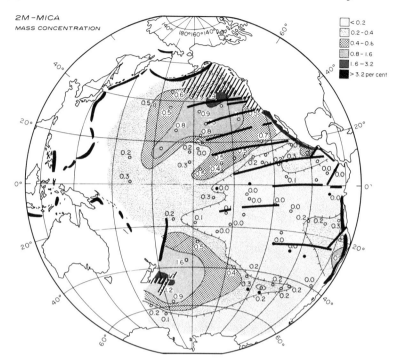

FIG. 5. Concentration distribution of mica at the present sediment surface of the Pacific Ocean floor. Features of particular interest are the zones of wind-transported mica from the Australian continent and from the Atacama Desert, and the apparent absence of a similar transport from the Asian continent in intermediate northern latitudes. Similar distributions are found for other detrital minerals like quartz and plagioclase. The distribution of mica and associated minerals off the North American continent is complex; south of the Murray Fracture Zone eolian transport prevails. North of that area turbidite transport appears to be responsible for much of the sediment distribution. Such geomorphological features, which are effective in protecting the oceanic basins against inflow of turbidites (fracture zones, trenches, and troughs) are indicated in black in the map; also shown by white striping are the flat bottom areas in the northeast Pacific and off New Zealand, indicating extensive accumulation of turbidites. Mica distribution from Bonatti and Arrhenius [in preparation]; geomorphological features compiled from Menard [1964].

as indicators because of the unusual sensitivity with which they can be determined and because of the flaky grain shape which favors wind transport over large distances.

Figure 5 shows the distribution of $2M$-mica in the present sediment surface of the Pacific Ocean. The distribution of windborne dust is particularly well recorded in the South Pacific, where confusion with river-transported continental minerals does not occur, and where deep trenches prevent turbidity currents, loaded with such minerals from the continental slope, to spread over the ocean floor.

Perhaps the most remarkable feature of the distribution in the Southern Hemisphere is the trace of high mica concentration left in the roaring forties, where dust is carried from Australia almost all the way across the South Pacific. Another quantitatively less important source of eolian dust is the Atacama Desert on the South American coast.

In the northern part of the Pacific Ocean, the region of extensive

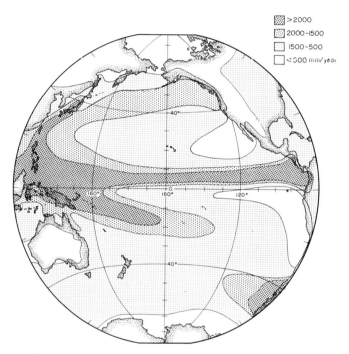

FIG. 6. Rate of rainfall over the Pacific Ocean and adjacent areas, demonstrating the control of eolian tropospheric dust transport exerted by coprecipitation with rain. The rate of rainfall over open oceanic areas is uncertain and largely extrapolated from qualitative observations at sea and from measurements on islands and continental stations. Features of particular interest are the coincidences of low precipitation with transport zones of tropospheric dust, as indicated in Figure 5, extending downwind from the Sonoran, Atacama, and Australian deserts. Furthermore it is interesting to note the existence of a meridional zone of high rainfall off the Asian coast in the middle northern latitudes. Precipitation in this zone is probably responsible for the removal from the atmosphere of the large amounts of tropospheric dust which are transported from the Asian continent by strong westerly winds around 40°N. Modified from Schott [1935].

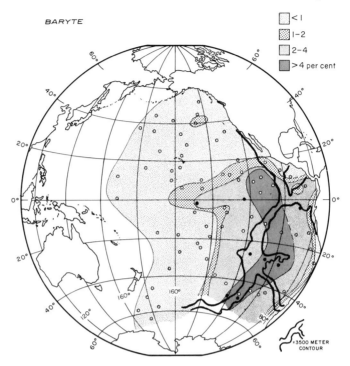

FIG. 7(a). Distribution of baryte in the carbonate-free component at the present sediment surface of the Pacific Ocean floor, showing the high baryte concentration on and near the East Pacific Rise (supposedly due to volcanic injection) and below the equatorial productivity zone, where precipitation is due to planktonic and benthonic organisms. From Boström *et al*. [in preparation].

eolian deposition overlaps the area off the northern United States and Canada, which, because of the absence of shielding trenches, receives considerable inflow of continental supensions rich in mica, chlorite, plagioclase, and quartz presumably spreading along the bottom in turbidity currents. From this transport results a characteristic basin-fill topography of extreme flatness [Menard, 1964, p. 195], contrasting with the rolling hill topography typical of the Pacific pelagic sedimentation basins [Arrhenius, 1952, 2.57; 3.4.1]. The flat bottom area, indicative of turbidity deposition, is delineated in Figure 5. The East Pacific deep ocean floor, south of the Murray Fracture Zone [Menard, 1964, p. 15], is protected against such influx from the continents by ridges, basins, or trenches parallel to the coast; spreading of turbidites from the North is prevented by the topographic barriers provided by the fracture zones. It is only south of the turbidite area that the concentration of minerals such as mica, chlorite, quartz, and plagioclase can be assumed to be mainly windborne and where their distribution is indicative of the intensity and direction of easterly winds.

Another remarkable feature of the present-time mica distribution is the absence of a zone of high concentration in the latitudinal zone extending off the coast of Japan, in the strong westerlies, which are known to carry large amounts of dust from the Asian continent over the China Sea [Futi, 1939]. It is probable that the failure of this aerosol to reach far out in the ocean is due to precipitation by rain in the meridional zone of high rainfall over and east of Japan (Figure 6) in contrast to the relatively low rainfall over the New Zealand area. Still lower rainfall characterizes the oceanic areas with eolian deposition off the coasts of lower California and Chile, confirming the observation by Wilkins [1958] that scrubbing by atmospheric precipitation is an important factor controlling the distribution of aerosols. This has to be taken into account also in interpreting the eolian mineral distribution in the past.

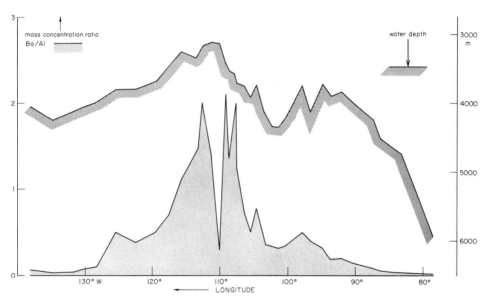

FIG. 7(b). Distribution of barium in a profile across the East Pacific Rise at latitude 12°S. The graph illustrates the high abundance of barium in relation to aluminum on the Rise. Both elements were determined by X-ray spectroscopy. The elemental distribution of barium shown in this figure parallels that of crystalline baryte ($BaSO_4$) determined by X-ray diffraction and illustrated in Figure 7a. Although a minor fraction of the barium occurs in the form of the barium zeolite harmotome, the major amount of the element is present as baryte. From Boström et al. [in preparation].

The changes in time of the distribution and transport rate of the eolian component is now being investigated; present evidence indicates a marked intensification of the wind transport during a Pleistocene stage, older than 30,000 years [Bonatti and Arrhenius, 1965; Arrhenius and Bonatti, 1965].

One of the memory devices which appears promising is the control of crystal structure and substitution exerted by the sea water itself on the solids that crystallize in equilibrium with it. A notable example is barium sulfate (baryte), which in solid solution harbors strontium, lead, and radium among other cations, and complexes such as chromate and possibly borotetrafluoride substituting for sulfate. These crystal species show a remarkable distribution (Figure 7a) with a very high concentration roughly coinciding with the East Pacific Rise (shown by the heavy contour in Figure 7a), which is also characterized by extensive volcanism and high heat flow [von Herzen and Uyeda, 1963].

Elemental analysis of barium (Figure 7b) indicates a distribution analogous to that of baryte. (The barium concentration is given in relation to unit mass of aluminum in order to eliminate the effect of varying dilution with biogenic minerals.)

At the present time sufficient information does not exist to determine conclusively the reason for the high relative barium concentration on the East Pacific Rise. In light of available evidence, however, we favor the suggestion that this is a source area where barium is introduced in the deep ocean water by volcanic processes and precipitates as baryte due to supersaturation. The tongue of high baryte content extending below the Equator is probably again the effect of the production mechanism, maintaining by its rain of organic solids over the ocean floor an abundant bottom fauna. This includes a most remarkable group of primitive organisms, the Xenophyophora, which are known for their habit of secreting baryte into the protoplast [Arrhenius and Bonatti, 1964]. The distribution of these organisms (Figure 8) illustrates the supporting effect of the surface productivity in the Equatorial Zone, also on the bottom living organisms.

Besides the benthic Xenophyophora, a number of planktonic organisms are known to concentrate the heavy elements of group 2a. The rain of organic detritus, which is continuously falling from the surface toward the bottom of the ocean in the plane of the Equator, transports these katabolites vertically into the deep water and to the ocean floor at a much more rapid rate and in a considerably more localized fashion than would be possible solely by hydrodynamic convection processes, which typically have turnover times of the order

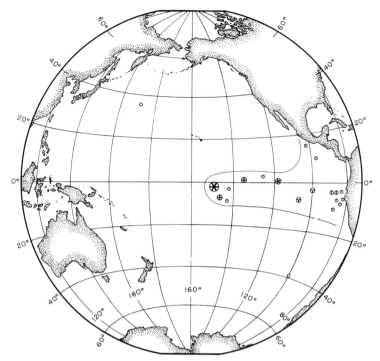

FIG. 8. Occurrence of Xenophyophora in dredge hauls from the Pacific Ocean floor. The number and length of radii drawn in each circle indicate the number of species found on the corresponding stations. In general, the abundance of species is known to be related to the production rate of the organism group in question. Data from Schulze [1906].

of $10^3$ years [Wooster and Volkman, 1960; Bien, Rakestraw, and Suess, 1960].

Some of the biogenic precipitates, such as crystalline strontium sulfate (celestite), secreted as skeletal spicules by a group of planktonic protozoa, dissolve rapidly on their way to the bottom; this is the case also with the silica of the majority of diatoms. For this reason the equatorial productivity zone is probably manifest also by a curtain of relatively high concentration of the ensuing dissolved chemical species. Ionic and convective diffusion would only slowly dissipate such concentrations. Under these circumstances it would be interesting to attempt tracing of the meridional flow in the intermediate and deep water by departure of the solution curtain from the equatorial plane —we have not yet had the opportunity to try this in practice. Strontium ion would appear to be an ideal tracer for this purpose.

The baryte distribution mentioned previously is only one of the many that indicate the importance of volcanic processes on the ocean floor—the extent and distribution (Figure 9) of this volcanic activity gives interesting aspects on the history of the ocean. One of these is related to the fact that volcanic effusion provides the raw material for a group of structures which show considerable promise as memory matter, namely, the aluminosilicates. The crust of the earth consists essentially of close-packed oxygen atoms held together by mixed covalent-ionic bonds to small, highly charged silicon and aluminum ions. The major part of the other elements simply serve as interstitial cations, preserving the electroneutrality of the structure. Aluminum is particularly interesting in this context in view of its ambivalent and indicative behavior. As is well known, the coordination of large ligands,

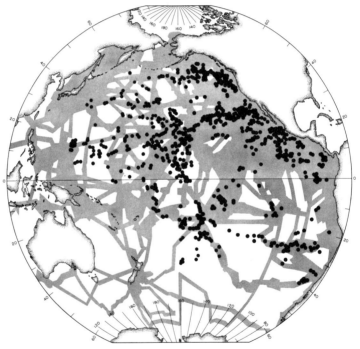

FIG. 9. Distribution of volcanoes in explored areas of the Pacific Ocean. Light surface tone indicates the distribution of all bottom exploration tracks on which information was available to the present author. (The area within 60 nautical miles from each track is included in the area indicated as explored.) White areas are unexplored, or explored by spot soundings only. Centers of black circles indicate the position of features which are believed or known to be volcanoes. The map is based on information from Menard [1964, and personal communication].

like oxygen, around small core atoms can be achieved by several different configurations (Figure 10), for example, in a trigonal fashion such as in carbonates, borates, and nitrates; tetrahedrally such as around silicon in silicates, or octahedrally, with six ligands, such as in the transition metal oxides. The actual choice is determined by two factors, one steric and simply depending on the size of the core ion relative to the ligand. The larger the effective size of the core ion, the more room there is on its surface for associated ligands, or with constant size of the core, higher coordination is possible with smaller ligands. The other factor is the directional interaction of the wave

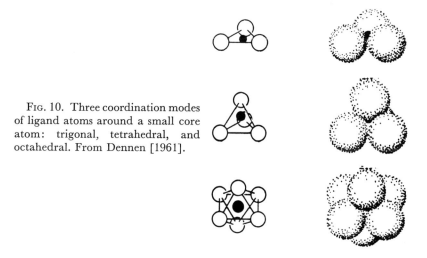

FIG. 10. Three coordination modes of ligand atoms around a small core atom: trigonal, tetrahedral, and octahedral. From Dennen [1961].

functions of the valence electrons, minimizing the energy by different hybrid orbital configurations. The trigonal and tetrahedral coordinations need only $s$-$p$ hybrids, whereas the octahedral coordination with six vectors involves also $d$ states.

Aluminum, which ties oxygen in a half-ionic, half-covalent bond, has such a size range that it can accommodate a surrounding cluster of either four or six oxygen ligands. Furthermore, although aluminum in the ground state has no $d$ electrons, the excitation of $d$ states within the energy range of the bond is becoming possible in this part of the periodic system, and the six-coordination can supposedly be stabilized by a certain amount of $sp^3d^2$-hybrid character of the molecular orbital [Hägg, 1963; 24-3f]. As a result, aluminum-oxygen structures of both tetrahedral and octahedral coordination types are stable at room temperature and pressure. The former is widely represented by the tetrahedral network in feldspars and zeolites; the latter, also frequent in nature, occurs particularly in the octahedral sheets of mica and the micalike clay minerals.

It has been suggested [Arrhenius, 1959b] that with other chemical potentials constant, as is largely the case in sea water, the hydrogen ion activity controls the coordination in aluminosilicates condensing from solution. With increasing hydrogen ion activity in the solution, hydroxyl bonds are formed between the hydrogen ions and the tetrahedrally coordinated oxygen ligands in the aluminosilicate polymer; this leads to change in coordination from four to six around the aluminum core.

It is not *a priori* clear if this transformation is due primarily to steric effects or to hybrid orbital effects of the type mentioned above. If steric effects are assumed to be responsible, it would at first seem surprising that a change of ligand from oxygen to hydroxyl ion, corresponding to a slight increase in effective size of the ligand, could lead to an increase in the coordination number. To account for the phenomenon one could argue that the formation of the hydroxyl bond leads to a decrease in the polarizing power of the ligand, and that electrons are for this reason released back onto the aluminum core atom, which as a result increases its effective size and consequently would be able to accommodate a large number of ligands around it even in spite of their increased size. It is interesting to confront such an explanation with the information on the valence band structure that can be obtained from X-ray emission spectra.

It is well known from early work in X-ray spectroscopy that changes in the valence state measurably affect the transition energies between electron shells inside the valence shell. White, McKinstry, and Bates [1959] were the first to demonstrate that changes in coordination also result in energy shifts of this nature. These effects have been used extensively on an empirical basis for assessment of the valence state of atoms, particularly by Faessler and, more recently, for determination of the coordination state. In contrast to the X-ray absorption phenomena used for similar purposes, particularly by Cauchois and by Faessler, no theoretical explanation appears to have been given for the emission energy shift phenomena, which could justify the generalized interpretation in terms of coordination- or valence-state changes.

It appears reasonable that the explanation of these phenomena has to be sought in the fact that valence orbitals with low azimuthal quantum states penetrate near the nucleus and consequently contribute to screening also between inner orbitals. This is qualitatively illustrated by Figure 11 which demonstrates that primarily the $s$ and to a lesser extent $p$ electrons in the valence shell can effectively interpose charge between the 2*p* and 1*s* states. Any process in the valence band, resulting in the removal of electrons with a low azimuthal quantum number, or in a change in the effective azimuthal quantum number would consequently be expected to lead to a shift in the emission lines from the inner level transitions. Such changes may

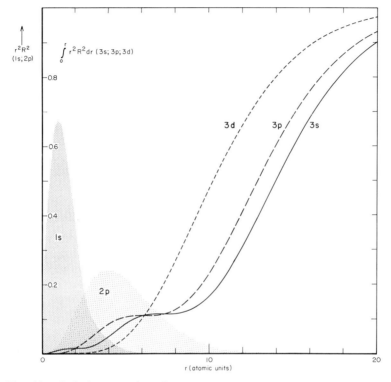

FIG. 11. Relative screening effect of outer electrons on the $2p \rightarrow 1s$ transition (resulting in $K\alpha$ emission). Hydrogenlike orbitals are assumed, and the deformation caused by excitation of one of the 1*s* states is neglected. The charge density distributions $r^2R^2$ for 2*p* and 1*s* are indicated by surface tone. To demonstrate the interposition between these levels of charge, originating from outer electrons (valence electrons in the case of period-3 elements), the charge density distribution functions for 3*s*, 3*p*, and 3*d* have been integrated, and the radial distribution of the integrals are plotted. The graph indicates the large screening effect of the outer *s* and *p* orbitals compared with *d* and higher states, and qualitatively explains the observed emission frequency shift resulting from changes in oxidation number or ligand type, and, according to the present analysis, also from changes in orbital hybridization. A numerical evaluation of these effects is being attempted at the present time, using a self-consistent field approach and incorporating the fact that only one of the 1*s* orbitals actually is occupied in the X-ray experiment.

be achieved in at least three different ways: by change in the oxidation number of the emitting atom, or by change in the electronegativity of the ligand atom, both resulting in removal or addition of electrons with respect to the emitting atom, and finally also by change in hybridization of the valence orbitals resulting in an alteration of the radial distribution of charge in the emitting atom. In many cases, two, or all of these three permutations occur simultaneously but in widely varying proportions.

It is interesting then to note that the measurements by White, McKinstry, and Bates [1959] show that the $2p$–$1s$ transition in aluminum is considerably more energetic when the ion is octahedrally coordinated than in the case of tetrahedral coordination. This is not consistent with the assumption that the cause of the coordination change is the return to the core of electrons due to depolarization; this would lead to the opposite effect by increased screening between the $2p$ and $1s$ levels. Instead, it is suggested that the impression of hydrogen bonds on the ligands leads to an increased amount of $sp^3d^2$-hybrid character of the molecular orbital and that this is the cause for the stabilization of the octahedral coordination. This would be in accord with the observed energy shift in the aluminum $K\alpha$-emission line ($2p$–$1s$) since the change in character from mainly $sp^3$ in the tetrahedral case to at least partly $sp^3d^2$ associated with octahedral coordination corresponds to an effective decrease in the density of $s$ states resulting in decrease in screening between the inner orbitals.

In principle then, the balance between the two structure types constitutes a record of the hydrogen ion concentration in the solution from which the equilibrium solids grew. The resulting information contained in the structure appears potentially useful in the interpretation of the sedimentary record, both from the ocean and from continental water bodies. The ocean today is a buffered alkaline system and the tetrahedral structure appears to prevail in all situations where aluminosilicates crystallize in equilibrium with sea water at low temperature. But in the primeval ocean or in closed interstitial systems, the activities might well have been different enough to change this balance. The experiments indicate that the transformation falls in the range of pH 7.5 to 8.5, near or within the hydrogen ion activity range now prevailing in the ocean.

I conclude with these examples of investigations in process in order to indicate some of the possible pathways toward a quantitative treatment of the ocean's memory mechanisms. I mention them also to emphasize the eagerness with which everybody active in this field is looking forward to the day when it will be politically expedient to make use of existing intermediate drilling technology for the purpose of penetrating the total preserved record of the evolution of the earth. This record might well be contained in the ocean floor in a higher state of fidelity than anywhere else. The present interrogation of the most superficial layers of the ocean's memory should be only a beginning.

## Acknowledgments

Many of the results discussed above are new, and previously unpublished. They are the outcome of research, sponsored by the U.S. Atomic Energy Commission (AT11-1-34) and by the American Chemical Society's PRF-fund (875-C6). The generous support from these agencies is gratefully acknowledged.

The author also wishes to thank Drs. Kurt Boström, Sigurd Burckhardt, Joseph Mayer, Jason Saunders, and Captain John Sinkankas for helpful comments and criticism.

## References

Arrhenius, G., Sediment cores from the East Pacific, *Rep. Swed. Deep Sea Exp.*, 5, Göteborg, 1952.
Arrhenius, G., Climatic records on the ocean floor, Rossby Memorial Volume, Rockefeller Institute Press, New York, 1959a.
Arrhenius, G., Crystallization of zeolites on the ocean floor, in *Internat. Oceanogr. Congr. Preprints*, p. 447, edited by M. Sears, A.A.A.S., Washington, D.C., 1959b.
Arrhenius, G., Pelagic sediments, *The Sea*, 3, p. 655, edited by M. Hill, Interscience Publishers, Inc., New York, 1963.
Arrhenius, G., and E. Bonatti, Neptunism and vulcanism in the ocean, in *Progress in Oceanography*, edited by M. Sears, Pergamon Press, New York, 1964.
Arrhenius, G., and E. Bonatti, The Mesa Verde loess, *Soc. Am. Archeol. Mem.*, 18, 1965.
Bien, G., N. Rakestraw, and H. Suess, Radiocarbon concentration in Pacific Ocean water, *Tellus*, 12, 436–443, 1960.
Bonatti, E., and G. Arrhenius, Pleistocene eolian sedimentation in the Pacific off the coast of Mexico, *Marine Geology*, in press.
Boström, K., E. Bonatti, B. Holm, and G. Arrhenius, Oceanic baryte, in preparation, 1965.
Cromwell, T., R. Montgomery, and E. Stroup, Equatorial undercurrent in Pacific Ocean, revealed by new methods, *Science*, 119, 648–649, 1954.
Dennen, W., Principles of mineralogy, Ronald Press Co., New York, 1960.
Futi, H., On dust storms in China and Manchukuo, *J. Met. Soc. Japan*, Ser. 2, 17, 473–486, 1939.
Hägg, G., *Allmän och oorganisk kemi*, Almquist och Wiksell, Uppsala, 1963.

Jousé, A., Les diatomées des dépôts de fond de la partie nord-ouest de l'Océan Pacifique, *Deep-Sea Res.*, *6*, 187–192, 1960.

Jousé, A., Diatomovye vodorosli i ikh rol v vyiasnenii istorii okeanov, *Izv. Akad. Nauk SSSR Ser. Geogr.* no. 2, 13–20, 1961.

Knauss, J., Measurements of the Cromwell Current, *Deep-Sea Res.*, *6*, 265–286, 1960.

Menard, H. W., *Marine Geology of the Pacific*, McGraw-Hill Book Company, New York, 1964.

Schott, G., *Geographie des Indischen und Stillen Ozeans*, Boysen, Hamburg, 1935.

Schulze, F. E., Die Xenophyophoren der amerikanischen Albatross-Expedition 1904/1905, *Sitzber. Ges. Naturforsch. Freunde Bln.*, no. 8, 205–229, 1906.

Sverdrup, H., The place of physical oceanography in oceanographic research, *J. Marine Res.*, *14*, 292, 1955.

Von Herzen, R., and S. Uyeda, Heat flow through the eastern Pacific ocean floor, *J. Geophys. Res.*, *68*, 4219–4250, 1963.

White, E., H. McKinstry, and T. Bates, Crystal chemical studies by x-ray fluorescence, *Proc. Ann. Conf. Ind. Appl. X-Ray Analy.*, *1958*, Denver Research Institute, Denver, Colo., 1959.

Wilkins, E. M., Precipitation scavenging from atomic bomb clouds at distances of one thousand to two thousand miles, *Trans. Am. Geophys. Union*, *39*, 60–62, 1958.

Wooster, W., and G. Volkman, Indications of deep Pacific circulation from the distribution of properties at five kilometers, *J. Geophys. Res.*, *65*, 1239–1249, 1960.

# THE LARGE-SCALE OCEANIC CIRCULATION

*Henry M. Stommel*

Massachusetts Institute of Technology, Cambridge, Massachusetts

Our information about the general circulation of the ocean is rather incomplete. We do not have systematic data on currents, velocity, temperature, and salinity at frequent intervals from a permanent world network of stations, such as the meteorologist has in the air; we do not even have techniques developed for obtaining such data in limited areas. Fortunately certain fields such as temperature and salinity in the deep waters of the ocean seem to be only slightly time-dependent, so that mapping of these properties over large volumes of the ocean can be done using data collected at very irregular times. On the other hand, studies of time variability of properties are extremely scarce, and it is impossible to estimate even the magnitude of turbulent fluxes of properties on most time scales.

So in trying to think about the large-scale oceanic circulation, we rely on certain mappings of temperature and salinity data which we suppose depict the mean state and upon certain hypotheses about the nature of the dynamical processes at play in the ocean. Let me expand a little on this last clause. The ocean is a fluid and is described by the Navier-Stokes equations of hydrodynamics plus other equations expressing the conservation of heat and salt in the presence of advection and diffusion. The structure of these equations is too rich to yield completely to analytical or even numerical methods. In all branches of science that deal with fluid media it is found that simplified systems of equations can be postulated and justified which seem to describe certain limited subsets of the set of all possible fluid processes; thus we have the subsets of incompressible flow, of perfect fluids, of low Reynolds number laminar viscous flow, etc. And in thinking about various engineering works or natural fluid-dynamical phenomena, such as flow in turbines or rivers or waterfalls or hurricanes, we are concerned largely with establishing the dominant interplays of forces and of mechanism for transfer of properties. Stated in words corresponding to the mathematical formulation, what are the dominant terms in the hydrodynamical equation in different parts of the ocean? Are there regions within the ocean distinguished from others by having a different dynamics?

I want, therefore, to try to present an over-all picture of our present-day theoretical interpretation of the general oceanic circulation in terms of a composite of different types of dynamical regimes. It is basically a mathematical-hydrodynamical model in which many important details of rigor have not been logically worked out yet and which involve elements that the observational side of oceanography is not yet able to test or measure because of lack of suitable observing techniques. When I emphasize the imperfection of observing techniques perhaps I should say that I wrote this chapter during a succession of midnight-to-dawn watches during an attempt to survey the Somali Current neat Socotra in the heart of the Southwest Monsoon. It is rather quixotic to try to get the measure of so large a phenomenon armed only with a 12-knot vessel and some reversing bottles and thermometers. Clearly some important phenomena slip through the observational net, and nothing makes one more convinced of the inadequacy of present-day observing techniques than the tedious experience of garnering a slender harvest of thermometer readings and water samples from a rather unpleasant little ship at sea. A few good and determined engineers could revolutionize this backward field.

Basically our task of theoretical interpretation is to show how, using the hydrodynamical equations and the equation of heat transfer, we can construct a mathematical model of circulation which bears some resemblance to what we dimly perceive by very imperfect observational techniques to be the real ocean's circulation.

As is well known, the general qualitative features of any hydrodynamical process depend essentially on certain nondimensional numbers—for example, in ordinary hydraulic and low-speed aerodynamics, the Reynolds number plays a decisive role: its magnitude determines whether the flow will be turbulent, whether boundary layers will detach and form waves, etc. The ocean is on a rotating globe, and this introduces another number, the Rossby number, which is a ratio of local relative vorticity to the angular velocity of the earth. Thermal processes introduce still another dimensionless combination. Thus, depending on the choice of these dimensionless numbers, there are many different physical regimes, each corresponding to different limited regions in the total number space. To explore all of the associated parameter space would carry us through most of the range of hydrodynamical phenomena. We do not know enough about turbulence in the ocean to specify appropriate values for these parameters *a priori*.

What we have done is to search around in this parameter space for a region in which the theoretical qualitative behavior of the fluid dynamical regime seems to correspond to that of the real ocean, and this is what we have found: First, that the stress field produced at the surface of the ocean by the wind does not penetrate beyond a hundred meters depth in the ocean. This is a most remarkable result of the interaction of viscous shearing stress and Coriolis force, first rigorously demonstrated by Ekman many years ago, and many times confirmed in the atmosphere and in laboratory tanks. A refined set of measurements in the ocean has never been achieved, but there is such evidence that the concept of a thin Ekman layer is applicable to the ocean.

In order to obtain quantitative agreement, it appears necessary to make use of an "eddy viscosity," in place of the much smaller molecular stirring and mixing processes in the surface of the sea which are not defined as part of the simple Ekman theory. The transport of water in the Ekman layer, fortunately, does not depend upon the value of the eddy viscosity and is at right angles to the direction of the applied stress (to the right of it in the Northern Hemisphere). This is a very useful quantitative result which holds in quite a reasonably large volume of nondimensional parameter space. The chief weakness of application of the Ekman theory to calculating transports in the upper viscous boundary layer of the ocean is our inability to produce reliable estimates of the mean wind-stress from observed winds over the sea. This is a micrometeorological problem of considerable importance and difficulty and needs to be pursued vigorously. It poses very difficult technological and engineering problems.

In the idealized ocean which we have set up, the Ekman transports are all meridional, so the meridional coasts are not barriers to the Ekman flow. However, the Ekman transport is a function of latitude. Mass conservation within the Ekman layer therefore requires a vertical flux into and out of the Ekman layer. This evidently must occur mostly at the bottom of the layer because only a small amount of water flows through the sea surface by evaporation precipitation. The

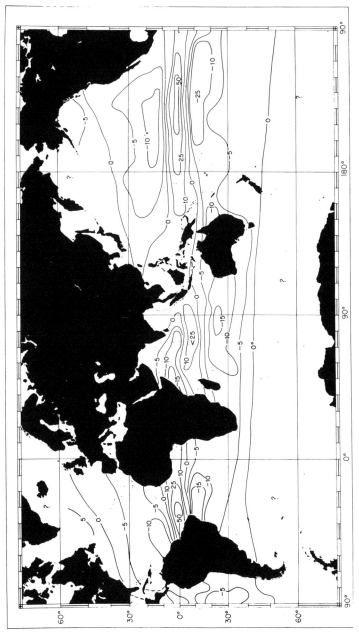

Fig. 1. 5° means of annual mean vertical component of velocity beneath the Ekman layer in cm/day computed from wind stress and Ekman theory.

Ekman layer in our model therefore imposes a vertical mass flux on the deeper regions of the ocean below. Figure 1 shows the mean vertical component of velocity beneath the Ekman layer in cm/day, positive upward.

The neighborhood of the equator presents us with a singularity because here the calculated transport of the Ekman layer increases without limit. Evidently some higher order dynamical regime must prevail here, the simple Ekman theory being locally inapplicable.

Beneath the Ekman layer—from a depth of about 50 meters to the neighborhood of the bottom where there may be another much weaker Ekman layer of little importance—the flow is nearly geostrophic. This is a miliar type of flow to anyone who has studied weather maps. The horizontal components of the flow are parallel to the isobars: Dynamically, the forces balancing in the horizontal are Coriolis forces and horizontal pressure gradients, and the vertical balance of forces is purely hydrostatic. Geostrophy of the deep interior ocean is an approximation: We do not know precisely how good an approximation because we are unable to undertake the type of observational program that would be needed to achieve statistical significance to the results. However, we have lived with the concept of geostrophy in the deep ocean interior for many years and have not found it unreasonable.

The geostrophic flow fields in the atmosphere are incessantly changing with time. In the ocean we visualize them as more nearly stationary.

Figure 2 shows the pressure on a level surface just beneath the sea surface, expressed in terms of meters of water, or in other words, this chart is a topographic map of the physical sea surface above the geoid. Around Antarctica the level is low, less than 0.4 meter above an arbitrary level. As we proceed northward we encounter large high-pressure ridges running from West to East. These ridges are highest at the western sides of the oceans. The ridges lie on opposite sides of the Equator which is where the topography is generally low and fairly flat. The highest point of all is in the North Pacific near Japan—a height of 2.4 meters above the arbitrary level. So all in all, this topography of the sea surface is quite gentle. How do we obtain this map? We obtain it not by direct leveling determinations, certainly, but by computation from the observed density field and the assumption, which we cannot fully justify, that deep pressure surfaces are nearly level, i.e., that the thin low-density surface layers of the ocean float isostatically on top of the large volume of water of nearly uniform density which fills most of the ocean basins. Under this assumption the less dense and thicker the upper layers, the higher their top

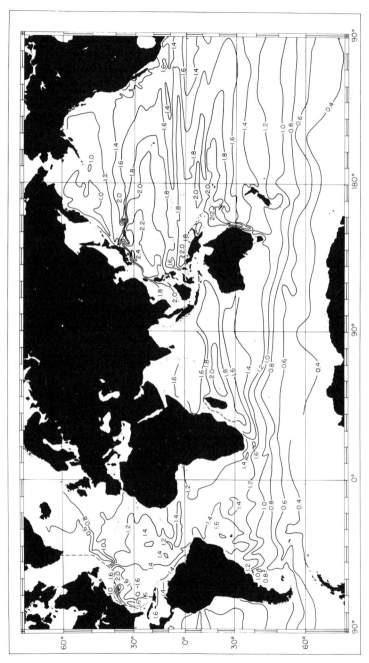

Fig. 2. Levels of the sea surface (in meters) above an arbitrary level surface.

surface floats. Moreover, the effects of seasonal variability of density are mostly limited to very shallow depths so this topographic map is not seriously modified by season. In constructing it we find that a choice of 20-cm contour interval permits us to ignore seasonal differences.

If the ocean circulation were completely geostrophic, this chart would give the direction and magnitude of the currents just beneath the Ekman layer: velocity parallel to the contours, with high level to right in the Northern Hemisphere, and, in the open ocean, characteristic magnitudes of from 1 to 3 cm/sec. Since the low-density water is as deep as 1000 meters in the ocean (the total oceanic depth being about 4000 meters), the horizontal pressure gradients extend to 1000 meters depth with diminishing magnitude but similar direction so that even slow velocities like 1 cm/sec do transport considerable water. Application of the geostrophic principle to the high-pressure ridges in mid-latitudes produces a remarkable result: The low-density water is found to be flowing toward the Equator in most of the width of the ocean instead of toward the poles as we might have expected from simplest intuition, but this result does appear to be in accord with the facts.

As a result of the fact that the Coriolis parameter is a function of latitude, equatorward components of geostrophic motion are horizontally convergent; water must be supplied to or taken away from a meridionally directed geostrophic stream for it to remain steady. This occurs by vertical flows from other layers above or below the geostrophic layer. This is of course the link between the viscous wind-driven Ekman layer on the top of the ocean and the underlying geostrophic regime. The meridional component of the vertically integrated geostrophic flow is determined by the vertical flux from above by the Ekman layer.

The other component of the geostrophic flow must be determined from the continuity (or mass conservation) equation. This leads us to a differential equation of first-order for the mass-transport function which can satisfy only one lateral boundary condition, but there are two boundaries. This means that, even outside the Ekman layer, the flow cannot be geostrophic everywhere. Indeed, we then find that under certain circumstances, narrow boundary layers can logically develop at either boundary, but in nature the main nongeostrophic boundary currents seem to be confined to the western boundaries of oceans. Thus we have strong currents like the Gulf Stream, the Kuroshio, the Somali Current, as local regions where pure geostrophy breaks down. With the introduction of these regions, which can be incorporated most easily into the formal theory by means of singular perturbation theory, the theory of the vertically integrated mass transport is in principle completed. But there are some important logical points about the existence of such boundary layers when viscosity is small and of possibly more stable quasi-periodic solutions such as limit cycles, which make it impossible to declare that the formal mathematical problem is completely and satisfactorily resolved by this steady solution.

We must now consider the role of density structure (due to the temperature field) in the ocean and how that is related to the circulation. In the simplified model, which we have been discussing, we might at first suppose that there was no imposed temperature field at the surface: that the temperature is uniform. Then the water everywhere would be homogeneous, and the geostrophic currents would not vary with depth. For this case the vertical component of velocity within the geostrophic layer decreases from its value imposed by the Ekman layer at about 50 meters to zero at the bottom.

The real interior of the ocean has a marked density structure. At depths of from 200 to 900 meters in tropical and sub-tropical regions, the warm surface water is separated from the much more voluminous cold deep water by a layer of rapid temperature change called the main thermocline. From a mathematical point of view, what we need to do is to couple the linear equation for the geostrophic dynamics with the nonlinear equation describing the advection and diffusion of heat. Nonlinear partial differential equations are difficult to solve in general, but fairly general forms of similarity transformation have been found that reduce the equation to a single nonlinear ordinary differential equation, and this, when solved numerically, has been shown to yield many of the main features of the main density distribution of the ocean. Nearly everywhere in the ocean the deep water flows upward, and the sinking regions which supply the cold water are evidently quite limited in geographical extent. The theory of the sinking areas is now receiving attention.

In summary, then, let us review how the model works, guided on one hand by hints for observation that tell us in general what kind of physical regime occurs in the ocean, and on the other by partial solution of an idealized theoretical model which gives us quantitative results consistent with general physical conservation principles (Figure 3).

We start with a parcel of water at (1) which by being strongly cooled sinks to the bottom (2) and then flows southward in a western boundary current in which it spends only a few weeks before reaching

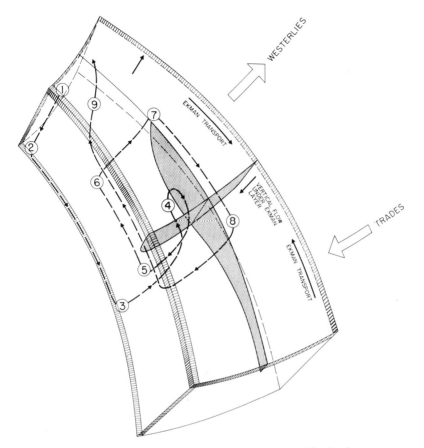

Fig. 3. Diagram showing circulation pattern in an idealized ocean.

a lower latitude (3), where it enters the geostrophic interior, flowing northeasterly and gradually ascending over a period of perhaps 200 to 600 years until it presses against the warm water being forced down from above by the Ekman layer at (4). As it becomes warmed it starts to flow southward at a higher rate of flow and in perhaps 1 to 3 years enters the western boundary current again at a higher level (5), this time flowing northward until it reaches a point (6) where it leaves the boundary current and ascends to the surface into a divergent Ekman layer. Once in the Ekman layer it is quickly (a month or two) passed southward until it reaches the region (8) of convergent Ekman layers, where it sinks into the geostrophic region, to meet the water ascending from (4), and is then passed through the western boundary current to the north (9) again—this cycle occurring 10 or more times until finally it reaches the sinking point (1) again and the cycle repeats itself.

As I have just sketched it, the motion is laminar; there were no irregularities in the motion. The picture is actually meant to illustrate a state of mean motion: Actual water particles do not stay on one of the paths long but execute partly random jumps from one path to another. In most of the ocean these jumps seem to occur 6 or 10 times a year and with a lateral displacement of perhaps 10 km. Vertical turbulent exchange of particles also occurs on a smaller scale and more frequently. The amplitude of the vertical exchange enters as an important parameter in the theory of the thermocline, but the lateral exchange, important in the dynamics of the oceanic circulation, is somewhat uncertain and will remain so until some definitive programs of measurements have been carried out in the deep sea to evaluate their influence.

# COUPLING OF WAVES, TIDES, AND THE CLIMATIC FLUCTUATION OF SEA LEVEL

*Walter H. Munk*

*Institute of Geophysics and Planetary Physics, University of California, La Jolla, California*

## Introduction

Some progress has been made in the last decade in studying the coupling of waves of different scales, particularly by Phillips, Hasselmann, and Longuet-Higgins. In the linear (uncoupled) approximation, each of the topics (local seas, distant swells, tides, etc.) can be independently discussed, and I would deserve your scorn in reviewing waves *and* tides when I could so easily saturate you with waves *or* tides. But the world is cruel and nonlinear; the "and/or" option may not always be available.

I will confine this discussion to three observed features that owe their existence to nonlinear coupling. Since water waves (unlike turbulences) are weakly coupled, theory is easy and observations difficult.

## Red Shift of Swell

Last summer a group from the Institute of Geophysics installed and maintained six wave stations between New Zealand and Alaska (Figure 1) to study the propagation of ocean swell over very large

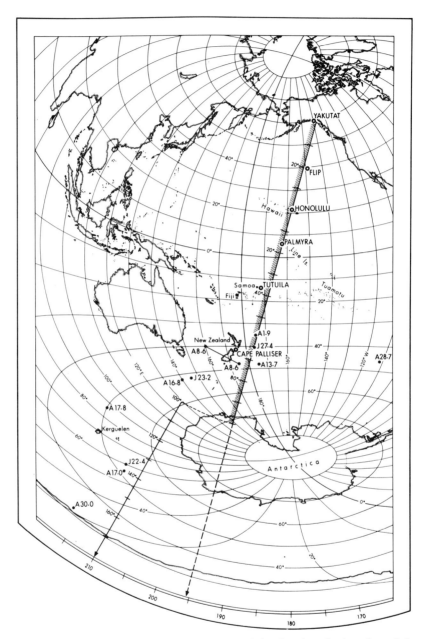

FIG. 1. Great-circle chart based on Honolulu showing the location of the six wave instruments and of the principal storm sources. Each storm is marked by a dot and its fractional date (J27.4 means July 24, 9.6h GMT).

distances [Snodgrass *et al.*, in preparation]. Wave records were taken twice daily for 100 days each station (30 days for FLIP), and power spectra computed. Swell generated by intense antarctic storms could be tracked from station to station for a dozen events. The power spectrum at the end of the storm fetch and at the nearest and farthest stations (typical distances from storms are 1000 km and 10,000 km, respectively) are somewhat schematically portrayed in Figure 2 for a typical event. The storm spectrum is inferred from Pierson's work; the station spectra are the composite of successive spectra so combined as to remove the

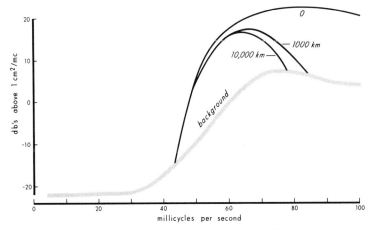

FIG. 2. Typical wave spectra at 0, 1000 km, and 10,000 km from an intense storm in the South Pacific, and the background spectrum at Honolulu, Hawaii. The effect of dispersion is removed from the storm spectra.

dispersive delay of high frequencies. Let $E(f, t)$ designate the spectral density at frequency $f$ and time $t$. The plotted spectra are $E(f, t_0 + \Delta/V)$ where $t_0$ is the time of the storm, $\Delta$ its great-circle distance from some station, and $V(f) = g/(4\pi f)$ the group velocity appropriate to frequency $f$.

We note a pronounced attenuation of high frequencies (and resultant "red shift") between storm and first station, and a small subsequent attenuation across the entire Pacific (typically 40,000 wavelengths). The early attenuation must be essentially complete after the first few hundred miles. Subsequent attenuation (apart from geometric spreading) is less than 0.05 db/deg below 70 mc/sec and of the order 0.1 db/deg at 80 mc (1 degree is 60 nautical miles = 111 km). At higher frequencies the swell train from a given event is lost beneath the background, which we attribute to a steady radiation from the entire storm belts of the southern oceans. Thus at low frequencies the attenuation is too small to be measured, and at high frequencies the swell is too small to be measured.

Nevertheless the observations pose some questions which are definitely beyond the scope of linear wave theory. Dissipation by molecular viscosity is utterly negligible; at 80 mc it is $2.5 \times 10^{-4}$ db/deg as compared to the observed amount 0.1 db/deg. The abrupt decay of the decay (a feature we had not properly anticipated) has an

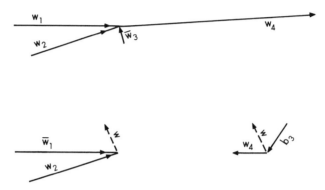

FIG. 3. Feynman-Hasselmann diagrams for triple interactions. Top: production of high wave numbers (shorter waves) by three interacting wave trains, a process ultimately leading to whitecapping. Bottom: production (in two steps) of low wave numbers by two gravity wave trains interacting with a "bottom wave."

all-or-nothing aspect so typical of nonlinear processes. How can we inquire whether and in what manner such nonlinear processes can account for the observed decay?

Hasselmann (in preparation) has exploited a geometric display introduced by Feynman to discuss particle interaction. Examples of F-H diagrams are shown in Figure 3. Any elementary wave train **w** is designated by an arrow pointing in the direction of wave propagation with length proportional to the reciprocal wavelength (or wave number) **κ**. This wave train is associated with three frequencies $\omega$, $\kappa_x$, $\kappa_y$: cycles per unit time, and the $x, y$ components of cycles per unit distance. In the particle analogy $\omega$ is energy and **κ** is momentum. It is now useful to introduce as a convenient abstraction the antiwave (antiparticle) **w̄** having a negative frequency (energy) $\bar{\omega}$ and a negative wave number (momentum) **κ̄**; but the wave slowness (reciprocal wave velocity) **κ̄**/$\bar{\omega}$ is still positive and in the direction of the arrow, as previously.

We consider the triple interaction

$$(\mathbf{w}_1\, \mathbf{w}_2\, \bar{\mathbf{w}}_3) \rightarrow \mathbf{w}_4$$

of two waves and an antiwave in producing a wave $\mathbf{w}_4$, whose frequency and wave number are determined by the rules

$$\varkappa_1 + \varkappa_2 + \bar{\varkappa}_3 = \varkappa_4$$

$$\omega_1 + \omega_2 + \bar{\omega}_3 = \omega_4$$

Thus $\mathbf{w}_1$ and $\mathbf{w}_2$ are added vectorially and $\bar{\mathbf{w}}_3$ antivectorially* (Figure 3). A solid vector represents $\mathbf{w}_4$ if the values of $\varkappa_4$ and $\omega_4$ are appropriate to a free gravity wave (particle), and a dashed vector if they are not (these are called *virtual* particles in the analogy). For deep gravity waves we have the relation $\omega^2 = g|\varkappa|$, and it can be shown that there is no doublet interaction

$$\varkappa_1 + \bar{\varkappa}_2 = \varkappa_3, \qquad \omega_1 + \bar{\omega}_2 = \omega_3$$

such that $\omega_i^2 = g|\varkappa_i|$, $i = 1, 2, 3$. But under certain conditions first formulated by Phillips [1960] the relation $\omega_i^2 = g|\varkappa_i|$ can be fulfilled for triple interactions leading to a "resonance transfer" portrayed in Figure 3. A single interaction (collision) may lead approximately to a doubling of frequency. Repeated interactions lead to a cascade of energy into high frequencies for which energy can be dissipated in whitecaps. The efficiency of this process can be determined only by rather detailed calculations of the coupling coefficients (collision cross sections). Hasselmann [1963] has shown that in a fully developed "Neumann Sea" the characteristic time scale for transferring energy from low to high frequencies is of the order of a few hours. In the storm area the lower frequencies are continuously regenerated. Just beyond the storm the regeneration stops but the nonlinear energy flux continues, and this leads within a few hours to a substantial reduction of the medium and high frequencies. Calculations show that the differences between the upper two curves in Figure 2 can be interpreted in this fashion.

The subsequent attenuation is more difficult to explain. Interaction of the existing swell with the *average* background is too weak by a factor of 100. But the background is sometimes temporarily raised by earlier winds associated with main event, or by winds from other storms, and it is under just these circumstances that attenuation

* The diagram would be simpler if we used $\varkappa$ vectors to represent interacting wave trains, for then the resultant wave train is obtained by straightforward vector addition. The disadvantage of this scheme is that $\mathbf{w}_3$ is then represented by a vector pointing opposite to the direction of wave propagation.

appears to be most pronounced. It turns out that the triple interaction (which may be regarded as "forward scattering" of $\mathbf{w}_1$ and $\mathbf{w}_2$ with the help of $\bar{\mathbf{w}}_3$) can then lead to the observed order of attenuation, provided some of the scattered energy cannot reach the station because of island shadowing. Under other circumstances the scattering may lead to an enhancement of waves with increasing distance. Both have been observed. It is then improper to speak of an *attenuation coefficient*. Attenuation depends on the meteorological and geographical circumstances. For wave prediction purposes it will be necessary to compute the collision cross sections per unit time per unit area for each situation. This is not a cheerful prospect.

**Surf Beat**

The previous discussion dealt with an interaction leading to high (or sum) frequencies. Next we consider the formation of low (or difference) frequencies. The interaction $\bar{\mathbf{w}}_1$ and $\mathbf{w}_2$ leads to a virtual wave $\mathbf{w}$ (Figure 3); over a flat bottom this would be of no consequence because of the absence of resonance; over a rough bottom there is a further interaction with each Fourier component of the bottom, and for the appropriate combinations of wave trains we obtain a triplet resonant interaction

$$(\bar{\mathbf{w}}_1 \ \mathbf{w}_2 \ \mathbf{b}_3) \to \mathbf{w}_4$$

$$\bar{\varkappa}_1 + \varkappa_2 + \varkappa_3 = \varkappa_4$$

$$\bar{\omega}_1 + \omega_2 + 0 = \omega_4$$

where the bottom is represented by a "bottom wave" of zero frequency. By taking $\bar{\omega}_1$ nearly equal to $-\omega_2$ we can generate oscillations of very low frequencies; in the extreme case of $\bar{\omega}_1 = -\omega_2$ we have a dc change in sea level due to radiation pressures [Longuet-Higgins and Stewart, 1962].

There is some evidence for the occurrence of this interaction over the shallow margins of the ocean basins. Between 5 and 30 mc the wave spectrum is remarkably smooth (Figure 2). The spectral density is typically 40 db below the peak spectrum of the waves, and is observed to be high on days of high waves and vice versa. It has long been suspected that these low-frequency oscillations are the result of the beating of the waves against one another; hence the name surf beats.

We have subjected the first part of the interaction leading to the virtual wave $\mathbf{w}$ (but not yet the second part) to experimental tests [Hasselmann *et al.*, 1963] with the aid of a bispectral analysis (first proposed by John Tukey). Suppose we inquire as to whether the oscillations at some frequency $\omega = \omega_2 - \omega_1$ is the result of an interaction between frequencies $\omega_1$ and $\omega_2$ in producing $\omega_2 - \omega_1$. The record is played through band-pass filters centered at $\omega$, $\omega_1$, and $\omega_2$, producing the filtered records $x(t)$, $x_1(t)$, and $x_2(t)$. The triple product (or bispectrum*)

$$B(\omega, \omega_1) = <x(t)x_1(t)x_2(t)>$$

is a measure of the interaction. The computation is repeated with other frequency pairs whose difference is again $\omega$, and in this manner the interaction of $\omega$ with all possible frequency pairs differing by $\omega$ is evaluated. Note the experimental maximum value $B(0.225, 0.075) = -5 \times 10^4$ cm³ sec² in the lower triangle of Figure 4, indicating an interaction between 0.075, 0.225, and 0.330 Nyquist frequency units. The power spectra show peaks at 0.225 and 0.300 Nyquist and one suspects then that the weak low-frequency peak at 0.075 Nyquist is the *result* of an interaction of the two strong high-frequency peaks in producing a difference frequency. A theoretical calculation based on the nonlinear Navier-Stokes equation and the observed power spectrum yields $B(0.075, 0.225) = -3 \times 10^4$ cm³sec², and this supports the hypothesis. The theoretical value $B(0.025, 0.225) = -2 \times 10^5$ cm³sec² and corresponding experimental values $B(0.225, 0.025) = -4 \times 10^4$ is probably the result of difference frequencies produced by interactions *within* the main peak. The negative sign implies that a group of high waves leads to a temporary *lowering* of sea level near shore; in the asymptotic case of vanishing difference frequency this leads to a "dc" lowering of sea level in the presence of waves. Figure 4 has the appearance of Rorschach inkblots, thus indicating symmetry about the 45° axis arising from a resemblance of the theoretical and experimental octants.

The work pertains only to the first part of the triple interaction $(\bar{\mathbf{w}}_1, \mathbf{w}_2, \mathbf{b}_3)$. The conclusion is that irregular wave trains are associated with low-frequency oscillations related to the envelope of the wave train. But these are virtual waves, confined in space and time to the space-time configuration of the irregular wave trains. In the presence of an irregular bottom there will be an interaction with that particular Fourier component of the bottom which yields $\omega_4$ and $\varkappa_4$ related according to gravity wave theory. Some of the low-frequency energy is radiated seaward; most is trapped in the continental wave

---

* The notation $B(\omega, \omega_1)$ infers interaction between $\omega$, $\omega_1$ and $\omega + \omega_1$ (or $\omega_2$) without implication as to what causes what. Cause and effect depends on the relative cross sections.

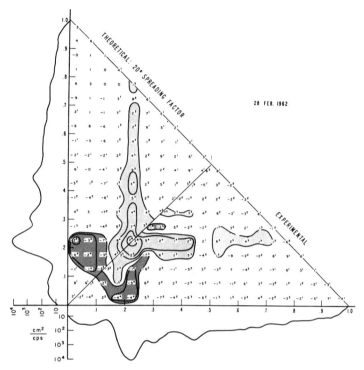

Fig. 4. Bispectra of ocean wave record [Hasselmann et al., 1963]. The numbers give the contributions towards the mean-cubed record (in cm³) per unit frequency band squared (in cps²) and are thus in units cm³ sec². The number $-7^4$ denotes $-7 \times 10^4$. Contours are drawn for $-10^3$, $-10^4$, $-10^5$, $-10^6$. In the case of perfect agreement between theory and experiment the pattern would be symmetrical about the 45° line. The two axes give frequencies in Nyquist units; 1 Nyquist is 0.25 cps. The (identical) plots along the two axes are the power spectra in cm²/cps.

guide. Attenuation is very slow, and the entire ocean basin is filled with low-frequency energy that expresses a radiative balance of the nonlinear processes along the world's coast lines.

### Tidal Cusps

During the last century sea level has risen by something like 10 cm, and this can be associated, at least in part, with the melting of ice at high latitudes. The geologic record over the last 10,000 years indicates variations by as much as 1 meter per century. If we had a time series of sea level over the last million years, I suspect we would find its spectrum to have a broad maximum near $10^{-4}$ cycle per year, the time scale being governed by the thermal inertia inherent in the melting of ice caps.

Tidal observations extending over something like a century are available for a dozen stations. The spectra consist of the tidal line spectrum superimposed on a continuous spectrum [Munk *et al.*, 1965]. The continuum rises markedly and monotonically with decreasing frequency down to the lowest frequencies to which the analysis can be carried with existing records (Figure 5). The peak at "zero frequency" is an indication of the climatic fluctuation of sea level. In addition, the continuum "cusps" in the vicinity of the strong lines.

The physical basis of these cusps is quite simple. Tides at a given harbor are affected by water depth over the adjoining shelf. In years of relatively high "mean sea level," the tide wave propagates shoreward somewhat more rapidly, and the phase of high and low tide is advanced (and the amplitude slightly reduced, I believe). The phase modulation in the time domain is reflected by the spectral cusps in the frequency domain.

In the language of communication engineering, a carrier (the tides) is modulated by a noise (the climatic fluctuation). Demodulation would provide some information concerning climatic variations during the last century. It is amusing to contemplate that the daily tide records have some of this information "built in." If the instrumental response to low frequencies were unfavorable, one might still learn

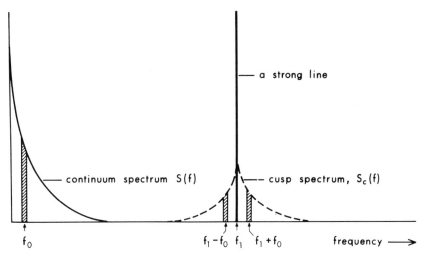

FIG. 5. A strong line interacting with a continuum peaked at "zero frequency" forms a cusp.

something about climatic fluctuations by examining the cusps. But as far as standard tide gauges are concerned, I believe nothing is gained over examining the low-frequency record directly.

The cusps have some bearing on the precision of tide prediction.

**Climatic Fluctuations**

In the previous section we noted some effects of variations in sea level of long period. The long time scale is imposed by the thermal inertia of the ice caps. The detailed examinations of Ewing and Donn make it plausible how long-period oscillations can arise in this manner.

Now suppose that there would be no processes with very long time scales. Does this imply that the "mean sea level" would remain precisely fixed? The answer must be "no," because oscillation of arbitrarily low frequency can be produced by nonlinear interactions between neighboring frequencies in various spectral bands. We have here an extension to climatic frequencies of the processes that lead to formation of surf beats from interactions in the main wave peak. What happens there is that the *direct* generation of waves in the cycle-per-minute band is lacking, and nonlinear interactions fill up the vacuum.

There are then two distinct ways for generating low frequencies: ($i$) by processes which have inherently long time constants, and ($ii$) by nonlinearities in the medium and the resulting coupling of neighboring frequencies.

Whether the nonlinear energy flux is up or down the frequency scale depends on the cross sections and relative intensities. In the case of surf beats the flux is downward from the intense wave frequencies near 0.1 cps to the low region at 0.01 cps. In the case of tidal cusps it is upward from the relatively intense climatic frequencies to the low spectral region surrounding the tidal lines. Thus the low-frequency oscillations of sea level are *not* the result of interferences at higher frequencies. But in the case of atmospheric pressure the reverse situation may hold. At southern Alaska the rms variations below 1 cycle per year are typically 2 mb, whereas short-term changes by as much as 96 mb have been observed [Roden, 1965]. Veronis and others working at Woods Hole are looking into the possibility that long-term variations in the drift of ocean water are the result of interference between variable currents of higher frequency. Perhaps the famed Cromwell current which flows beneath the surface along the Equator is the result of interactions between equatorially trapped waves (if I may add a hypothesis to five existing theories). In fact, the problem of the general circulation (as the dc problem is called in meteorology and oceanography) is a case in point. It was long held that the solution should be formed in terms of the appropriate dc forces. Subsequently Jeffreys suggested and Starr demonstrated that interactions between cyclonic waves having typical frequencies of 2 cycles per week leads to the prerequisite flux of momentum to maintain the westerly winds.

Dr. Klaus Hasselmann has permitted me to use some of his unpublished work and has otherwise been most helpful in the preparation of this note. The National Science Foundation has supported this work under grant NSF-G-13575.

### References

Hasselmann, K., On the non-linear energy transfer in a gravity-wave spectrum, Part 3. Evaluation of the energy flux and swell-sea interaction for a Neumann spectrum, *J. Fluid Mech.*, 15, No. 3, 385–398, 1963.

Hasselmann, K., W. H. Munk, and G. J. F. MacDonald, Bispectra of ocean waves, in *Time Series Analysis*, edited by M. Rosenblatt, John Wiley & Sons, New York, 1963.

Longuet-Higgins, M. S., and R. W. Stewart, Radiation stress and mass transport in gravity waves, with application to "surf beats," *J. Fluid Mech.*, in press, 1962.

Munk, W. H., B. Zetler, and G. W. Groves, Tidal cusps, *Geophys. J.*, in press, 1965.

Phillips, O, M., On the dynamics of unsteady gravity waves of finite amplitude, Part 1. The elementary interactions, *J. Fluid Mech.*, 9, No. 2, 193–217, 1960.

Roden, G. I., On atmospheric pressure oscillations along the Pacific Coast of North America 1873–1963, *J. Atmospheric Sci.*, in press, 1965.

Snodgrass, F. E., G. W. Groves, K. F. Hasselmann, G. R. Miller, W. H. Munk, and W. H. Powers, Propagation of ocean swell across the Pacific, *Phil. Trans. Roy. Soc. London*, in preparation.

**PART IV**

# THE "SOLID" EARTH I

# THE FIGURE AND LONG-TERM MECHANICAL PROPERTIES OF THE EARTH

*Gordon J. F. MacDonald*

Institute of Geophysics and Planetary Physics, University of California,
Los Angeles, California

## Introduction

The hypothesis of continental drift seems to provide an elegant solution to many questions of the evolution of the earth's crust. Advocates of the theory affirm its ability to explain puzzling problems raised by tectonics, ice ages, animal and plant migration, and paleomagnetism. While the hypothesis appears to solve many problems connected with the surface of the earth, it also raises substantial questions about the earth's interior. The only plausible mechanism that has been suggested for the breakup and movement of continents is that thermal convection occurs within the solid mantle of the earth. The convection theory has taken many forms, but in all of them a column of mantle material is supposed to rise toward the surface, spread out horizontally, and sink again. The driving force for this convective motion is the buoyancy of the hotter and lighter rock. Near the surface, where the postulated crustal current flows horizontally, it produces a drag on the overlying material; this drag is the force which separates continental blocks. The velocities required for the currents are on the order of 1 to 10 cm per year in order to account for the presumed continental motions.

Controversy arises in two ways: from the ambiguity in the empirical evidence for continental drift, and from the doubtful theoretical treatment of convection. With regard to the latter, one may ask the question: how does mantle material respond to small stresses imposed over long periods, say a hundred million years? We lack the detailed knowledge of the structure of the interior of the earth, and of the properties of matter at extreme pressures. The mechanical properties of the material under physical conditions within the mantle must be inferred either from geophysical observation or from laboratory studies. The latter are most unsure guides because of the time scale required for continental drift. No theory exists which allows an extrapolation of laboratory phenomena over geologic time scales.

The surest guide to the mechanical behavior of materials at high pressures and high temperatures over long periods of time can be found in the geophysical observations themselves. Of these the most pertinent are those obtained from studies of the motion of artificial satellites. These investigations yield a description of the large-scale features of the earth's gravitational field so detailed and accurate that they must be accounted for in any theory of continental drift. It is my purpose to explore this connection thoroughly. I have previously reviewed the limitations placed on continental drift by observations of continent and mantle structure [MacDonald, 1963] and will not discuss these matters.

The following section reviews the observations used in support of continental drift. Bullard [1964] has provided a more extended, admirable, and balanced discussion of this evidence. I then discuss those properties of matter essential for convection to occur. The geophysical evidence is of two kinds: that pertaining to small-scale regional uplift following loading, and that concerned with the large-scale features of the earth's figure. It is in this latter subject where satellite observations provide a wealth of new material. In order to illustrate clearly the importance of satellite observations, a development of the theory of the earth's internal gravitational field is presented. This discussion leads to the application of the tensor virial equations introduced by Chandrasekhar [1961] in discussions of the figures of rotating fluids. These powerful global relations, similar to the equations of conservation of mass and energy, give information, without going into details, of great importance regarding the distribution of internal stresses. The paper concludes with a brief summary of the important problems which must be resolved if we are to describe the mechanical properties of the materials making up the earth's interior.

## Review of Observations Supporting Large Surface Displacements

*Paleomagnetic Evidence*

The results obtained in the studies of the magnetization of ancient rocks constitute the main reason for the revival of interest in continental drift. The direction of the magnetization of the rock is presumed to determine the position of the magnetic pole at the time the rock acquired its magnetization. In application to geodynamics, it is further supposed that the magnetic and rotational poles coincide on the average and that the magnetic field in the past has maintained its dipolar character. The interpretation of the paleomagnetic results is complicated by the possibility of polar wandering and is further confused by magnetic reversals. The possibility that the direction of the earth's magnetic field in the past has reversed its polarity is used to explain oppositely magnetized rocks, though there is a well-established process by which certain rocks become magnetized in a direction opposite to that of the magnetic field in which they are placed. Excellent reviews of the paleomagnetic data from different points of view have been given by Cox and Doell [1960], Runcorn [1962], and in a series of papers by Irving [1959, 1960–1963].

The strongest evidence advanced in support of the paleomagnetic method comes from determinations of the position of the poles during the Tertiary and more recent times. In rocks ranging from the Pleistocene to the recent the pole positions are closely clustered about the present axis of rotation [see Cox and Doell, 1960]. This implies that for times on the order of $10^5$ years the average magnetic axis coincides with the axis of rotation. The period of averaging is uncertain, but it is on the order of $10^3$ years. The paleomagnetic results from older Tertiary rocks show a wider scatter about the present axis of rotation, but there are no preferred groupings of pole positions from the separate continents.

For rocks earlier than Tertiary, the Permian and Carboniferous are the best studied. The inferred pole positions in these rocks do not cluster about the present pole position but rather about a number of distinct positions. The pole positions for an individual continent group more closely than pole positions from different continents. Thus, the European results show a different mean position from those of North America, which are, in turn, distinct from the Australian poles. It is this different clustering that proponents of continental drift use as evidence that the continents have moved relative to each other.

This evidence can be questioned on two grounds. A single paleomagnetic observation determines only the latitude of the magnetic pole; the longitude of the magnetic pole is arbitrary. When this restriction of the paleomagnetic data is taken into account, discrepancies between the different clusterings of the pole are reduced. For example, the reader can interpret Runcorn [1962], Figures 7, 11, and 14. The interpretation of the paleomagnetic results is further based on the presumption that the field prior to the Tertiary was dipolar in character and aligned with the current axis of rotation. As discussed earlier, the dipolar character of the field through the Tertiary is strong presumptive evidence for a dipolar field in more remote times but one admits this as no more than an extrapolation. There is no theoretical reason why the field could not have been multipolar. The resolution of this question can be obtained by a determination of the paleomagnetic pole position of relatively closely spaced rocks of a given geologic period, but covering a wide geographic area.

The data for rocks older than the Permian are limited. For Pre-Cambrian rocks, poles for a given continent group to a far lesser extent, and the most that can be said is that since these times parts of continents have moved relative to each other.

In summary, the consistency of results from a given locality for any geologic time and from different localities for the Tertiary strongly supports the methods of paleomagnetism. A continuing increase in the degrees of freedom required to account for paleomagnetic results argues against their validity. Paleomagnetism, however, is just one of a number of lines of evidence used in support of continental drift.

*Faults with a Large Horizontal Displacement*

Many transcurrent faults with large displacements are known. Among the faults with largest displacement are the San Andreas Fault where movements of several hundred kilometers have taken place [Hamilton, 1961; Crowell, 1962], the Great Glen Fault with displacements on the order of 100 km [Kennedy, 1946], the Alpine Fault of New Zealand, the Philippine Fault, and the Atacama Fault of Chile [see Allen, 1962, and Benioff, 1962 for excellent discussions of the circum-Pacific faults]. In the case of the San Andreas Fault, there can be no question that at present the coastward side of the San Andreas fault is moving northward at an average rate of about 5 cm a year.

Further evidence for large transcurrent displacements may be found in the fracture zones in the eastern Pacific. These zones, which range in length from 2000 to 3000 km and average from 100 to 250 km in width, consist of a broad ridge broken into individual ridges and troughs [Menard, 1964]. Magnetic surveys have established that the north-south trends of the magnetic field are cut off at the fracture zones. Vacquier *et al.* [1961] estimate that the displacement of the magnetic trends are 265 km for the Pioneer Fracture Zone and about 1200 km for the Mendocino Zone. The reality of these displacements can be questioned, since the fracture zones are not correlated with any feature on land; in fact, they die out before reaching the California coast.

There can be little doubt that strike-slip motion is important in major earthquake zones at present. There is, however, little agreement among geologists as to the importance of strike-slip motion over geologic time. Some geologists claim evidence for large-scale horizontal displacements, while others emphasize the importance of relative vertical displacements with only minor horizontal motions. The fact that blocks within continents move relative to each other is often used to support the hypothesis that entire continents shift.

*Ocean Floors*

The early discussions of continental drift [Wegner, 1924] were hampered by a lack of information regarding the structure of the ocean floor. In the last 15 years there has been marked advance in our knowledge of this region [Hill, 1963; Menard, 1964]. These studies clearly show that the ocean floor differs greatly from the continental regions in topography and structure as well as in petrology. The topography is characterized by oceanic rises, broad featureless plains, trenches, and, in certain regions, abundant seamounts or submarine volcanoes.

Some authors interpret the existence of the oceanic rises as supporting the notion of continental drift [Wilson, 1960]; the mid-ocean ridges are taken as the locus of the rising convective currents, the currents then sinking under the continental mountain systems. Such a hypothesis is used to explain the topographic features seen in the central portions of the ridges. The ridges are characterized by a faulted topography, normal faulting parallel to the rise, and some transverse faulting. Needless to say, these interpretations are speculative.

The thickness of the sediments on the sea floor is also considered evidence for convection and continental motion [Hess, 1962]. The sedimentation since the Pleistocene has been at a rate of one centimeter per thousand years and the Mohole test drilling indicates that this rate may have held since the Miocene. The present thickness of the sediments on the ocean floor would then accumulate in something like 200 to 300 million years. Because of the short time scale, Hess argues that the ocean floor must be renewed from time to time. Convection currents are presumed to provide the mechanism by which the ocean floor could be periodically rejuvenated.

The argument from the rate of sedimentation however, is reminiscent of attempts to date the earth from the salt content of the oceans. Rather than requiring a clean ocean floor once every few hundred million years, the present rate of sedimentation and the observed thickness of ocean sediments suggest, as does the salt content of the ocean, that the rate of erosion has varied widely in geologic times.

*The Classical Arguments for Continental Drift*

The principal types of evidence used in support of continental drift have been:

1. The matching of continental borders.
2. The matching of geologic structure on opposite sides of ocean basins.
3. Paleoclimatological arguments.
4. Biological data.

These have all been reviewed in a book edited by Runcorn [1962]. In recent years, the classical arguments have been sharpened by the application of new methods. For example, Everett [see Bullard, 1964] has reexamined the fit of the continental margins using an electronic computer. He finds a remarkable alignment between Europe and North America and between South America and Africa.

Biological data have given rise to great controversies in the past and continue do do so. For example, Axelrod [1964] argues that paleofloras are arranged in latitudinal zones and indicate continental stability. Hamilton [1964] suggests that the paleobotanical evidence of subtropical floras in the present Arctic and of cold-climate and temperate floras in the present Tropics argue strongly for continental drift. Axelrod emphasizes the importance of topography in controlling paleofloras. This complication, together with uncertainties of the relation between the organisms and their environment make the subject one of continuing uncertainty.

In summary, any single line of evidence for drift can be attacked. Taken together the observations provide a strong case for motion of blocks within continents. The observations are not sufficiently precise to distinguish between the organized motion of continental blocks or the relative motions of pieces of continents. In the latter case, the mean relative position of the continental masses, may have remained constant.

## Mechanical Properties Required for Drift

Material of the mantle behaves as an elastic solid with a slight degree of imperfection for deformations of frequency 10 cps to $10^{-7}$ cps [Knopoff and MacDonald, 1958]. At these frequencies there can be no doubt that material could sustain a small-amplitude stress difference without rupturing or flowing. The theory of continental drift driven by convection requires the solid crystalline mantle material to flow over time scales measured in $10^8$ to $10^9$ years ($10^{-16}$ cps). A description of the mechanical behavior of the mantle material for long-term deformations is central to the problem of the construction of an adequate theory of continental drift.

It has become conventional in discussions of the mechanical properties of the mantle to assume a Newtonian viscosity. In a Newtonian fluid, flow takes place at all stress differences with a resistance to flow proportional to the gradient of velocity. This type of behavior can be described in terms of a single parameter, viscosity, if the fluid is taken as incompressible. Such a model for mantle material has the great advantage of yielding readily to mathematical analysis.

A crystalline solid subjected to stresses applied for a long period undergoes nonrecoverable deformation at stress differences well below those required to produce rupture or flow. The time-dependent deformation of solids under stress is termed creep. The most detailed studies of creep have been made with metals. The applicability of these results to crystalline silicates remains unknown.

The investigation of Andrade and Aboav [1964] on cadmium illustrates the complexities of creep in crystalline materials and furthermore, clearly demonstrates the irrelevancy of Newtonian viscosity to discussions of flow in solids. Andrade and Aboav find that at stresses below a critical stress $\sigma_c$ the shear strain, $\gamma$, dependence on time is represented by

$$\gamma = M(1 - e^{-mt}) + K_M t$$

Above the critical stress, $\gamma$ is given by

$$\gamma = A + Bt^{1/3} + Kt$$

The rate of deformation for long times is thus independent of time in both cases. The critical stress $\sigma_c$ depends both on temperature $T$ and grain size $d$ with

$$\sigma_c = \frac{\text{const}}{d^{1/2}} \left( \frac{1}{T} - \frac{1}{T_M} \right)$$

where $T_M$ is the melting temperature.

In the long time regime the flow takes place by grain boundary migration. For stresses below $\sigma_c$, $K_M$ varies as $\sigma^2$, while for stresses above $\sigma_c$, $K$ depends exponentially on

$$K = Q e^{q\sigma}$$

where $Q$ is independent of stress. The rate of deformation at constant stress thus varies as the square of the stress for low stresses and exponentially with the stress at high applied stresses. This behavior contrasts with that of Newtonian viscosity in which the rate of deformation varies linearly with the stress. Both $K$ and $K_M$ are found to depend on temperature, and for temperatures below the critical temperature $T_c$ this dependence involves $[1 - (T/T_c)]^{-3/2}$, where $T_c$ is the temperature above which grain boundaries associated with the flow are unstable. Thus $T_c$ depends on the grain size.

Certain of the details of Andrade and Aboav results may be related to the structural characteristics of cadmium. The nonlinear dependence of the rate of deformation on stress is, however, characteristic of a wide range of material [Kennedy, 1963]. A further characteristic of creep is that the type of nonlinearity depends on the relation of the applied stress to a critical stress. The critical stress in turn depends on the nearness of the temperature to that of melting.

In the upper mantle the temperatures closely approach the melting temperature [for example see MacDonald, 1963]. The deformation for all but very small stresses may be such that the rate of deformation varies exponentially with the stress. Deeper within the mantle, the temperatures may lie well below melting and the deformation would have a different dependence on stress. Furthermore, the temperatures under oceans are greater than under continents and the mechanical behavior of the mantle material may thus vary horizontally.

**Response of the Earth to Loads**

Numerous studies on the rebound of glaciated regions in Fenno-Scandia [Kääriäinen, 1953; Niskanen, 1948] may provide evidence on the mechanical properties of the crust and mantle. More recently, a detailed investigation of the uplift of Pleistocene Lake Bonneville has been presented by Crittenden [1963]. Like the Fenno-Scandian uplift, the time scale is on the order of $5 \times 10^3$ years, while the length scale is of the order of 200 km as compared with 2000 km for Fenno-Scandia. These observations can be readily interpreted in terms of viscosity and we proceed to do so, clearly recognizing that a simple viscous model is inappropriate for crystalline solids.

Suppose that the earth is distorted by a disturbing potential $U_n$ of degree $n$. The earth deforms as a result of finite elasticity, and the additional gravitational potential at the displaced surface, arising solely from the redistribution of mass, is $k_n U_n$, where $k_n$ is the dimensionless Love number of order $n$.

Kelvin showed that for an incompressible homogeneous sphere of rigidity $\mu'$, density $\rho$, surface gravity $g$, and radius $a$, the Love number $k_2$ is

$$k_2 = \tfrac{3}{2}(1 + \mu)^{-1} \tag{1}$$

where

$$\mu = \frac{19}{2} \frac{\mu'}{\rho g a} \tag{2}$$

The dimensionless rigidity $\mu$ in Equation 1 can be replaced by a value which yields a $k_2$ in agreement with observations of the Chandler wobble and bodily tides. Thus a substitution of $\mu = 2.3$ takes into account the radial variation of density and elasticity.

The Love number $k_2$ defined by Equation 1 is appropriate in considering the large-scale deformations imposed by an external potential. A surface load has two effects. There is a potential arising from the gravitational attraction of the load, and there is a stress which depresses the surface and redistributes the mass. Munk and MacDonald [1960] show that the Love number $k_2'$ due to the surface load is

$$k_2' = -\tfrac{2}{3}k_2 \tag{3}$$

In the general case of an external potential of degree $n$ the appropriate value for $k_n$ is

$$k_n = \frac{\tfrac{3}{2}}{n-1} \frac{1}{1 + N\mu} \tag{4}$$

where

$$N = \frac{2(2n^2 + 4n + 3)}{19n} \tag{5}$$

and, for a load, we have

$$k_n' = -\frac{1}{1 + N\mu} \tag{6}$$

The formalisms sketched in Equations 1 through 6 are appropriate provided the response of the material is elastic. For an elasto-viscous body, the elastic formalism can be applied provided the dimensionless rigidity $\mu$ (see Equation 2) is replaced by the operator $\hat{\mu}$

$$\hat{\mu} = \frac{\mu D}{D + \tau^{-1}}; \quad D = \frac{d}{dt}, \quad \tau = \frac{\eta'}{\mu'} \tag{7}$$

[Jeffreys, 1959]; $\eta'$ is the dynamic viscosity, the single parameter describing the anelastic deformation of the incompressible solid. The load Love number for an elastoviscous earth becomes

$$\hat{k}_n' = -\frac{1}{1 + N\hat{\mu}} \tag{8}$$

Equations 7 and 8 can be combined to determine the time history of a load suddenly applied at time zero. That history is given by

$$(1 + \hat{k}_n')H(t) = \frac{N\mu}{1 + N\mu} \frac{D}{D + \gamma_n} H(t)$$

$$= \frac{N\mu}{1 + N\mu} e^{-\gamma_n t} \tag{9}$$

where $H(t)$ is the Heaviside step function; $\gamma_n^{-1}$ is the compensation time and determined by

$$\gamma_n^{-1} = (1 + N\mu)\tau \tag{10}$$

The factor $N\mu/(1 + N\mu)$ accounts for the immediate effects due to elastic deformation.

The uplift in Lake Bonneville has a characteristic wave number $n$ of about 200. The compensation time is about 4000 years [Crittenden, 1963], so that the corresponding time constant $\tau$ is $2.3 \times 10^9$ sec. The characteristic time $\tau$ is related to the viscosity and rigidity so that

$$\eta' = 2.3 \times 10^9 \mu' \tag{11}$$

A rigidity of $8.3 \times 10^{11}$ dynes cm$^{-2}$, consistent with the value of $\mu$ of 2.31 yields a dynamic viscosity of $1.9 \times 10^{21}$ dynes sec cm$^{-2}$. A lesser rigidity appropriate to the upper mantle further lowers the viscosity. The length scale of the Fenno-Scandian uplift corresponds to an $n$ of 20. The time scale is on the order of 5000 years. The resulting dynamic viscosity is $2.3 \times 10^{22}$ dynes sec cm$^{-2}$. A large-scale deformation thus yields a higher value for the viscosity.

The order of magnitude difference between the values obtained from Lake Bonneville and Fenno-Scandia suggest the complexity in interpreting the mechanical behavior of the mantle in terms of a single-parameter system. The discrepancies can be accounted for either by assuming that the viscosity varies with depth, by using a more complicated model for the anelastic response of the material, or by assuming the creep parameters are depth-dependent. Since the length scales for the two loads differ by an order of magnitude in dimension, it might be expected that the Lake Bonneville load response yields information regarding shallower mantle material than does the larger-scale Fenno-Scandian uplift. In neither case do the viscosities relate to the properties of the mantle as a whole.

## Figure of the Earth

Satellite observations lead to the determination of the lower-order harmonics of the earth's gravitational field. These harmonics can then be compared with those expected for an earth in which hydrostatic equilibrium prevails. The theory of the figure of the earth in fluid equilibrium is based on two related assumptions. In the hydrostatic theory it is assumed that, throughout the earth, stress differences vanish and further that the equation of state is barotropic, i.e., the density is a function solely of pressure. It is quite clear that horizontal compositional and thermal differences exist within the mantle [MacDonald, 1963] and the assumption of the barotropic equation of state must fail. Indeed, the satellite observations provide a measure of this failure, and thus determine the degree to which the mantle is baroclinic.

We write the external gravitational potential of the earth as

$$U = \frac{GM}{r}\left[1 + \sum_{n=1}^{\infty}\sum_{m=0}^{n}\left(\frac{R}{r}\right)^{n}P_{n}^{m}(\sin\beta)(C_{n}^{m}\cos m\lambda + S_{n}^{m}\sin m\lambda)\right] \quad (12)$$

where

$G$ = gravitational constant
$M$ = mass of earth
$r$ = distance from center of mass
$R$ = mean equatorial radius
$P_{n}^{m}$ = associated Legendre polynomial
$\beta$ = latitude
$\lambda$ = longitude

The coefficient $C_{n}^{m}$ and $S_{n}^{m}$ are dimensionless quantities specifying the contribution of each harmonic to the external potential. An alternative and useful notation is

$$J_n = -C_n^0 \tag{13}$$

Table 1 lists the values for certain of the zonal and tesseral harmonics.

### Table 1

COEFFICIENTS IN THE EXTERNAL POTENTIAL
(dimensionless, unit = $10^{-6}$)

|  | Observed | Hydrostatic | Isostatic Continent (1000 fathom) |
|---|---|---|---|
| $J_2$ | 1082.7 | 1072.1 | −0.42 |
| $J_3$ | −2.54 | 0 | −0.17 |
| $J_4$ | −1.40 | −2.3 | −0.34 |
| $J_5$ | −0.01 | 0 | 0.56 |
| $J_6$ | 0.37 | 0.00 | 0.18 |
| $J_7$ | −0.45 | 0 | 0.34 |
| $J_8$ | 0.07 | — | 0.03 |
| $J_{10}$ | −0.50 | — |  |
| $J_{12}$ | 0.31 | — |  |
| $C_2^2$ | 1.68 | 0 | 0.20 |
| $S_2^2$ | −0.64 | 0 | 0.12 |
| $C_3^1$ | 1.77 | 0 | 0.02 |
| $S_3^1$ | 0.19 | 0 | 0.01 |
| $C_4^1$ | −0.57 | 0 | 0.02 |
| $S_4^1$ | −0.46 | 0 | −0.01 |

The zonal harmonics are due to King-Hele et al. [1964] who derive new values for the even-order zonal harmonics using the measurements of seven satellites. The tesseral harmonics are taken from the results of Guier [1963].

Part of the observed nonhydrostatic contribution to the external gravitational potential could arise from an isostatically compensated crust. We recall that the vacuum gravitational field can be interpreted in terms of the number of lines of gravitational force per unit area. An isostatically compensated continent will give rise to a second-order anomaly because of the relatively large area through which the same number of lines of force pass, since the number of lines per steradian is constant. Table 1 lists the gravitational anomaly expected for an isostatically compensated crust where the density model of the crust is that of Worzel and Shurbet [1954]. The anomalous potential due to the continents is calculated on the assumption that the continental crust extends to the 1000-fathom line. The fact that the potential expected from the isostatically compensated crust is smaller and, in general, of opposite sign to that observed, demonstrates without question that the mass anomalies giving rise to the external potential lie within the earth's mantle. A study of these should then provide information regarding the mechanical properties of mantle material.

In examining the numerical magnitude of the coefficients in the expansion we note that the low-order zonal harmonics are largest, even if account is taken of the normalization of the associate Legendre functions. In the subsequent discussion, I will discuss the zonal harmonics and their interpretation. The arguments can be extended with some algebraic complications to the tesseral harmonics.

*Interpretation of the Coefficients in the External Potential in Terms of the Moments of the Mass Distribution*

We consider a mass distribution given by $\rho(\mathbf{r})$. The potential of the mass distribution is

$$U(\mathbf{r}) = G \int_V \frac{\rho(\mathbf{r}')}{|\mathbf{r} - \mathbf{r}'|} d\tau \qquad (14)$$

where the integration is over the volume $V$ bounded by the surface $S$. Expanding the denominator in the integral in the usual form in terms of zonal harmonics $P_n$, we obtain for $\mathbf{r}$ external to $V$

$$U(\mathbf{r}) = \frac{G}{r} \int_V \rho(r') \left[ 1 + P_1(\mu) \frac{r'}{r} + P_2(\mu) \frac{r'^2}{r^2} + \cdots \right] d\tau \qquad (15)$$

where $\mu$ is

$$\mu = \frac{x_i x_i'}{rr'} \qquad (16)$$

$x_i$ are the Cartesian coordinates of the point $\mathbf{r}$, and $x_i'$ are the coordinates of the element of matter at $\mathbf{r}'$. The potential $U$ can also be written in the form

$$U = U_0 + U_1 + U_2 + U_3 + U_4 + \cdots \qquad (17)$$

where the individual terms are given by a comparison with Equation 15. The zeroth-order term is determined by the zeroth-order moment of the mass distribution:

$$U_0 = \frac{G}{r} \int \rho(\mathbf{r}') d\tau \qquad (18)$$

$U_1$ is

$$U_1 = \frac{G}{r^3} \left( x_i \int \rho(\mathbf{r}') x_i' d\tau \right) \qquad (19)$$

Thus, if the coordinate system has its origin at the center of mass, $U_1$ vanishes. Similarly, $U_2$ becomes

$$U_2 = \frac{G}{r^3}\left[\mathscr{I}_{;ii} - \frac{3}{2r^2}\mathscr{I}_{;ij}(r^2\delta_{ij} - x_ix_j)\right] \quad (20)$$

provided the coordinate system is a principal axis system and where $\mathscr{I}_{;ij}$ are the second-order moments of the mass distribution

$$\mathscr{I}_{;ij} = \int_V \rho(\mathbf{r}')x_i'x_j'\,d\tau \quad (21)$$

The summation convention is used throughout the paper. For an axially symmetric density distribution $U_3$ is

$$U_3(\mathbf{r}) = \frac{G}{2r^4}\frac{x_3}{r}\left[\left(\frac{5x_3^2}{r^2} - 3\right)\mathscr{I}_{;333}\right.$$
$$\left. + \left(15\frac{x_1^2}{r^2} - 3\right)\mathscr{I}_{;311} + \left(15\frac{x_2^2}{r^2} - 3\right)\mathscr{I}_{;322}\right] \quad (22)$$

$$\mathscr{I}_{;ijk} = \int_V \rho(\mathbf{r}')x_i'x_j'x_k'\,d\tau \quad (23)$$

$$\mathscr{I}_{;311} = \mathscr{I}_{;322}$$

In terms of $\mu$ Equation 22 becomes

$$U_3 = \frac{5}{2}\frac{G}{r^4}\mathscr{I}_{;333}P_3(\mu) \quad (24)$$

By a similar reduction, $U_4$ can be written as

$$U_4(\mathbf{r}) = \frac{G}{r^5}[\mathscr{I}_{;1111} + \mathscr{I}_{;3333} - 6\mathscr{I}_{;1133}]P_4(\mu) \quad (25)$$

$$\mathscr{I}_{;ijkl} = \int_V \rho(\mathbf{r}')x_i'x_j'x_k'x_l'\,d\tau$$

for an axially symmetric density distribution.

If Expressions 20, 24, and 25 are compared with 12, we obtain the following identities:

$$J_2 = \frac{\mathscr{I}_{;11} - \mathscr{I}_{;33}}{MR^2} = \frac{C - A}{MR^2}$$

$$J_3 = -\frac{5}{2}\frac{\mathscr{I}_{;333}}{MR^3} \quad (26)$$

$$J_4 = -\frac{1}{MR^4}[\mathscr{I}_{;1111} + \mathscr{I}_{;3333} - 6\mathscr{I}_{;1133}]$$

where

$$C = \mathscr{I}_{;11} + \mathscr{I}_{;22}$$
$$A = \mathscr{I}_{;33} + \mathscr{I}_{;22} \quad (27)$$

Here $J_2$ determines the difference between two second moments of the density distribution, $J_3$ determines the third-order moment of the density distribution about the $x_1 - x_2$ plane, while $J_4$ is fixed by the difference between fourth-order moments of the mass distribution.

It is sometimes stated that the improved knowledge of the terms in the external potential will lead to a better determination of the radial variation of the density. In order that the radial variation of density be determined, we require integrals of the form

$$\int_V \rho(\mathbf{r}')r'^2 \, d\tau, \quad \int_V \rho(\mathbf{r}')r'^4 \, d\tau, \cdots \qquad (28)$$

The even-order zonal harmonics provide only the difference between the moments of the density and thus do not determine the integrals in Expression 28 uniquely. Additional information is required before the coefficients $J_2$, $J_4$, etc., can be used to limit the radial variation density.

*The Interpretation of $J_n$ in Terms of Surfaces of Equal Density*

Let

$$r = b\{1 + g_0(b) + g_2(b)P_2(\cos\theta) + g_3(b)P_3(\cos\theta) + \cdots\} \qquad (29)$$

denote the equation of a surface of constant density, where $\theta$ is the geocentric colatitude. The origin of the coordinate system is taken coincident with the center of mass. We interpret $b$ as the mean radius of the surface of equal density. This interpretation requires that the mass contained within a sphere of radius $b$ equals the mass contained within the surface $r(b,\theta)$

$$g_0 = -\tfrac{1}{5}g_2^2 \qquad (30)$$

provided

$$|g_2(b)| \ll 1$$
$$|g_3(b)| \ll |g_2(b)|; \quad |g_4(b)| \ll |g_2(b)| \qquad (31)$$

and terms of second order in the small quantity $g_2$ and terms of first order in $g_3, g_4$ are kept. The bounding surface of the earth is taken to be $r(b_0,\theta)$, where $b_0$ represents the smallest root of the equation

$$\rho(b) = 0 \qquad (32)$$

In the following, we shall suppose that the volume $V$ over which the integrations are extended includes the whole system, so that all the variables, including the density, may be assumed to vanish on the bounding surface $S$. The actual physical surface of the earth is, of course, not a surface of equal density, but is enclosed by a surface on which the density is vanishingly small compared with the density of the underlying material. For numerical calculations, the mean radius is assigned the value of $6.3710 \times 10^8$ cm [Kaula, 1963a; Clarke, 1962].

For the integrations determining the different moments of the mass distribution, $\rho$ can be regarded as a function of a single variable $b$. We thus obtain

$$C = \frac{8\pi}{15} \int_0^{b_0} \rho(b) \frac{d}{db} \{b^5(1 - g_2 + \tfrac{3}{7}g_2^2)\} \, db$$
$$A = \frac{8\pi}{15} \int_0^{b_0} \rho(b) \frac{d}{db} \{b^5(1 + \tfrac{1}{2}g_2 + \tfrac{9}{7}g_2^2)\} \, db \qquad (33)$$

for the principal moments of inertia $C$ and $A$ about the polar axis and about an axis in the equatorial plane, respectively. The mean moment of inertia $I$

$$I = \tfrac{1}{3}(2A + C) \qquad (34)$$

is then

$$I = \frac{8\pi}{15} \int_0^{b_0} \rho(b) \frac{d}{db} \{b^5(1 + g_2^2)\} \, db \qquad (35)$$

To terms of the second order in the small quantity $g_2$, $I$ is

$$I = \frac{8\pi}{3} \int_0^{b_0} \rho(b) b^4 \, db \qquad (36)$$

Since

$$C - A = -\frac{4\pi}{5} \int_0^{b_0} \rho \frac{d}{db} \{(g_2 + \tfrac{4}{7}g_2^2)b^5\} \, db \qquad (37)$$

we have

$$J_2 = -\frac{4}{5} \frac{\pi}{MR^2} \int_0^{b_0} \rho \frac{d}{db} \{b^5(g_2 + \tfrac{4}{7}g_2^2)\} \, db \qquad (38)$$

The observational determination of $J_2$ thus places a single integral constraint on function $g_2(b)$.

Similarly, we have that

$$\mathscr{I}_{;333} = \frac{8\pi}{35} \int_0^{b_0} \rho \frac{d}{db} (g_3 b^6) \, db \qquad (39)$$

provided the mass distribution is axially symmetric. If the center of mass is chosen as the origin of the coordinate system, then

$$\mathscr{I}_{;311} = -\tfrac{1}{2}\mathscr{I}_{;333} \qquad (40)$$

From Equation 39 and the definition of $J_3$, we obtain

$$J_3 = -\frac{4\pi}{7MR^3} \int_0^{b_0} \rho \frac{d}{db} (g_3 b^6) \, db \qquad (41)$$

The fourth-order moments entering into $J_4$ are given by

$$\mathscr{I}_{;1111} = \frac{4\pi}{35}\int_0^{b_0} \rho\,\frac{d}{db}\{(1 - 2g_2 + \tfrac{8}{5}g_2^2 + \tfrac{1}{3}g_4)b^7\}\,db$$

$$\mathscr{I}_{;3333} = \frac{4\pi}{35}\int_0^{b_0} \rho\,\frac{d}{db}\{b^7(1 + 4g_2 + \tfrac{3.5}{8}g_2^2 + \tfrac{8}{9}g_4)\}\,db \quad (42)$$

$$\mathscr{I}_{;1133} = \frac{4\pi}{105}\int_0^{b_0} \rho\,\frac{d}{db}\{b^7(1 + g_2 + \tfrac{8}{5}g_2^2 - \tfrac{4}{3}g_4)\}\,db$$

so that

$$J_4 = -\frac{4\pi}{9MR^4}\int_0^{b_0} \frac{d}{db}\{b^7(g_4 + \tfrac{54}{35}g_2^2)\}\,db \quad (43)$$

Equation 41 provides an integral constraint on the radial variation of $g_3$. Equation 43 provides a similar constraint on $g_4$, provided $g_2$ is determined.

*Equipotential Surfaces*

The total potential $\psi$ is

$$\psi(\mathbf{r}') = U(\mathbf{r}') + \tfrac{1}{3}\Omega^2 r'^2[1 - P_2(\mu)] \quad (44)$$

where $U$ is determined by Equation 14 and is due to the distribution of mass, while the second term in Equation 44 is the contribution from the centrifugal potential; $\Omega$ is the angular velocity. The density distribution is taken to be axisymmetric with $\mu$ given by

$$\mu = \cos\theta \quad (45)$$

The potential due to the mass distribution can be expanded in terms of the Legendre polynomials as

$$U(\mathbf{r}') = U_0(r') + U_2(r')P_2(\mu) + U_3(r')P_3(\mu) \\ + U_4(r')P_4(\mu) + \cdots \quad (46)$$

where, provided the departures from spherical symmetry are small, we have that

$$|U_2| \ll |U_0|;\quad |U_3| \ll |U_2|;\quad |U_4| \ll |U_2| \quad (47)$$

Let the surface of equal geopotential be

$$r(b,\mu) = b[1 + f_2(b)P_2(\mu) + f_3(b)P_3(\mu) + f_4(b)P_4(\mu) + \cdots] \quad (48)$$

As before, we take $b$ to be the mean radius of the surface of equipotential passing through $\mathbf{r}'$. In general, the surface will not coincide with a surface of equal density and the coefficients $f_n$ are not equal to $g_n$.

Combining Equations 46 and 48 with 44, the equation for the geopotential becomes

$$\psi(r, \mu) = U_0(r) + \tfrac{1}{3}\Omega^2 b^2 (1 - \tfrac{2}{5}f_2)$$
$$+ [U_2(r) - \tfrac{1}{3}\Omega^2 b^2 (1 - \tfrac{10}{7}f_2)]P_2(\mu)$$
$$+ U_3(r)P_3(\mu) + [U_4(r) - \tfrac{12}{35}\Omega^2 b^2 f_2]P_4(\mu) + \cdots \quad (49)$$

where $r$ is now understood to denote the point interior to the surface $S$ bounding the earth. The conditions that the geopotential $\psi$ does not vary on the surfaces fixed by Equation 48 requires that

$$b f_2 \left.\frac{dU_0}{dr}\right|_b + \frac{b^2}{7} f_2^2 \left.\frac{d^2 U_0}{dr^2}\right|_b + U_2(b)$$
$$+ \tfrac{2}{7} b f_2 \left.\frac{dU_2}{dr}\right|_b - \tfrac{1}{3}\Omega^2 b^2 (1 - \tfrac{10}{7} f_2) = 0$$

$$b \left.\frac{dU_0}{dr}\right|_b f_3 + U_3(b) = 0 \quad (50)$$

$$b \left.\frac{dU_0}{dr}\right|_b f_4 + \frac{9 b^2 f_2^2}{35} \left.\frac{d^2 U_0}{dr^2}\right|_b + U_4(b)$$
$$+ \tfrac{18}{35} b f_2 \left.\frac{dU_2}{dr}\right|_b - \tfrac{12}{35}\Omega^2 a^2 f_2 = 0$$

where terms of the second order have been kept and

$$b^2 \Omega^2 = O(U_2) \quad (51)$$

Equations 50 can be transformed into integral relationships connecting the coefficients $f_n$ governing the equipotential surfaces with the coefficients determining the surfaces of equal density $g_n$. To do this we require expressions for $U_n$ in terms of $g_n$. Returning to Equation 14 and expanding $|\mathbf{r} - \mathbf{r}'|$ in spherical harmonics in the usual manner, we find

$$U_0(r) = \frac{4\pi G}{r} \int_0^b \rho s^2 \, ds + 2\pi G \int_b^{b_0} \rho \frac{d}{ds}\{s^2(1 - \tfrac{1}{5}g_2^2)\} \, ds$$

$$U_2(r) = \frac{4\pi G}{r^3} \int_0^b \rho \frac{d}{ds}\{s^5(g_2 + \tfrac{4}{7}g_2^2)\} \, ds$$
$$+ \frac{4\pi G}{4} r^2 \int_b^{b_0} \rho \frac{d}{ds}(g_2 - \tfrac{1}{7}g_2^2) \, ds$$

$$U_3(r) = \frac{4\pi G}{7r^4} \int_0^b \rho \frac{d}{ds}(g_3 s^6) \, ds + \frac{4\pi G}{7} r^3 \int_b^{b_0} \rho \frac{d}{ds}(g_3 s^{-1}) \, ds \quad (52)$$

$$U_4(r) = \frac{4\pi G}{9 r^5} \int_0^b \rho \frac{d}{ds}\{s^7(g_4 + \tfrac{54}{35}g_2^2)\} \, ds$$
$$+ \frac{4\pi G}{9} r^4 \int_b^{b_0} \rho \frac{d}{ds}\{s^{-2}(g_4 - \tfrac{27}{35}g_2^2)\} \, ds$$

where the terms of second order in $g_2$ and of first order in $g_3$ and $g_4$ have been kept.

Substituting from Equations 52 into 50, we obtain the needed relations between the coefficients describing the equipotential surfaces and those determining the surfaces of equal density:

$$f_2(b) = -\tfrac{1}{3}m(b)(1 - \tfrac{10}{7}f_2) + \tfrac{2}{7}f_2^2(b)$$

$$+ \frac{4\pi}{5M(b)b^2}(1 - \tfrac{6}{7}f_2)\int_0^b \rho\frac{d}{ds}\{s^5(g_2 + \tfrac{4}{7}g_2^2)\}\,ds$$

$$+ \frac{4\pi b^3}{5M(b)}(1 + \tfrac{4}{7}f_2)\int_0^{b_0} \rho\frac{d}{ds}(g_2 - \tfrac{1}{7}g_2^2)\,ds$$

$$f_3(b) = \frac{4\pi}{7b^3}\int_0^b \rho\frac{d}{ds}(g_3 s^6)\,ds + \frac{4\pi}{7}b^4\int_b^{b_0} \rho\frac{d}{ds}(g_3 s^{-1})\,ds \quad (53)$$

$$f_4(b) = \frac{54}{35}f_2^2(b) - \frac{72}{35}\frac{\pi}{M(b)}\frac{f_2}{b^2}\int_0^b \rho\frac{d}{ds}(s^5 g_2)\,ds$$

$$+ \frac{4\pi}{9M(b)}\left[\frac{1}{b^4}\int_0^b \rho\frac{d}{ds}\{s^7(g_4 + \tfrac{54}{35}g_2^2)\}\,ds\right.$$

$$\left. + b^5\int_b^{b_0} \rho\frac{d}{ds}\{s^{-2}(g_4 - \tfrac{27}{35}g_2^2)\}\,ds\right]$$

where

$$m(b) = \frac{\Omega^2 b^3}{GM(b)}$$

$$M(b) = 4\pi\int_0^b \rho s^2\,ds \quad (54)$$

Equation 53 can be used to calculate $f_n$ provided $g_n$ are determined in some way.

Combining Equation 53 with 38, 41, and 43, we obtain the surface values for $f_n$:

$$f_2(b_0) = -\tfrac{1}{3}m(b_0)[1 - \tfrac{10}{7}f_2(b_0)] + \tfrac{2}{7}f_2^2(b_0) - \left(\frac{R}{b_0}\right)^2(1 - \tfrac{6}{7}f_2)J_2$$

$$f_3(b_0) = -\left(\frac{R}{b_0}\right)^3 J_3 \quad (55)$$

$$f_4(b_0) = \frac{54}{35}f_2^2(b_0) + \frac{18}{7}\left(\frac{R}{b_0}\right)^2 f_2 J_2 - \left(\frac{R}{b_0}\right)^4 J_4$$

The value of the coefficients in the external gravitational potential thus determine the shape of the equipotential surface having a mean radius $b_0$. The shape of this surface can be obtained independently of any assumptions regarding hydrostatic equilibrium. The shape of the surfaces internal to $b_0$, however, depend on the detailed knowledge of the density surfaces. Equations 53, when specialized for hydrostatic equilibrium, are identical to the integral relations obtained by Jeffreys [1953] when account is taken of the differences in notation.

The numerical values for $f_2$, $f_3$, and $f_4$ are listed in Table 2. These values have been obtained using the values of $J_n$ given in Table 1;

Table 2

NUMERICAL VALUES FOR THE COEFFICIENTS DESCRIBING THE EQUIPOTENTIAL SURFACE HAVING MEAN RADIUS $b_0$

| | |
|---|---|
| $f_2(b_0)$ | $-2.2383 \times 10^{-3}$ |
| $f_3(b_0)$ | $2.55 \times 10^{-6}$ |
| $f_4(b_0)$ | $2.89 \times 10^{-6}$ |

equations given in 55 have been solved by iteration. In discussions of the earth's figure, it is traditional to describe the equipotential surfaces in terms of an ellipsoid of revolution with semimajor axis $a$ and ellipticity $e$. The ellipticity $e$, corresponding to the value $f_2$ listed in Table 2 is

$$e^{-1} = 298.25 \tag{56}$$

since

$$f_2 = -\tfrac{2}{3}e - \tfrac{23}{63}e^2 \tag{57}$$

*The Figure of a Hydrostatic Earth*

In the theory of fluid equilibrium it is assumed that the stress is hydrostatic throughout. A further important restriction is that the equation of state is barotropic with the density depending solely on the pressure. The hydrostatic theory can be developed from Equations 53 provided the coefficients $g_n$ are set identically equal to $f_n$; i.e., the surfaces of equal potential are assumed to coincide with the surfaces of equal density. The identity between Equations 53 and the usual formulation of the theory of the earth's internal gravitational field is easily established. For simplicity, we neglect terms of order $f_2^2$. The hydrostatic theory then requires that

$$f_2(b) = -\tfrac{1}{3}m(b) + \frac{4\pi}{5M(b)b^2} \int_0^b \rho \frac{d}{ds}(s^5 f_2)\,ds + \frac{4\pi}{5M(b)} b^3 \int_b^{b_0} \rho \frac{d}{ds} f_2\,ds \tag{58}$$

This is an integral equation for $f_2$ which can be solved with the appropriate boundary conditions. We integrate both of the integrals in Equation 58 by parts, so as to obtain

$$M(b)b^2 f_2(b) = -\frac{1}{3}\frac{\Omega^2 b^5}{G} - \frac{4\pi}{5}\int_0^b s^5 f_2 \frac{d\rho}{ds} ds - \frac{4\pi}{5} b^5 \int_b^{b_0} f_2 \frac{d\rho}{ds} ds \quad (59)$$

Differentiation of Equation 59 yields

$$\frac{d}{db}\{M(b)b^2 f_2(b)\} = -\frac{5}{3}\frac{\Omega^2 b^4}{G} - 4\pi b^4 \int_b^{b_0} f_2 \frac{d\rho}{ds} ds \quad (60)$$

while a second differentiation gives

$$\frac{d}{db}\left[\frac{1}{b^4}\frac{d}{db}\{M(b)b^2 f_2(b)\}\right] = 4\pi f_2 \frac{d\rho}{db} \quad (61)$$

If Equation 61 is expanded, we obtain the first-order Clairaut equation [Jeffreys, 1959; Chandrasekhar and Roberts, 1963].

Equation 60 yields a useful relation involving the surface values of the parameters. At the bounding surface, Equation 60 reduces to

$$2b_0 M f_2 + b^2 M \frac{df_2}{db} = -\frac{5}{3}\frac{\Omega^2 b^4}{G} \quad (62)$$

Dividing by $b_0 M f_2$ and rearranging, we obtain

$$\eta(b_0) = \frac{b}{f_2}\frac{df_2}{db} = -\frac{5}{3}\frac{m}{f_2} - 2 \quad (63)$$

In the usual theory [Jeffreys, 1959], this relation follows only after the use of Radau's approximation. It should be noted that an additional assumption in the hydrostatic theory is that the density stratification has triplanar symmetry. For a nonvanishing angular velocity I know of no way of proving this assertion.

If we adopt the Radau approximation, we have to first order in $f_2$ that

$$\frac{1(b)}{M(b)b^2} = 1 - \tfrac{2}{5}(1 + \eta)^{1/2} \quad (64)$$

Equations 63 and 64 combine to yield a relation determining the surface value of $f_2$ in terms of two observables, the mean moment of inertia $I$ and $m$. Henriksen [1960] first noted that satellite observations could yield an improved estimate of the flattening of the earth.

The mean moment of inertia is determined by $J_2$ and the precessional constant $H$:

$$\frac{1}{MR^2} = \frac{J_2}{H} - \tfrac{2}{3}J_2; \quad H = \frac{C - A}{C} \quad (65)$$

For $H$, I adopt the value $3.273 \times 10^{-3}$ [Jeffreys, 1963]. The resulting value for the mean moment of inertia is

$$\frac{1}{Mb_0{}^2} = 0.3308 \tag{66}$$

Table 3 lists the value of $f_2(b_0)$ and $e^{-1}(b_0)$ including terms of order

**Table 3**

VALUES OF $f_2(b_0)$ AND $e^{-1}(b_0)$ FOR HYDROSTATIC EARTH HAVING VARIOUS RADIAL DENSITY DISTRIBUTIONS

| Model | $f_2(b_0)$ (unit = $10^{-3}$) | $e^{-1}(b_0)$ |
|---|---|---|
| Bullard I  | 2.2531 | 296.41 |
| Bullard II | 2.2534 | 296.47 |
| Bullen A   | 2.2489 | 296.94 |
| Bullen B   | 2.2592 | 295.61 |
| M 1        | 2.2286 | 299.69 |
| M 3        | 2.2292 | 299.60 |

$f_2{}^2$ if the earth were in hydrostatic equilibrium and had various density distributions; the density distributions are listed in Table 4. The values for Bullen A listed in Table 3 agree with those derived by Jeffreys [1963].

Table 4 lists the mean moment of inertia, mass, and mean density distribution for various density models of the earth. In addition, the gravitational potential energy as a function of radius is also given. The final column lists the values of $f_2$, calculated on the assumption that the earth is in hydrostatic equilibrium. The calculations in Table 4 are to first order in $f_2$ (see Equation 58). As can be seen in Table 4, small variations in the internal density distribution lead to rather large variations in $f_2(b)$.

If the hydrostatic values of $f_n$ are adopted, then Equation 55 can be used to determine the hydrostatic values for $J_n$. The numerical values are listed in Table 1. A comparison of the hydrostatic and observed values of $J_2$ shows that the difference is larger by a factor of about 3 than any other terms of the gravitational expansion, if account is taken of the normalization. The earth thus deviates in a major way from a body in fluid equilibrium and this deviation is largest in the $J_2$ term.

## Table 4

### Parameters for Density Models of the Earth

Bullard I (Bullard, 1957)

| Depth (km) | Density (gm cm$^{-3}$) | $I(b)$ = mean moment of inertia (unit = $Mb_0^2 = 2.426 \times 10^{45}$ gm cm$^2$) | $M(b)$ = mass enclosed within sphere of radius $b$ (unit = $M(b_0)$ = $5.977 \times 10^{27}$ gm) | $\bar{\rho}(b)$ = mean density of matter enclosed in sphere of radius $b$ (unit = $\bar{\rho}(b_0)$ = $5.517$ gm cm$^{-3}$) | $|W(b)|$ = absolute value of gravitational potential energy of matter enclosed in sphere of radius $b$ (unit = $\frac{3}{5}GM^2(b_0)/b_0$ = $3.74 \times 10^{39}$ ergs) | $f_2(b)$ (unit = surface value of $f_2$) |
|---|---|---|---|---|---|---|
| 0 | 2.84 | 0.3336 | 1.0000 | 1.000 | 1.166 | 1.000 |
| 32 | 2.84 | 0.3285 | 0.9923 | 1.007 | 1.153 | 0.997 |
| 32 | 3.67 | 0.3285 | 0.9923 | 1.007 | 1.153 | 0.997 |
| 100 | 3.74 | 0.3147 | 0.9708 | 1.019 | 1.118 | 0.991 |
| 200 | 3.83 | 0.2952 | 0.9400 | 1.035 | 1.067 | 0.983 |
| 300 | 3.92 | 0.2764 | 0.9095 | 1.052 | 1.017 | 0.974 |
| 400 | 4.07 | 0.2561 | 0.8754 | 1.071 | 0.963 | 0.964 |
| 500 | 4.08 | 0.2413 | 0.8495 | 1.086 | 0.922 | 0.956 |
| 700 | 4.21 | 0.2095 | 0.7914 | 1.123 | 0.833 | 0.938 |
| 1000 | 4.40 | 0.1681 | 0.7087 | 1.184 | 0.711 | 0.910 |
| 1300 | 4.67 | 0.1335 | 0.6245 | 1.254 | 0.603 | 0.881 |
| 1600 | 4.72 | 0.1053 | 0.5609 | 1.336 | 0.507 | 0.852 |
| 2000 | 4.92 | 0.0762 | 0.4762 | 1.475 | 0.401 | 0.813 |
| 2400 | 5.11 | 0.0055 | 0.4028 | 1.665 | 0.316 | 0.776 |
| 2800 | 5.30 | 0.0041 | 0.3405 | 1.935 | 0.248 | 0.749 |
| 2900 | 5.36 | 0.0038 | 0.3266 | 2.021 | 0.234 | 0.746 |
| 2900 | 10.06 | 0.0038 | 0.3266 | 2.021 | 0.234 | 0.746 |
| 3600 | 11.06 | 0.0038 | 0.1758 | 2.125 | 0.087 | 0.734 |
| 5120 | 12.35 | 0.0003 | 0.0170 | 2.259 | 0.002 | 0.720 |
| 6370 | 12.63 | — | — | — | — | 0.718 |

## Table 4 (continued)
Bullard II (Bullard, 1957)

| Depth (km) | Density (gm cm$^{-3}$) | $I(b)$ = mean moment of inertia (unit = $Mb_0^2 = 2.426 \times 10^{45}$ gm cm$^2$) | $M(b)$ = mass enclosed within sphere of radius $b$ (unit = $M(b_0)$ = $5.977 \times 10^{27}$ gm) | $\bar{\rho}(b)$ = mean density of matter enclosed in sphere of radius $b$ (unit = $\bar{\rho}(b_0)$ = $5.517$ gm cm$^{-3}$) | $\|W(b)\|$ = absolute value of gravitational potential energy of matter enclosed in sphere of radius $b$ (unit = $\frac{3}{5}GM^2(b_0)/b$ = $3.74 \times 10^{39}$ ergs) | $f_2(b)$ (unit = surface value of $f_2$) |
|---|---|---|---|---|---|---|
| 0    | 2.84  | 0.3336 | 1.0000 | 1.000 | 1.161  | 1.000 |
| 32   | 2.84  | 0.3285 | 0.9923 | 1.007 | 1.148  | 0.997 |
| 32   | 3.32  | 0.3285 | 0.9923 | 1.007 | 1.148  | 0.997 |
| 100  | 3.38  | 0.3160 | 0.9727 | 1.021 | 1.116  | 0.991 |
| 200  | 3.46  | 0.2984 | 0.9449 | 1.040 | 1.071  | 0.983 |
| 300  | 3.55  | 0.2814 | 0.9173 | 1.061 | 1.025  | 0.974 |
| 400  | 3.63  | 0.2630 | 0.8864 | 1.084 | 0.975  | 0.965 |
| 500  | 3.70  | 0.2495 | 0.8630 | 1.103 | 0.938  | 0.958 |
| 700  | 3.81  | 0.2208 | 0.8103 | 1.150 | 0.856  | 0.942 |
| 1000 | 4.97  | 0.1852 | 0.7395 | 1.223 | 0.749  | 0.920 |
| 1300 | 5.18  | 0.1451 | 0.6468 | 1.284 | 0.613  | 0.859 |
| 1600 | 5.36  | 0.1115 | 0.5664 | 1.350 | 0.503  | 0.876 |
| 2000 | 5.59  | 0.0784 | 0.4703 | 1.457 | 0.382  | 0.846 |
| 2400 | 5.80  | 0.0546 | 0.3870 | 1.599 | 0.286  | 0.818 |
| 2800 | 6.00  | 0.0380 | 0.3164 | 1.798 | 0.213  | 0.795 |
| 2900 | 6.06  | 0.0348 | 0.3007 | 1.861 | 0.198  | 0.793 |
| 2900 | 9.34  | 0.0348 | 0.3007 | 1.861 | 0.198  | 0.793 |
| 3600 | 10.19 | 0.0121 | 0.1612 | 1.949 | 0.0734 | 0.781 |
| 5120 | 11.27 | 0.0062 | 0.0155 | 2.061 | 0.0014 | 0.768 |
| 6370 | 11.51 | —      | —      | —     | —      | 0.765 |

Table 4 (continued)
Bullen A (Bullen, 1953)

| Depth (km) | Density (gm cm$^{-3}$) | $I(b)$ = mean moment of inertia (unit = $Mb_0^2 = 2.426 \times 10^{45}$ gm cm$^2$) | $M(b)$ = mass enclosed within sphere of radius $b$ (unit = $M(b_0) = 5.977 \times 10^{27}$ gm) | $\bar{\rho}(b)$ = mean density of matter enclosed in sphere of radius $b$ (unit = $\bar{\rho}(b_0) = 5.517$ gm cm$^{-3}$) | $\|W(b)\|$ = absolute value of gravitational potential energy of matter enclosed in sphere of radius $b$ (unit = $\tfrac{3}{5}GM^2(b_0)/b = 3.74 \times 10^{39}$ ergs) | $f_2(b)$ (unit = surface value of $f_2$) |
|---|---|---|---|---|---|---|
| 0 | 2.84 | 0.3334 | 1.0000 | 1.000 | 4.166 | 1.000 |
| 32 | 2.84 | 0.3283 | 0.9928 | 1.007 | 1.153 | 0.997 |
| 32 | 3.32 | 0.3283 | 0.9928 | 1.007 | 1.153 | 0.997 |
| 100 | 3.38 | 0.3159 | 0.9739 | 1.022 | 1.122 | 0.991 |
| 200 | 3.47 | 0.2982 | 0.9460 | 1.042 | 1.076 | 0.983 |
| 300 | 3.55 | 0.2812 | 0.9184 | 1.062 | 1.030 | 0.974 |
| 400 | 3.65 | 0.2628 | 0.8875 | 1.086 | 0.980 | 0.964 |
| 500 | 3.87 | 0.2490 | 0.8635 | 1.104 | 0.942 | 0.957 |
| 700 | 4.30 | 0.2177 | 0.8062 | 1.144 | 0.852 | 0.941 |
| 1000 | 4.65 | 0.1744 | 0.7917 | 1.202 | 0.723 | 0.916 |
| 1300 | 4.83 | 0.1382 | 0.6384 | 1.275 | 0.605 | 0.890 |
| 1600 | 5.00 | 0.1080 | 0.5631 | 0.342 | 0.505 | 0.864 |
| 2000 | 5.22 | 0.0771 | 0.4733 | 1.467 | 0.392 | 0.829 |
| 2400 | 5.42 | 0.0548 | 0.3955 | 1.635 | 0.302 | 0.786 |
| 2800 | 5.61 | 0.0393 | 0.3295 | 1.873 | 0.232 | 0.770 |
| 2900 | 5.66 | 0.0363 | 0.3148 | 1.988 | 0.217 | 0.767 |
| 2900 | 9.70 | 0.0363 | 0.3148 | 1.984 | 0.217 | 0.767 |
| 3600 | 10.62 | 0.0127 | 0.1696 | 2.050 | 0.0813 | 0.755 |
| 5120 | 11.97 | 0.0002 | 0.0165 | 2.191 | 0.0016 | 0.739 |
| 6370 | 12.30 | — | — | — | — | 0.735 |

## Table 4 (continued)
Bullen B (Bullen, 1953)

| Depth (km) | Density (gm cm$^{-3}$) | $I(b)$ = mean moment of inertia (unit = $Mb_0^2 = 2.426 \times 10^{45}$ gm cm$^2$) | $M(b)$ = mass enclosed within sphere of radius $b$ (unit = $M(b_0) = 5.977 \times 10^{27}$ gm) | $\bar{\rho}(b)$ = mean density of matter enclosed in sphere of radius $b$ (unit = $\bar{\rho}(b_0) = 5.517$ gm cm$^{-3}$) | $\|W(b)\|$ = absolute value of gravitational potential energy of matter enclosed in sphere of radius $b$ (unit = $\tfrac{3}{5}GM^2(b_0)/b = 3.74 \times 10^{39}$ ergs) | $f_2(b)$ (unit = surface value of $f_2$) |
|---|---|---|---|---|---|---|
| 0    | 2.84  | 0.3344 | 1.0000 | 1.000 | 1.153 | 1.000 |
| 32   | 2.84  | 0.3292 | 0.9923 | 1.008 | 1.140 | 0.997 |
| 32   | 3.32  | 0.3292 | 0.9923 | 1.008 | 1.140 | 0.997 |
| 100  | 3.88  | 0.3163 | 0.9722 | 1.020 | 1.107 | 0.991 |
| 200  | 3.94  | 0.2961 | 0.9403 | 1.035 | 1.054 | 0.983 |
| 300  | 4.00  | 0.2768 | 0.9091 | 1.051 | 1.003 | 0.974 |
| 400  | 4.06  | 0.2585 | 0.8784 | 1.067 | 0.954 | 0.965 |
| 500  | 4.07  | 0.2412 | 0.8482 | 1.084 | 0.906 | 0.957 |
| 600  | 4.18  | 0.2247 | 0.8087 | 1.102 | 0.861 | 0.948 |
| 1000 | 4.41  | 0.1675 | 0.7065 | 1.180 | 0.692 | 0.911 |
| 1600 | 4.74  | 0.1046 | 0.5582 | 1.130 | 0.488 | 0.852 |
| 2000 | 4.94  | 0.0753 | 0.4731 | 1.466 | 0.382 | 0.812 |
| 2400 | 5.13  | 0.0542 | 0.3995 | 1.651 | 0.297 | 0.775 |
| 2800 | 5.42  | 0.0395 | 0.3369 | 1.915 | 0.230 | 0.746 |
| 2900 | 5.57  | 0.0366 | 0.3226 | 1.996 | 0.216 | 0.742 |
| 2900 | 9.74  | 0.0366 | 0.3226 | 1.996 | 0.216 | 0.742 |
| 3600 | 10.72 | 0.0127 | 0.1755 | 2.135 | 0.083 | 0.721 |
| 5120 | 15.00 | 0.0003 | 0.0218 | 2.883 | 0.004 | 0.641 |
| 6370 | 17.90 | —      | —      | —     | —     | 0.620 |

**Table 4 (continued)**

M 1 (Landisman et al., 1964)

| Depth (km) | Density (gm cm$^{-3}$) | $I(b)$ = mean moment of inertia (unit = $Mb_0^2 = 2.426 \times 10^{45}$ gm cm$^2$) | $M(b)$ = mass enclosed within sphere of radius $b$ (unit = $M(b_0)$ = $5.977 \times 10^{27}$ gm) | $\bar{\rho}(b)$ = mean density of matter enclosed in sphere of radius $b$ (unit = $\bar{\rho}(b_0)$ = $5.517$ gm cm$^{-3}$) | $|W(b)|$ = absolute value of gravitational potential energy of matter enclosed in sphere of radius $b$ (unit = $\tfrac{3}{5}GM^2(b_0)/b_0$ = $3.74 \times 10^{39}$ ergs) | $f_2 b$ (unit = surface value of $f_2$) |
|---|---|---|---|---|---|---|
| 0 | 2.84 | 0.3308 | 1.0000 | 1.000 | 1.164 | 1.000 |
| 32 | 2.84 | 0.3258 | 0.9929 | 1.009 | 1.151 | 0.997 |
| 32 | 3.32 | 0.3258 | 0.9929 | 1.009 | 1.151 | 0.997 |
| 100 | 3.34 | 0.3135 | 0.9740 | 1.022 | 1.120 | 0.991 |
| 200 | 3.42 | 0.2960 | 0.9466 | 1.042 | 1.074 | 0.982 |
| 300 | 3.53 | 0.2792 | 0.9192 | 1.063 | 1.029 | 0.973 |
| 400 | 3.69 | 0.2628 | 0.8917 | 1.084 | 0.985 | 0.964 |
| 500 | 3.85 | 0.2468 | 0.8639 | 1.105 | 0.940 | 0.955 |
| 700 | 4.18 | 0.2160 | 0.8076 | 1.146 | 0.852 | 0.938 |
| 1000 | 4.61 | 0.1737 | 0.7230 | 1.207 | 0.725 | 0.912 |
| 1400 | 4.98 | 0.1259 | 0.6143 | 1.294 | 0.572 | 0.876 |
| 1600 | 5.08 | 0.1063 | 0.5641 | 1.344 | 0.505 | 0.857 |
| 2000 | 5.08 | 0.0726 | 0.4748 | 1.471 | 0.392 | 0.818 |
| 2400 | 5.08 | 0.0543 | 0.4005 | 1.655 | 0.306 | 0.729 |
| 2800 | 5.08 | 0.0400 | 0.3397 | 1.931 | 0.240 | 0.749 |
| 2900 | 5.26 | 0.0373 | 0.3262 | 2.019 | 0.227 | 0.745 |
| 2900 | 9.92 | 0.0373 | 0.3262 | 2.019 | 0.227 | 0.745 |
| 3600 | 10.90 | 0.0130 | 0.1776 | 2.147 | 0.100 | 0.728 |
| 4710 | 12.27 | 0.0011 | 0.0433 | 2.450 | 0.010 | 0.692 |
| 5160 | 13.72 | 0.0002 | 0.0178 | 2.600 | 0.002 | 0.679 |
| 6370 | 15.42 | — | — | — | — | 0.639 |

**Table 4 (continued)**

M 3 (Landisman et al., 1964)

| Depth (km) | Density (gm cm$^{-3}$) | $I(b)$ = mean moment of inertia (unit = $Mb_0^2 = 2.426 \times 10^{45}$ gm cm$^2$) | $M(b)$ = mass enclosed within sphere of radius $b$ (unit = $M(b_0) = 5.977 \times 10^{27}$ gm) | $\bar{\rho}(b)$ = mean density of matter enclosed in sphere of radius $b$ (unit = $\bar{\rho}(b_0) = 5.517$ gm cm$^{-3}$) | $|W(b)|$ = absolute value of gravitational potential energy of matter enclosed in sphere of radius $b$ (unit = $\tfrac{3}{5}GM^2(b_0)/b_0 = 3.74 \times 10^{39}$ ergs) | $f_2b$ (unit = surface value of $g_2$) |
|---|---|---|---|---|---|---|
| 0 | 2.84 | 0.3308 | 1.0000 | 1.000 | 1.169 | 1.000 |
| 32 | 2.84 | 0.3258 | 0.9926 | 1.008 | 1.156 | 0.997 |
| 32 | 3.32 | 0.3258 | 0.9926 | 1.008 | 1.156 | 0.997 |
| 100 | 3.34 | 0.3135 | 0.9737 | 1.021 | 1.124 | 0.991 |
| 200 | 3.40 | 0.2961 | 0.9463 | 1.042 | 1.079 | 0.982 |
| 300 | 3.49 | 0.2794 | 0.9192 | 1.063 | 1.035 | 0.973 |
| 400 | 3.65 | 0.2632 | 0.8920 | 1.084 | 0.991 | 0.984 |
| 500 | 3.85 | 0.2473 | 0.8644 | 1.105 | 0.946 | 0.955 |
| 700 | 4.18 | 0.2165 | 0.8081 | 1.146 | 0.858 | 0.938 |
| 1000 | 4.61 | 0.1741 | 0.7234 | 1.208 | 0.731 | 0.912 |
| 1400 | 4.98 | 0.1264 | 0.6148 | 1.295 | 0.577 | 0.877 |
| 1600 | 5.08 | 0.1068 | 0.5646 | 1.345 | 0.510 | 0.858 |
| 2000 | 5.08 | 0.0761 | 0.4753 | 1.473 | 0.398 | 0.820 |
| 2400 | 5.08 | 0.0547 | 0.4009 | 1.657 | 0.312 | 0.783 |
| 2800 | 5.08 | 0.0405 | 0.3402 | 1.933 | 0.246 | 0.755 |
| 2900 | 5.24 | 0.0377 | 0.3267 | 2.022 | 0.232 | 0.752 |
| 2900 | 10.06 | 0.0377 | 0.3267 | 2.022 | 0.232 | 0.752 |
| 3600 | 11.06 | 0.0131 | 0.1758 | 2.126 | 0.085 | 0.740 |
| 5120 | 12.35 | 0.0003 | 0.0170 | 2.260 | 0.002 | 0.726 |
| 6370 | 12.63 | — | — | — | — | 0.723 |

Munk and MacDonald [1960] attribute the deviation from equilibrium in the $J_2$ term to the gradual slowing down of the earth. The equatorial bulge, at present, is too large, and this is consistent with the hypothesis that, as the earth slows down due to tidal friction, the mantle lags in its response through some anelastic processes. The characteristic time scale associated with the lag is easily determined since

$$\frac{\delta f_2}{f_2} = \frac{2\delta\Omega}{\Omega} \qquad (67)$$

If the present rate of tidal deceleration remains constant $\dot{\Omega} = -5.5 \times 10^{-22}$ radian sec$^{-1}$ [Munk and MacDonald, 1960], the present deviation of the earth's figure from equilibrium would develop in $9.5 \times 10^6$ years. If this value is to be interpreted in terms of a single parameter, viscosity, then the dynamic viscosity corresponding to a rigidity of $8.38 \times 10^{11}$ dynes cm$^{-2}$ is $7.9 \times 10^{25}$ dynes sec cm$^{-2}$. This value is about $4 \times 10^3$ larger than the value obtained from the Fenno-Scandian uplift and about $4 \times 10^4$ larger than that derived from the Lake Bonneville studies. The consistent association of a high viscosity with a low value of wave number for deformation may be taken to indicate that, on the average, the mantle is quite viscous, but that there may be a near-surface layer of low viscosity [Takeuchi, 1963]. This interpretation comes into difficulty in that one must suppose that all small-scale deviations from isostasy, such as the association of large gravity anomalies with mountain chains, must be maintained by some dynamic process.

An alternative interpretation has been discussed previously and that is that the mantle material cannot be described in terms of Newtonian viscosity, but that it undergoes creep at low stress differences. Thus, the number of mechanical parameters that must be assigned to describe the behavior of the material is greater than one and, furthermore, there are the complexities introduced by the undoubted variation of properties with depth.

The deviations in the coefficients of the external potential from those expected for a fluid earth indicate that, within the earth, surfaces of equal density are not surfaces of equal potential. Due to temperature and/or chemical inhomogeneities, equal potential and equal density surfaces intersect; the mantle is baroclinic. Since the mantle is baroclinic, stress differences exist. An equilibrium theory is powerless to give an estimate of the magnitude of the stress differences. Kaula [1963b] has numerically determined the stress differences corresponding to the observed gravitational potential, assuming an elastic behavior for the mantle and, furthermore, taking the elastic strains within the mantle to be such as to minimize the total strain energy. An alternative and highly instructive fashion of determining large-scale features of the earth's stress field is through the application of the tensor virial equations.

## The Tensor Virial Equations

*Equations for an Anelastic Solid*

The equations of motion in Cartesian coordinates for an anelastic solid, self-gravitating and rotating at a constant angular velocity $\mathbf{\Omega}$ are

$$\rho \frac{dv_i}{dt} + 2\rho\epsilon_{ijk}\Omega_j v_k = \frac{\partial p_{ij}}{\partial x_j} + \rho \frac{\partial}{\partial x_i}(U + \tfrac{1}{2}|\mathbf{\Omega} \times \mathbf{r}|^2) \qquad (68)$$

where $U$ is the potential of the self-gravitation, $\mathbf{v}$ is the velocity, and the operator $d/dt$ is, as usual,

$$\frac{d}{dt} = \frac{\partial}{\partial t} + v_j \frac{\partial}{\partial x_j} \qquad (69)$$

$\epsilon_{ijk}$ is a completely antisymmetric unit tensor of rank three, and $p_{ij}$ is the stress tensor. For an inviscid fluid the stress tensor reduces to $-p\delta_{ij}$, where $p$ is the hydrostatic pressure. For a viscous fluid the stress tensor will deviate from that appropriate for hydrostatic stress with terms dependent on the velocity field. For an elastic or anelastic solid the stress tensor will, in general, be of a more complicated form.

The tensor virial equation follows from expression 68 if one multiplies expression 68 by $x_k$ and integrates over the volume $V$ occupied by the self-gravitating matter:

$$\int_V \rho x_k \frac{dv_i}{dt} d\tau + 2\epsilon_{ijl}\Omega_j \int_V \rho x_k v_l \, d\tau$$
$$= \int_V \frac{\partial p_{ij}}{\partial x_j} x_k \, d\tau + \int_V \rho x_k \frac{\partial U}{\partial x_i} d\tau + \frac{1}{2}\int_V \rho x_k \frac{\partial}{\partial x_i}|\mathbf{\Omega} \times \mathbf{r}|^2 \, d\tau \qquad (70)$$

Over the bounding surface $S$, the stress tensor $p_{ij}$, the density $\rho$, and the velocity $\mathbf{v}$ vanish.

With the aid of the law of conservation of mass, the first term on the left-hand side of Equation 70 becomes

$$\int_V \rho x_k \frac{dv_i}{dt} d\tau = \frac{d}{dt}\int_V \rho x_k v_i \, d\tau - \int_V \rho v_i v_k \, d\tau \qquad (71)$$

# MECHANICAL PROPERTIES OF THE EARTH

The first term on the right-hand side of Equation 71 can be further reduced since

$$\int_V \rho \frac{d}{dt}(x_k v_i)\, d\tau = \frac{1}{2}\int_V \rho \frac{d}{dt}\left(x_i \frac{dx_j}{dt} + x_j \frac{dx_i}{dt}\right) d\tau$$

$$= \frac{1}{2}\frac{d^2}{dt^2}\int_V \rho x_i x_j\, d\tau$$

$$= \frac{1}{2}\frac{d^2}{dt^2}\mathscr{I}_{;ij} \qquad (72)$$

where $\mathscr{I}_{;ij}$ is the second-order moment tensor of the mass distribution. The second term on the right-hand side of Equation 71 is clearly related to the kinetic energy. Introducing the notation

$$\mathscr{K}_{ij} = \frac{1}{2}\int_V \rho v_i v_j\, d\tau \qquad (73)$$

we note that

$$\mathscr{K}_{ii} = \frac{1}{2}\int_V \rho |\mathbf{v}|^2\, d\tau \qquad (74)$$

The first term on the right-hand side of Equation 70 is

$$\int_V \frac{\partial p_{ij}}{\partial x_j} x_k\, d\tau = \int_V \frac{\partial}{\partial x_j}(p_{ij} x_k)\, d\tau - \int_V p_{ij}\delta_{jk}\, d\tau \qquad (75)$$

which, on application of the divergence theorem, yields

$$\int_V \frac{\partial p_{ij}}{\partial x_j} x_k\, d\tau = \int_S p_{ij} x_k n_j\, d\sigma - \int_V p_{ik}\, d\tau \qquad (76)$$

In Equation 76, $\mathbf{n}$ is the unit vector directed outward and associated with the element of surface $d\sigma$. The surface integral vanishes, since on the bounding surface enclosing the matter the stress vanishes:

$$\int_V \frac{\partial p_{ij}}{\partial x_j} x_k\, d\tau = -\int_V p_{ik}\, d\tau$$

The potential $U$ entering into the second term on the right-hand side of Equation 70 is determined by Equation 14. We thus have

$$\int_V \rho x_k \frac{\partial U}{\partial x_i}\, d\tau = G\int_V \int_V \rho(r)\rho(r') x_k \frac{(x_i - x_i')}{|\mathbf{r}-\mathbf{r}'|^3}\, d\tau\, d\tau' \qquad (77)$$

On interchanging primed and unprimed integration variables in the last integral, we find

$$\int_V \rho x_k \frac{\partial U}{\partial x_i}\, d\tau = -G\int_V \int_V \rho(\mathbf{r})\rho(\mathbf{r}') x_k' \frac{(x_i - x_i')}{|\mathbf{r}-\mathbf{r}'|^3}\, d\tau'\, d\tau \qquad (78)$$

Averaging Equations 77 and 78, we obtain

$$\int_V \rho x_k \frac{\partial U}{\partial x_i} d\tau = -\frac{1}{2} \int_V \rho u_{ik} d\tau = \mathscr{W}_{ik} \tag{79}$$

where $\mathscr{U}_{ik}$

$$\mathscr{U}_{ik}(\mathbf{r},t) = G \int_V \rho(\mathbf{r}') \frac{(x_i - x_i')(x_k - x_k')}{|\mathbf{r} - \mathbf{r}'|^3} d\tau' \tag{80}$$

is the symmetric potential tensor introduced by Chandrasekhar [1961]. We note that the contracted form of the tensor $\mathscr{W}_{ij}$ is the gravitational potential energy

$$\mathscr{W}_{ii} = -\tfrac{1}{2} G \int_V \int_V \frac{\rho(\mathbf{r})\rho(\mathbf{r}')}{|\mathbf{r} - \mathbf{r}'|} d\tau\, d\tau' \tag{81}$$

The last term on the right-hand side of Equation 70 may be rewritten in the form

$$\frac{1}{2} \int_V \rho x_k \frac{\partial}{\partial x_i} |\mathbf{\Omega} \times \mathbf{r}|^2 d\tau = \int_V \rho x_k (\Omega^2 x_i - \Omega_i \Omega_l x_l) d\tau \tag{82}$$
$$= \Omega^2 \mathscr{I}_{;ik} - \Omega_i \Omega_l \mathscr{I}_{;lk}$$

Substituting Equations 72, 73, 76, 79, and 82 into Equation 70, we obtain the tensor virial equation in the form

$$\frac{1}{2}\frac{d^2}{dt^2} \mathscr{I}_{;ij} = 2\mathscr{K}_{ij} + \mathscr{W}_{ij} + \Omega^2 \mathscr{I}_{;ij} - \Omega_i \Omega_l \mathscr{I}_{;lj} - \bar{p}_{ij} V$$
$$- 2\epsilon_{ikl}\Omega_k \int_V \rho x_j v_l\, d\tau \tag{83}$$

Equation 83 is identical to that obtained by Chandrasekhar and Lebovitz [1962] and Lebovitz [1961], except for the volume average of the stress tensor. In the applications considered by Chandrasekhar and his coworkers, the matter within $V$ is taken to be an inviscid fluid, so that the stress tensor reduces to a hydrostatic pressure. In the problem under discussion in this paper, we are concerned with the deviation from hydrostatic equilibrium. These deviations can result from finite values of the stress tensor and finite values of the velocity field. Later we will show that it is probable that the contributions of the velocity integrals are negligibly small compared with the other terms in the equations.

*Tensor Virial Equation for an Axisymmetric Solid*

We first examine the form of the tensor virial equation for the case

MECHANICAL PROPERTIES OF THE EARTH 231

where $\mathbf{v}$ is identically zero throughout the volume $V$. The $x_3$ axis is chosen in the direction of $\mathbf{\Omega}$. With these conditions, Equation 83 becomes

$$\mathscr{W}_{ij} + \Omega^2 \mathscr{I}_{;ij} - \Omega^2 \mathscr{I}_{;3j}\delta_{i3} = V\bar{p}_{ij} \qquad (84)$$

In applications to the earth we know from observations of the Chandler wobble that the average value of the off-diagonal components of the inertia tensor vanish [Munk and MacDonald, 1960]. Because of this, Equation 84 is equivalent to

$$\begin{aligned}
\mathscr{W}_{11} + \Omega^2 \mathscr{I}_{;11} &= V\bar{p}_{11} \\
\mathscr{W}_{22} + \Omega^2 \mathscr{I}_{;22} &= V\bar{p}_{22} \\
\mathscr{W}_{33} &= V\bar{p}_{33} \\
\mathscr{W}_{12} &= V\bar{p}_{12} \\
\mathscr{W}_{13} &= V\bar{p}_{13} \\
\mathscr{W}_{23} &= V\bar{p}_{23}
\end{aligned} \qquad (85)$$

We now proceed to investigate Equation 85 on the assumption that the mass distribution is axisymmetric. Equations 85 can be combined to yield

$$(\mathscr{W}_{33} - \mathscr{W}_{11}) - \Omega^2 \mathscr{I}_{;11} = V(\bar{p}_{33} - \bar{p}_{11}) \qquad (86)$$

The right-hand side of Equation 86 is a measure of the deviation from hydrostatic equilibrium. An estimate of the left-hand side of Equation 86 yields the average value of the stress differences within the earth.

Writing Equation 79 in terms of spherical coordinates, we have that

$$\mathscr{W}_{11} = \frac{1}{2} \int_V \rho r \left[ (1 - \mu^2) \frac{\partial U}{\partial r} - \frac{\mu(1 - \mu^2)}{r} \frac{\partial U}{\partial \mu} \right] d\tau \qquad (87)$$

$$\mathscr{W}_{33} = \int_V \rho r \left[ \mu^2 \frac{\partial U}{\partial r} + \frac{\mu(1 - \mu^2)}{r} \frac{\partial U}{\partial \mu} \right] d\tau$$

In evaluating the integrals of Equation 87, we express the potential $U$ according to Equation 46 and express the density in terms of the surfaces of equal density defined by Equation 29. If the term of order $g_2$ and $U_2$ is maintained, but all higher-order terms are discarded, Equation 87 takes the form

$$\mathcal{W}_{11} = -\frac{4\pi G}{3}\int_0^{b_0} \rho(s)sM(s)\,ds - \frac{4\pi}{15}\left[\int_0^{b_0} \rho \frac{d}{db}(s^3 U_2)\,ds\right.$$
$$\left. + G\int_0^{b_0} s^2 g_2 M(s)\frac{d\rho}{ds}\,ds\right] \quad (88)$$

and

$$\mathcal{W}_{33} = -\frac{4\pi G}{3}\int_0^{b_0} \rho s M(s)\,ds + \frac{8\pi}{15}\left[\int_0^{b_0} \rho \frac{d(s^3 U_2)}{ds}\,ds\right.$$
$$\left. + G\int_0^{b_0} s^2 g_2(s)\frac{d\rho}{ds}M(s)\,ds\right] \quad (89)$$

where Equation 52 has been used in evaluating $U_0$. The first term on the left-hand side of Equation 86 becomes

$$\mathcal{W}_{33} - \mathcal{W}_{11} = \frac{4\pi}{5}\left[\int_0^{b_0} \rho(s)\frac{d}{ds}[s^3 U_2(s)]\,ds\right.$$
$$\left. + G\int_0^{b_0} s^2 g_2(s)M(s)\frac{d\rho}{ds}\,ds\right] \quad (90)$$

From Equation 52, we have

$$\frac{d(b^3 U_2)}{db} = -4\pi G b^4 \int_0^{b_0} g_2 \frac{d\rho}{ds}\,ds \quad (91)$$

Introducing Equation 91 into Equation 90, and changing the order of integration in the first integral, we obtain

$$\mathcal{W}_{33} - \mathcal{W}_{11} = \frac{4\pi G}{5}\int_0^{b_0} g_2(s)\frac{d\rho}{ds}[s^2 M(s) - I(s)]\,ds \quad (92)$$

where

$$I(s) = \frac{8\pi}{3}\int_0^b \rho s^4\,ds \quad (93)$$

Substituting Equation 92 into 86 gives the virial equation appropriate to an axially symmetric self-gravitating body, provided the deviations from sphericity are not too great; i.e., terms of order $g_2{}^2$ are negligible. The virial equation

$$\frac{4\pi G}{5}\int_0^{b_0} g_2 \frac{d\rho}{ds}[s^2 M(s) - I(s)]\,ds = \tfrac{1}{2}\mathscr{I}_{;11}\Omega^2 + V(\bar{p}_{33} - \bar{p}_{11}) \quad (94)$$

places an integral condition on $g_2$. Unlike the condition imposed by $J_2$ (see Equation 38), the application of Equation 94 requires knowledge regarding the state of stress in the earth.

## Tensor Virial Equation for Hydrostatic Equilibrium

Equation 94 is satisfied identically, provided the state of stress is hydrostatic and the surfaces of equal potential correspond with the surfaces of equal density. To prove this, we note from Equation 50 that

$$\frac{d}{db}[b^3 U_2(b)] = \tfrac{5}{3}\Omega^2 b^4 + G\frac{d}{db}[f_2 b^2 M(b)] \qquad (95)$$

where $f_2$ refers to both the equal potential and equal density surfaces. Substituting Equations 95 into 90 and replacing $g_2$ by $f_2$, we obtain

$$\mathscr{W}_{33} - \mathscr{W}_{11} = \frac{\Omega^2 I(b_0)}{2} \qquad (96)$$

Substituting Equation 96 into Equation 86, we obtain

$$\frac{\Omega^2 I(b_0)}{2} = \Omega^2 \mathscr{I}_{;11} = \frac{\Omega^2}{2}[I + 0(g_2)] \qquad (97)$$

and Equation 94 is satisfied identically to terms of order $g_2^2$, since at hydrostatic equilibrium

$$p_{33} = p_{11}$$

## Deviations from Hydrostatic Equilibrium

We have previously obtained an estimate of the magnitude by which the surface value of $f_2$ differs from that to be expected if the earth were truly in hydrostatic equilibrium. In order to make use of this, we cast virial equation 86 into a form in which $f_2$ appears explicitly. Equations 91 and 90 can be combined in the form

$$\mathscr{W}_{33} - \mathscr{W}_{11} = \frac{4\pi G}{5}\left[-\int_0^{b_0}\left[4\pi\rho_0 b^4 \int_0^{b_0} g_2 \frac{d\rho}{ds} ds\right]db \right.$$
$$\left. + \int_0^{b_0}\left[4\pi b^2 g_2 \frac{d\rho}{db}\int_0^b \rho s^2 ds\right]db\right] \qquad (98)$$

If we change the order of integration in the second integral, Equation 98 becomes

$$\mathscr{W}_{33} - \mathscr{W}_{11} = \frac{4\pi G}{5}\left[-\int_0^{b_0}\left[\frac{dM(b)}{db} b^2 \int_0^{b_0} g_2 \frac{d\rho}{ds} ds\right] db \right.$$
$$\left. + \int_0^{b_0}\left[\frac{dM(b)}{db}\int_0^{b_0} s^2 g_2 \frac{d\rho}{ds} ds\right] db\right] \qquad (99)$$

An integration by parts yields

$$\mathscr{W}_{33} - \mathscr{W}_{11} = -\frac{2}{5}\int_0^{b_0} \frac{M(b)}{b^3}\frac{d}{db}(b^3 U)\, db \qquad (100)$$

The derivative in the integrand is given by Equation 50. We thus express $\mathscr{W}_{33} - \mathscr{W}_{11}$ in terms of $f_2$:

$$\mathscr{W}_{33} - \mathscr{W}_{11} = -\frac{2}{5}\int_0^{b_0} M(b)\left[\tfrac{5}{3}\Omega^2 b + \frac{G}{b^3}\frac{d}{db}[f_2 b^2 M(b)]\right] db \qquad (101)$$

The first integral on the right-hand side in Equation 101 can be integrated by parts to yield

$$-\tfrac{2}{3}\Omega^2\int_0^{b_0} M(b) b\, db = -\tfrac{1}{3}\Omega^2 M(b_0) b_0^2 + \frac{I\Omega^2}{2} \qquad (102)$$

Combining Equations 101 and 102 with Equation 86, we obtain

$$-\tfrac{2}{5}G\int_0^{b_0}\frac{M(b)}{b^3}\frac{d}{db}[f_2 b^2 M(b)]\, db - \tfrac{1}{3}\Omega^2 M(b_0) b_0^2 = V(\bar{p}_{33} - \bar{p}_{11}) \qquad (103)$$

If hydrostatic equilibrium prevails, Equation 103 would be satisfied, with a vanishing right-hand side. Let $h_2$ denote the second-order term in the expansion of the equipotential surfaces for hydrostatic equilibrium. We define the deviation from hydrostatic equilibrium by

$$\delta h_2 = h_2 - f_2 \qquad (104)$$

where, from results discussed in the earlier sections, $\delta h_2(b_0)$ is positive. Since the right-hand side is a measure of the deviation from hydrostatic equilibrium, Equation 103 takes the form

$$V(\bar{p}_{33} - \bar{p}_{11}) = \frac{2G}{5}\int_0^{b_0}\frac{M(b)}{b^3}\frac{d}{db}[\delta h_2 b^2 M(b)]\, db \qquad (105)$$

If $\delta h_2$ is positive and does not change sign within the earth, the integral on the right-hand side of Equation 105 is positive definite for any reasonable distribution of density. We integrate the right-hand side of Equation 105 by parts:

$$\tfrac{2}{5}G\int_0^{b_0}\frac{M(b)}{b^3}\frac{d}{db}[\delta h_2 b^2 M(b)]\, db = \tfrac{2}{5}G\frac{M^2(b_0)\delta h_2(b_0)}{b_0}$$
$$-\tfrac{2}{5}G\int_0^{b_0}\delta h_2 b^2 M(b)\frac{d[M(b)/b^3]}{db}\, db \qquad (106)$$

where, since

$$\frac{d[M(b)/b^3]}{db} = -\frac{4\pi}{b}[\bar{\rho}(b) - \rho(b)] \quad (107)$$

$$\bar{\rho}(b) = \frac{3}{b^3}\int_0^b s^2\rho(s)\,ds$$

Equation 105 becomes

$$V(\bar{p}_{33} - \bar{p}_{11}) = \tfrac{2}{5}G\frac{M^2(b_0)\delta h_2(b_0)}{b_0}$$

$$+ \frac{8\pi}{5}G\int_0^{b_0} \delta h_2 b M(b)[\bar{\rho}(b) - \rho(b)]\,db \quad (108)$$

Provided $\bar{\rho}$ is everywhere greater than $\rho$ and $\delta h$ does not change sign, the right-hand side of Equation 108 will have a sign determined by $\delta h$. The numerical values listed in Table 4 clearly show that the condition $\bar{\rho} > \rho$ on the density distribution is met by various models for the earth.

*Magnitude of $\bar{p}_{33} - \bar{p}_{11}$*

We can obtain an approximate value for $(\bar{p}_{33} - \bar{p}_{11})$ by assuming $\delta h_2$ is constant throughout the earth. If $\delta h_2$ is not constant, but decreases from its surface value, the calculation sets a maximum for the quantity $(\bar{p}_{33} - \bar{p}_{11})$ because of the positive definite character on the integral appearing on the right-hand side of Equation 105. However, there is no physical reason for the deviation of the surfaces of equal potential from those of a hydrostatic body to decrease within the earth.

We evaluate the derivative appearing in the second integral on the right-hand side of Equation 106 and integrate by parts to obtain

$$V(\bar{p}_{33} - \bar{p}_{11}) = \tfrac{2}{5}G\delta h_2(b_0)\left[5\int_0^{b_0} M(b)\frac{dM(b)}{db}\,db - 3\frac{M^2(b_0^2)}{b_0}\right] \quad (109)$$

The terms on the right-hand side of Equation 109 are related to the gravitational potential energies of the stratified earth and the homogeneous earth. Equation 109 can thus be rewritten as

$$\bar{p}_{33} - \bar{p}_{11} = \frac{2\delta h_2(b_0)}{V}[|W_0| - \tfrac{2}{3}|W_u|] \quad (110)$$

where $W_0$ is the gravitational potential energy determined by Equation 81 and where the terms of order $g_2$ have been neglected. Table 4 lists numerical values of $|W_0|$ for various density models of the earth;

$W_u$ is the gravitational potential energy for a uniform earth having the same mass and radius as the earth. The difference in stresses $p_{33}$ and $p_{11}$ averaged over the volume is proportional to the deviation of the equal potential surfaces from those of a hydrostatic body, and $\delta h_2$ fixes that fraction of the difference in potential energy of the earth and two-thirds the potential energy of an equivalent homogeneous earth that is associated with the disequilibrium $J_2$. Numerical values for the gravitational potential energy of various models of the earth in terms of the gravitational potential energy of an equivalent homogeneous earth are listed in Table 4. Table 5 lists the volume average

**Table 5**

Volume Average of $p_{33} - p_{11}$ on the Assumption $\delta h_2$ is Constant; $\delta h_2 = 9.1 \times 10^{-6}$

| Model | Stress Difference (unit = $10^6$ dynes cm$^{-2}$) |
|---|---|
| Bullard I | 18.8 |
| Bullard II | 18.6 |
| Bullen A | 18.8 |
| Bullen B | 18.3 |
| M 1 | 18.7 |
| M 3 | 18.9 |

of the stress difference, calculated on the assumption that $\delta h_2$ is constant throughout the earth. If it is assumed that $\delta h_2$ is constant over the mantle and vanishes in the core, the volume average stress difference is less because of the positive definite character of the integral appearing in the right-hand side of Equation 104. For the models listed in Table 5, the average value of $p_{33} - p_{11}$ is about 14 bars if $\delta h_2$ is constant over the mantle and vanishes elsewhere.

An estimate of the elastic energy associated with the deviations from a hydrostatic $J_2$ is

$$\text{elastic energy} = \frac{|p_{33} - p_{11}|^2}{4\bar{\mu}} \qquad (111)$$

where $\bar{\mu}$ is the average rigidity. Introducing numerical values we find that the total elastic energy associated with $\delta h_2(b_0)$ is $1.1 \times 10^{29}$ ergs if $\delta h_2$ is constant throughout the earth. The total energy $V(\bar{p}_{33} - \bar{p}_{11})$

is much greater, $2 \times 10^{34}$ ergs. The elastic energy can be compared with Kaula's [1963b] estimate of $2.2 \times 10^{29}$ ergs based on a numerical elastic model calculation. A discrepancy of about a factor of 2 is expected, and Kaula treated the crust as an elastically supported surface load. Kaula's assumption of minimum strain energy results in the load being compensated by a layer at the bottom of the crust. The energy stored within the crust due to the $J_2$ is about $10^{29}$ ergs.

In the discussion so far, we have neglected the velocity-dependent terms in the full virial equation 82. For a steady flow, the terms on the left-hand side of Equation 82 vanish. The term due to the Coriolis force will be of order

$$\omega \Omega | = 5.86 \times 10^{40} \omega$$

where $\omega$ is the characteristic frequency of convective motion. For motions having a time scale on the order of $10^7$ years, the last term in Equation 82 is of the magnitude of $10^{26}$ cm$^2$ gm sec$^{-2}$. This is negligibly small compared with $(\bar{p}_{33} - \bar{p}_{11})V$. The term in $\mathscr{K}$ is proportional to $\omega^2$, and thus even smaller than the Coriolis term.

**Third-Order Virial Equation**

The second-order virial equation yields relations involving the volume average of the stress differences. The relations involve the second-order moment of the mass distribution. The interpretation of the higher-order moments of the mass distribution requires higher-order virial equations. The third-order virial equation exhibits how third-order moments of the mass distribution are related to moments of the stresses within the earth.

The third-order virial equations are obtained by multiplying Equation 68 by $x_m x_n$ and integrating over the volume. The first term on the left-hand side becomes

$$\int_V \rho \frac{dv_i}{dt} x_m x_n \, d\tau = \frac{d\mathscr{V}_{i;mn}}{dt} - 2(\mathscr{K}_{im;n} + \mathscr{K}_{in;m}) \qquad (112)$$

where we use the notation

$$\mathscr{V}_{i;mn} = \int_V \rho v_i x_m x_n \, d\tau$$

$$\mathscr{K}_{im;n} = \frac{1}{2} \int_V \rho v_i x_m x_n \, d\tau \qquad (113)$$

$\mathscr{K}_{im;n}$, on contraction of the indices to the left of the semicolon, yields the first moment of the kinetic energy.

With the notation given in Equation 113, the term arising from the Coriolis force becomes

$$2\epsilon_{ijk}\Omega_j \int_V \rho v_k x_m x_n \, d\tau = 2\epsilon_{ijk}\Omega_j \mathscr{V}_{k;mn} \tag{114}$$

The first term on the right-hand side of Equation 68, after multiplication by $x_m x_n$ and integration over the volume, yields

$$\int_V \frac{\partial p_{ij}}{\partial x_j} x_m x_n \, d\tau = \int_S p_{ij} x_m x_n n_j \, d\sigma - \int_V p_{ij}(\delta_{ij}x_n + \delta_{jn}x_m) \, d\tau$$
$$= -\pi_{im;n} - \pi_{in;m} \tag{115}$$

where

$$\pi_{im;n} = \int_V p_{ij} x_m x_n \, d\tau \tag{116}$$

The term involving the gradient of the potential on the right-hand side of Equation 68, when multiplied by $x_m x_n$ and integrated over the volume, yields terms which are first moments of the potential energy tensor $\mathscr{W}_{ij}$

$$\int_V \rho \frac{\partial U}{\partial x_i} x_m x_n \, d\tau = G \int_V \int_V x_m x_n \rho(\mathbf{r})\rho(\mathbf{r}') \frac{\partial}{\partial x_i} \frac{1}{|\mathbf{r}-\mathbf{r}'|} \, d\tau \, d\tau'$$
$$= -\frac{G}{2} \int_V \int_V \rho\rho' x_m' x_n' \frac{(x_i'-x_i)}{|\mathbf{r}-\mathbf{r}'|^3} \, d\tau \, d\tau'$$
$$- \frac{G}{2} \int_V \int_V \rho\rho' x_m x_n \frac{(x_i-x_i')}{|\mathbf{r}-\mathbf{r}'|^3} \, d\tau \, d\tau'$$
$$= -\frac{G}{2} \int_V \int_V \rho\rho' x_n \frac{(x_m-x_m')(x_i-x_i')}{|\mathbf{r}-\mathbf{r}'|^3} \, d\tau \, d\tau'$$
$$- \frac{G}{2} \int_V \int_V \rho\rho' x_m \frac{(x_n-x_n')(x_i-x_i')}{|\mathbf{r}-\mathbf{r}'|^3} \, d\tau \, d\tau'$$
$$= \mathscr{W}_{im;n} + \mathscr{W}_{in;m} \tag{117}$$

where

$$\mathscr{W}_{im;n} = -\frac{1}{2} \int \rho \mathscr{U}_{im} x_n \, d\tau \tag{118}$$

and $\mathscr{U}_{im}$ is given by Equation 80. From contraction of the indices on the left-hand side of the semicolon, we see that $\mathscr{W}_{ii;m}$ is the first moment of the gravitational potential energy.

The terms derived from the centrifugal force can be rewritten in terms of the third-order mass moments:

$$\frac{1}{2}\int_V \rho \frac{\partial}{\partial x_i}[\epsilon_{ijk}\Omega_j x_k]^2 x_m x_n \, d\tau$$

$$= \Omega^2 \int_V \rho x_i x_m x_n \, d\tau - \Omega_i \Omega_j \int \rho x_j x_m x_n \, d\tau$$

$$= \Omega^2 \mathscr{I}_{;imn} - \Omega_i \Omega_j \mathscr{I}_{;jmn} \qquad (119)$$

Equations 112, 114, 115, 117, and 119 combine to yield the third-order virial equation

$$\frac{d}{dt}\mathscr{V}_{i,mn} = 2(\mathscr{K}_{im;n} + \mathscr{K}_{in;m}) - \pi_{im;n} - \pi_{in;m} + \mathscr{W}_{im;n}$$
$$+ \mathscr{W}_{in;m} + \Omega^2 \mathscr{I}_{;imn} - \Omega_i \Omega_j \mathscr{I}_{;jmn} - 2\epsilon_{ijk}\Omega_j \mathscr{V}_{k;mn} \qquad (120)$$

Equation 120, specialized to the case where the internal stress is hydrostatic, is identical with Equation 25 of Chandrasekhar [1962].

The number of terms in Equation 120 reduces markedly provided the density distribution contains some elements of symmetry. Suppose that the density distribution has biplanar symmetry with

$$\rho(x_1, x_2, x_3) = \rho(-x_1, x_2, x_3)$$
$$= \rho(x_1, -x_2, x_3) \qquad (121)$$

where, as before, the Cartesian axes are chosen with $x_3$ parallel to $\boldsymbol{\Omega}$. For biplanar symmetry the only nonvanishing integrals $\mathscr{I}_{;ijk}$ are

$$\mathscr{I}_{;333} \neq 0; \quad \mathscr{I}_{;311} \neq 0; \quad \mathscr{I}_{;322} \neq 0 \qquad (122)$$

For cylindrical symmetry

$$\mathscr{I}_{;311} = \mathscr{I}_{;322} \qquad (123)$$

and, if the origin of the coordinate system coincides with the center of mass, then

$$\mathscr{I}_{;311} = -\tfrac{1}{2}\mathscr{I}_{;333} \qquad (124)$$

From Equation 118 we note that

$$\mathscr{W}_{11;1} = \mathscr{W}_{22;2} = 0$$
$$\mathscr{W}_{33;3} \neq 0 \qquad (125)$$

for biplanar symmetry. Further, we have that $\mathscr{W}_{ij;k} = 0$ if $i \neq j$ and $i, j = 1, 2$, since the integrand is odd in $x_i$ and $x_j$, while the density is even in $x_1$ and $x_2$. Thus, for biplanar symmetry, only $\mathscr{W}_{13;1}$, $\mathscr{W}_{23;2}$, $\mathscr{W}_{22;3}$, and $\mathscr{W}_{33;3}$ are nonvanishing.

For the case of biplanar symmetry with the origin at the center of mass and $x_3$ along the axis of rotation, the third-order virial equations 120 reduce to

$$\mathscr{W}_{13;1} + \mathscr{W}_{11;3} + \Omega^2 \mathscr{I}_{;311} = \pi_{13;1} + \pi_{11;3}$$
$$\mathscr{W}_{23;2} + \mathscr{W}_{22;3} + \Omega^2 \mathscr{I}_{;322} = \pi_{23;2} + \pi_{22;3} \quad (126)$$
$$\mathscr{W}_{33;3} = \pi_{33;3}$$

Furthermore, if we assume axial symmetry, then a convenient form of the third-order virial equation is

$$\mathscr{W}_{33;3} - \mathscr{W}_{13;1} - \mathscr{W}_{11;3} = \mathscr{I}_{;311}\Omega^2 + \pi_{33;3} - \pi_{13;1} - \pi_{11;3} \quad (127)$$

where $\mathscr{I}_{;311}$ is known from observations of the earth's external potential. We can evaluate the left-hand side of Equation 127 in terms of the coefficients describing the equipotential surfaces and thus obtain an estimate of $\pi_{33;3} - \pi_{13;1} - \pi_{11;3}$.

*Third-Order Virial Equation for Axial Symmetry*

Introducing spherical coordinates, we have that

$$\mathscr{W}_{13;1} + \mathscr{W}_{11;3} = \frac{1}{2} \int_V \rho \left[ r^2 \mu (1 - \mu^2) \frac{\partial U}{\partial r} - r\mu^2(1 - \mu^2) \frac{\partial U}{\partial \mu} \right] d\tau \quad (128)$$

$$\mathscr{W}_{33;3} = \frac{1}{2} \int_V \rho \left[ r^2 \mu^3 \frac{\partial U}{\partial r} + r\mu^2(1 - \mu^2) \frac{\partial U}{\partial \mu} \right] d\tau$$

Expressing the density in terms of the surfaces of equal density and keeping the terms of order $g_3$, we obtain

$$\mathscr{W}_{13;1} + \mathscr{W}_{11;3} = -\frac{4\pi}{35} \int_0^{b_0} \left[ Gg_3 b^3 M(b) \frac{d\rho}{db} + \rho \frac{d}{db}(b^4 U_3) \right] db \quad (129)$$

$$\mathscr{W}_{33;3} = \frac{4\pi}{35} \int_0^{b_0} \left[ Gg_3 b^3 M(b) \frac{d\rho}{db} + \rho \frac{d(b^4 U_3)}{db} \right] db$$

With Equation 129, Equation 127 becomes

$$\frac{8\pi}{35} \int_0^{b_0} \left[ Gb^3 g_3 M(b) \frac{d\rho}{db} + \rho \frac{d}{db}(b^4 U_3) \right] db$$
$$= \pi_{33;3} - (\pi_{13;1} + \pi_{11;3}) - \Omega^2 \mathscr{I}_{;311} \quad (130)$$

From Equation 52, we have that

$$\frac{d}{db}(b^4 U_3) = -4\pi G b^6 \int_0^{b_0} g_3 b^{-1} \frac{d\rho}{db} db \quad (131)$$

# MECHANICAL PROPERTIES OF THE EARTH

Introducing Equation 131 into 130 and changing the order of integration in the first integral, we obtain

$$\pi_{33;3} - (\pi_{13;1} + \pi_{11;3}) - \Omega^2 \mathscr{I}_{;311}$$

$$= \frac{8\pi}{35} G \left[ \int_0^{b_0} \left[ \frac{dM}{db} \int_0^{b_0} s^3 g_3 \frac{d\rho}{ds} ds \right] db \right.$$

$$\left. - \int_0^{b_0} \left[ b^4 \frac{dM}{db} \int_b^{b_0} g_3 s^{-1} \frac{d\rho}{ds} ds \, db \right] \right]$$

$$= -\frac{8}{35} \int_0^{b_0} \frac{M(b)}{b^3} \frac{d}{db} (b^4 U_3) \, db \qquad (132)$$

In terms of the coefficients describing the surfaces of equal potential (see Equation 50), Equation 132 can be written as

$$\pi_{33;3} - (\pi_{13;1} + \pi_{11;3})$$

$$= -\Omega^2 \mathscr{I}_{;311} - \frac{8G}{35} \int_0^{b_0} \frac{M(b)}{b^3} \frac{d}{db} [b^3 f_3 M(b)] \, db \qquad (133)$$

Using Equations 41, 53, and 107, Equation 133 takes the form

$$\frac{32\pi G}{35} \int_0^{b_0} b^2 f_3(b) M(b) [\bar{\rho}(b) - \rho(b)] \, db$$

$$= \frac{MR^3}{5} J_3 \left[ \frac{8}{7} \frac{GM}{b_0^3} - \Omega^2 \right] + \pi_{13;1} + \pi_{11;3} - \pi_{33;3} \qquad (134)$$

Provided $f_3$ does not change sign within the earth, the left-hand side of Equation 134 has a sign determined by $f_3$, if $\bar{\rho} > \rho$ everywhere in the interior.

As in the case of the second-order virial equations, we can obtain an approximation by assuming that $f_3$ is constant throughout the earth. Using Equation 133 and integrating the integral on the right-hand twice by parts, we obtain that

$$\pi_{33;3} - \pi_{13;1} - \pi_{11;3}$$

$$= -\frac{8}{35} Gf_3(b_0) \left[ \frac{M^2(b_0)}{2} + 3 \int_0^{b_0} \frac{M^2(b)}{b} \, db \right] - \mathscr{I}_{;311} \Omega^2 \qquad (135)$$

for the case where $f_3$ is constant and maintains its surface value throughout the earth.

By using the numerical value of $J_3$ of $-2.54 \times 10^{-6}$, we can obtain that $\pi_{33;3} - \pi_{13;1} - \pi_{11;3}$ is of the order of $7 \times 10^{43}$ dynes cm$^{-1}$ with the values ranging from $6.9 \times 10^{43}$ dynes cm$^{-1}$ for Bullen's Model B to $7.6 \times 10^{43}$ dynes cm$^{-1}$ for Bullard's Model 1.

The evaluation of the virial integrals provides a means of estimating the average stress differences and moments of the stresses in terms of the deviations of the surface of equal potential from those for a body in fluid equilibrium. Calculations have been carried out through the third-order virial equation. Fourth-order calculations are possible, but here further information is required, since $J_4$ determines only the difference between the mass moments (see Equation 26). The calculations reviewed above have been made on the assumption that the density distribution is axially symmetric. Similar evaluations can be carried out for the real earth, which deviates from axial symmetry (see Table 1). In this way, expressions can be obtained for the means and moments of the various components of the stresses within the earth.

## Conclusions

It has long been known that on the small scale of a volcano or individual mountain range the surfaces of equal density and those of equipotential differ. The remarkable result of satellite geodesy is that on a global scale these surfaces stand apart. The implication is that there are large-scale temperature or chemical inhomogeneities within the earth. These inhomogeneities must be supported by stress differences. The existence and magnitude of these stress differences yield information about the large-scale mechanical properties of the earth. The full accounting of the derived stress differences remains a task for the future.

The geodetic results have a further significance in discussions of crustal evolution. The nonhydrostatic components of $C_n^m$ and $S_n^m$ provide a measure of the degree of mechanical disequilibrium within the earth. Energy stored within the disequilibrium figure can, in principle, drive those processes which shape the outer layers. Equation 110 provides an estimate of the nonequilibrium energy associated with the equatorial bulge of the earth. A fraction of the difference in gravitational potential energy between the earth and an equivalent homogeneous earth is available for tectonic processes. At present, about $2 \times 10^{34}$ ergs of energy are associated with the nonhydrostatic $J_2$. If the time scale for the development of the nonequilibrium bulge is $10^7$ years, as is suggested by the astronomical data, then, on the average, $2 \times 10^{27}$ erg year$^{-1}$ should be released. This rate of energy release is one-fifth the present heat flow and much larger than the $10^{25}$ erg year$^{-1}$ which is the estimated rate of release of seismic energy. The rate of conversion of kinetic energy of rotation into heat by tidal friction is about $8 \times 10^{26}$ ergs year$^{-1}$. Thus the gravitational energy being made available as a result of the change of shape of the earth is larger than the decrease in rotational kinetic energy.

Stored within the disequilibrium figure is an energy source comparable in magnitude with the radioactive heat sources. How is this energy released? Has this energy in the past been converted in such a way that continents have moved, ocean basins have opened, and mountains have formed? There are no simple answers to these questions. Much theory and experiment lies ahead. The nature of energy conversion in solids capable of creep must be explored. The detailed nature of flow processes in silicates must be described and understood in such terms that one can confidently extrapolate from the laboratory to the geologic time scale. The conditions within the mantle must be described far more precisely. Until these and other questions can be answered continental drift and convection will remain speculative subjects. The foregoing problems, however, illustrate the endless variety of challenges in geodynamics.

## References

Allen, C. R., Circum-Pacific faulting in the Philippines-Taiwan region, *J. Geophys. Res.*, *67*, 4795–4812, 1962.

Andrade, E. N., and D. A. Aboav, The flow of polycrystalline cadmium under simple shear, *Proc. Roy. Soc. London*, *A*, *280*, 353–382, 1964.

Axelrod, D. I., Reply to discussion of paper, "Fossil flora suggest stable, not drifting, continents," *J. Geophys. Res.*, *69*, 1669–1671, 1964.

Benioff, H., Movements on major transcurrent faults, in *Continental Drift*, edited by S. K. Runcorn, pp. 103–134, Academic Press, New York, 1962.

Bullard, E. C., The density within the earth, in *Verhandel. Ned. Geol. Mijnbouwk. Genoot.*, *18*, 23–41, 1957.

Bullard, E. C., Continental drift, *Quart. J. Geol. Soc. London*, *120*, 1–33, 1964.

Bullen, K. E., *An Introduction to the Theory of Seismology*, 2nd ed., Cambridge University Press, 1953.

Chandrasekhar, S., *Hydrodynamic and Hydromagnetic Stability*, pp. 557–581, Oxford at the Clarendon Press, 1961.

Chandrasekhar, S., On the point of bifurcation along the sequence of Jacobi ellipsoid, *Astrophys. J.*, *136*, 1048–1068, 1962.

Chandrasekhar, S., and N. R. Lebovitz, On the oscillations and stability of rotating gaseous masses, *Astrophys. J.*, *135*, 248–260, 1962.

Chandrasekhar, S., and P. H. Roberts, The ellipticity of a slowly rotating configuration, *Astrophys. J.*, *138*, 801–808, 1963.

Clarke, D. C., Constants and related data used in trajectory calculations at the Jet Propulsion Laboratory, *California Institute of Technology, Jet Propulsion Laboratory Technical Report*, 32–273, 1962.

Cox, A., and R. R. Doell, Review of paleomagnetism, *Bull. Geol. Soc. Am.*, *71*, 645–768, 1960.

Crittenden, M., Effective viscosity of the earth derived from isostatic loading of Pleistocene Lake Bonneville, *J. Geophys. Res.*, *68*, 5517–5530, 1963.
Crowell, J. C., Displacements along the San Andreas Fault, California, *Geol. Soc. Am. Spec. Papers*, *71*, 1962.
Guier, W. H., Determination of the non-zonal harmonics of the geopotential from satellite doppler data, *Nature*, *200*, 124–125, 1963.
Hamilton, W., Origin of the Gulf of California, *Bull. Geol. Soc. Am.*, *72*, 1307–1318, 1961.
Hamilton, W., Discussion of paper by D. I. Axelrod, "Fossil flora suggest stable, not drifting, continents," *J. Geophys. Res.*, *69*, 1666–1668, 1964.
Henriksen, S., The hydrostatic flattening of the earth, *Ann. Intern. Geophys. Year*, *12*, 192–198, 1960.
Hess, H. H., History of ocean basins, in *Petrologic Studies, A Volume in Honor of A. F. Buddington*, edited by A. E. J. Engle, H. L. James, and B. L. Lennard, pp. 599–620, Geol. Soc. Am., New York, 1962.
Hill, M. N., editor, *The Sea*, vol. III, Interscience Publishers, Inc., London, 1963.
Irving, E., Paleomagnetic pole positions: A survey and analysis, *Geophys. J.*, *2*, 51–79, 1959.
Irving, E., Paleomagnetic directions and pole positions, Parts II–V, *Geophys. J.*, *3*, 44–49; *5*, 70–79; *6*, 263–267; *7*, 263–274, 1960–63.
Jeffreys, H., The figures of rotating planets, *Monthly Notices Roy. Astron. Soc.*, *113*, 97–105, 1953.
Jeffreys, H., *The Earth*, 4th ed., Cambridge University Press, London, 1959.
Jeffreys, H., The hydrostatic theory of the figure of the earth, *Geophys. J.*, *8*, 196–202, 1963.
Kääriäinen, E., On the recent uplift of the earth's crust in Finland, *Publ. Finnish Geodetic Inst.*, *42*, 1953.
Kaula, W. M., A review of geodetic parameters, *NASA Technical Note D1847*, 1963a.
Kaula, W. M., Elastic models of the mantle corresponding to variations in the external gravity field, *J. Geophys. Res.*, *68*, 4967–4978, 1963b.
Kennedy, A. J., *Processes of Creep and Fatigue in Metal*, John Wiley & Sons, New York, 1963.
Kennedy, W. Q., The Great Glen Fault, *Quart. J. Geol. Soc. London*, *102*, 41–72, 1946.
King-Hele, D. C., G. E. Cook, and H. Margaret Watson, Even zonal harmonics in the earth's gravitational potential, *Nature*, *202*, 996, 1964.
Knopoff, L., and G. J. F. MacDonald, Attenuation of small-amplitude stress waves in solids, *Rev. Mod. Phys.*, *30*, 1178–1192, 1958.
Landisman, M., Y. Saito, and J. Nafe, The vibrations of the earth and the properties of the steep interior regions, Part 1, Density, *Geophys. J.*, in press, 1964.
Lebovitz, N. R., The virial tensor and its application to self-gravitating fluids, *Astrophys. J.*, *134*, 500–535, 1961.
MacDonald, G. J. F., Deep structure of continents, *Rev. Geophys.*, *1*, 587–665. 1963.
Menard, H. W., *Marine Geology of the Pacific*, McGraw-Hill Book Company, Inc., New York, 1964.
Munk, W. H., and G. J. F. MacDonald, *Rotation of the Earth*, Cambridge University Press, London, 1960.

Niskanen, E., On the viscosity of the earth's interior and crust, *Ann. Acad. Sci. Fennicae*, *15*, 1940.

Runcorn, S. K., Paleomagnetic evidence for continental drift and its geophysical cause, in *Continental Drift*, edited by S. K. Runcorn, pp. 1–40, Academic Press, New York, 1962.

Takeuchi, H., Time-scales of isostatic compensation, *J. Geophys. Res.*, *68*, 2357, 1963.

Vacquier, V., A. D. Raff, and R. E. Warren, Horizontal displacements in the floor of the north-eastern Pacific Ocean, *Bull. Geol. Soc. Am.*, *72*, 1251–1258, 1961.

Wegner, A., *The Origin of Continents and Oceans*, London, 1924.

Wilson, J. T., Hypothesis of earth's behavior, *Nature*, *189*, 925–929, 1960.

Worzel, J. L., and G. L. Shurbet, Gravity interpretations from standard oceanic and continental cross-section, in *Crust of the Earth*, edited by A. Poldervaart, *Geol. Soc. Am. Spec. Papers*, *62*, 87–100, 1954.

# SEISMOLOGICAL INFORMATION AND ADVANCES*

*Frank Press*

*California Institute of Technology, Pasadena, California*

## Introduction

I should like to begin with some disclaimers. This report is not a review of the entire field of seismology as the title implies. Rather, it is a discussion of what to me is new and exciting in the field. In a paper limited by time, choices must be made, and an author's taste, if not his prejudice, is involved in the selection of topics. I assume that I address an audience of physical scientists rather than specialists in seismology. In selecting topics, I take the point of view that seismology is not an end in itself. It is a significant subject because it is a principal tool for deducing composition and physical state of planetary interiors, for revealing internal thermal and mechanical processes. It is also significant because it may one day be concerned with the prediction of earthquakes and the identification of possible underground nuclear explosions.

It is a matter of much personal satisfaction to me to be able to participate in the dedication of the Cecil and Ida Green Building. Dr. Green is known to all of us for his philanthropy, particularly his interest in promoting eduction. Geophysicists also know him as a

---

* Contribution 1296, Division of Geological Sciences, California Institute of Technology, Pasadena, California.

contributing scientist, as a seismologist in fact. He has always been an innovator in geophysical exploration for petroleum. Many of the techniques and concepts of seismology which qualify as "new and exciting" were introduced to the oil business by Dr. Green and his colleagues. I have in mind such things as extending the spectral band of seismic waves, introducing digital techniques and array concepts, or advancing the physics of wave propagation to improve interpretation methods. If a modern history of geophysics is written, I hope that it will include exploration seismology so that we can recognize these developments and give due credit to our colleagues in the oil business for anticipating many of them.

## Instrumentation

Much of what will be reported was made possible by instruments and field methods introduced in recent years. The developments have

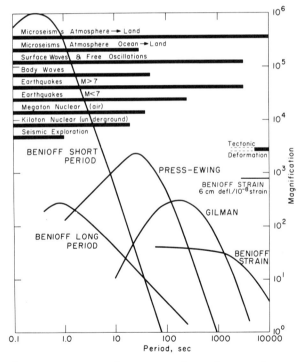

Fig. 1. Response curves of representative seismographs together with spectral bandwidth of seismic waves.

in common extended frequency response, use of digital data acquisition systems, and extended spatial coverage.

The several types of seismic waves which have been observed cover different spectral bands (Figure 1). This is not only a matter of relative excitation by the source, but also of absorption, scattering, and the limitations of instruments and analysis. In general, the larger the magnitude of the source, the larger is the areal extent of the source, and the greater the relative excitation of long waves. Low-mode free oscillations, for example, have not as yet been observed for earthquakes with magnitudes less than $8\frac{1}{4}$. Surface waves and free oscillations appear to have greater content of low-frequency components than body waves, again a matter of excitation and absorption, but also an artificiality of methods of analysis which separate surface waves and body waves. Strictly speaking, body waves result from the superposition of higher-mode surface waves or free oscillations. Also shown in Figure 1 are the response curves of representative old and new instruments. Instruments which respond well to long surface waves and free oscillations were not generally available until after 1952 when mechanically stable long-period pendulums and galvanometers and electrononically stable displacement transducers were introduced. Digital data acquisition systems provide geophysicists with the dynamic range, the precision, and the convenient entry to computers, which in turn make available the powerful methods of time series analysis, including correlation, general spectral analysis, and optimal filtering. Although seismic wave amplitudes can vary by more than 120 db, available dynamic range prior to the advent of digital seismographs was only 40 db. Figure 2 (courtesy of Professor Stewart Smith) shows one example of the use of a digital seismograph system developed at Caltech. Immediately following the earthquake, the automatically plotted seismogram was available in which the seismometers were rotated by the computer to correspond to radial, transverse, and vertical components with respect to the epicenter. Furthermore, the user could specify output to match most existing seismograph systems or he could specify a response curve, i.e., numerically design a seismograph, to best fit his needs. The advantage of rotation is evident in the clear distinction between modes with radial and transverse motion. The dynamic range can further be appreciated when it is observed that probably every conventional seismograph in North America was off scale in the period range of interest.

An example of a digital system to improve signal to noise is shown in Figure 3 [Shimshoni and Smith, 1964]. By performing the nonlinear operation of time averaging the cross products of vertical and

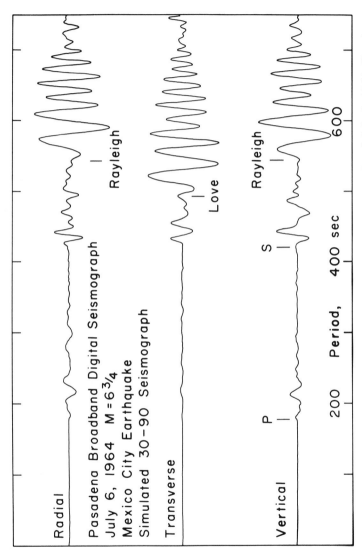

FIG. 2. Pasadena digital seismograph output for Mexican earthquake of July 6, 1964. Simulates Press-Ewing seismograph with rotated horizontal components.

horizontal components, a function of ground motion is produced in which linearly polarized motion is enhanced. Since orbital motion of noise is predominantly elliptical, significant signal-to-noise improvements were made, the indication of first motion was improved, and crustal reverberations following the $P$ group were discriminated against.

The use of expanded spatial coverage takes several forms, depending on the application. In explosion-seismic refraction studies it takes the form of jump correlation between linear arrays of detectors extending tens of kilometers with element spacing of a few hundred meters [Gamburtsev, 1952]. With such data, variations in pulse shape and amplitude, as well as travel times, can be used to infer the presence of discontinuities and gradients in the crust and upper mantle.

As a result of research programs to detect and identify small and distant explosions and earthquakes, several large arrays of detectors have been established. The arrays which have been installed are mostly less than 10 km in a linear dimension, a value which is small when measured in seismic wavelengths. However, the results which have been received are indeed impressive. Simple summation of elements have yielded magnification capabilities as high as 4 million. Theoretical studies show [Texas Instruments, 1962] that, with larger arrays of $N$ detectors, time shifting and summation together with optimal filtering can lead to signal/noise improvements somewhat better than $\sqrt{N}$ for realistic noise models, with no limitation on the value of $N$. We will discuss results from these arrays in a later section.

A brief word is in order to an audience like this because engineers and physicists frequently wonder whether seismologists have exploited advances in other fields in their design of instruments. The major problem in instrumental design is not sensitivity or instrument noise, since noise in the ground due to wind and atmospheric pressure changes limits magnification. Therefore, suggestions regarding the use, for example, of lasers or inertial guidance devices must offer advantages—such as increased dynamic range, transducers with direct digital output, mode discrimination, or long-term stability for frequencies like 1 cycle/day or 1 cycle/month.

Brief mention should be made of seismographs developed at Columbia University and Caltech for use on the moon. These special purpose devices are rugged, lightweight, broad-band, and self-stabilizing, and are probably much too fancy to be used on earth (except on the sea floor). One instrument can survive impact at velocity of several hundred miles per hour. Another can function as a gravimeter and record short-period body waves as well as low-mode lunar free oscillations.

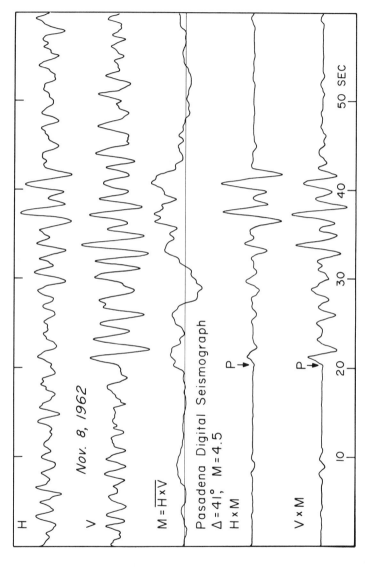

Fig. 3. Signal-to-noise improvement using digital seismograph correlation horizontal and vertical components.

## Seismic Rays

Although we now know the distinction to be an artificial one, seismologists have long dealt separately with rays and waves. Until recently ray methods were almost exclusively used to explore the earth's interior. Travel times and amplitudes of seismic rays generated by earthquakes were inverted to yield velocity-depth functions. Ray methods were adequate to discover the major features of the earth's interior, the crust-mantle discontinuity, some second-order discontinuities in the mantle, and the inner and outer core. However, the methods were severely limited in several ways—velocity reversals could not be easily recognized, and masked layers were always possible. Regional variations in the upper mantle were difficult to detect because measurements were restricted to those regions having both earthquakes and nearby seismograph networks.

Ray methods have been revitalized in recent years. For exploration to depths of 100 km, the impetus was given by Soviet scientists [Gamburtsev, 1952] who introduced the concept of massive field efforts in which moving linear arrays recorded explosive-generated seismic waves. In the United States, groups from the U.S. Geological Survey, the Carnegie Institution, and the University of Wisconsin combined this approach with multichannel magnetic recording and playback systems adapted from apparatus developed for oil prospecting. The results were unanticipated. Figure 4, modified from a recent summary by Pakiser [in press], shows at first glance an unexpected diversity in continental crustal and upper-mantle structure. From the usual isostatic point of view in which topographic loads are compensated by roots formed from crystal thickening, the thin crust in the intermount and plateau between the Rockies and the Sierra Nevada is puzzling. The mantle velocities are lower by some 5 per cent in this region, implying a similar decrease in mantle density. One interpretation would state that compensation through crustal thickening occurs where normal mantle (velocity 8 km/sec) is present and the depth of compensation is the depth to the $M$-discontinuity. The region from the Rocky Mountains to the east coast and from the Sierra Nevada to the Pacific coast falls in this category. In the region of low-velocity mantle, compensation occurs by the reduction of mantle density in the zone between the shallow $M$-discontinuity and a depth of about 80 km. Stated otherwise, a surface of constant density, say 3.45 gm/cm$^3$ would follow the $M$-discontinuity eastward to the Rocky Mountains where it would plunge to a depth of about 80 km (i.e., 50 km below Moho) and rise again to near the $M$-discontinuity west of the Sierra Nevada.

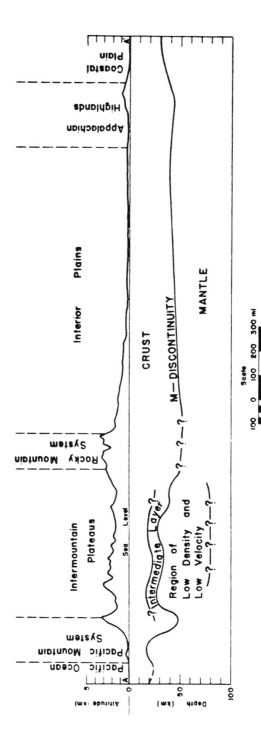

Fig. 4. Crustal and upper-mantle profile across United States derived from seismic refraction studies. After Pakiser.

To speculate still further with Pakiser, we can attempt to correlate these changes with recent laboratory studies on phase changes in rock-forming materials. In particular, the basalt-eclogite phase transformation occurs at temperatures and pressures corresponding

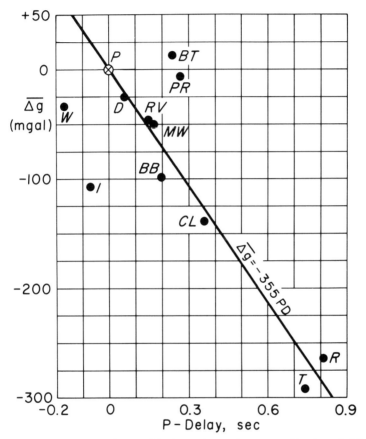

Fig. 5. Correlation of P residuals and gravity anomalies for California stations.

to the deep crust or upper mantle. The $M$-discontinuity probably represents a chemical change with rocks of basaltic composition in the crust and a mixture of periodite and basalt forming the mantle. (The diversity of depths and temperatures at which Moho is found under oceans, continental shields, and orogenic belts argues against a phase change.)

In the normal mantle basalt is in its dense phase—eclogite—the transformation to basalt occurring above Moho but near the bottom of the crust. Under the great basin the lower-mantle velocity implies that the basalt-eclogite phase change is not reached until depths like 80 km. This is consistent with the preliminary indications of higher heat flow in this region and surface wave studies.

Alternate explanations of the low mantle velocities would involve partial serpentinization, temperature effects, with or without partial melting.

Ray methods involving penetration of the deeper mantle have not led to major revisions of deep structures. The travel times have been made more precise using data from nuclear explosions, to the point that residuals now observed are due to variations in the upper mantle and crust immediately below the source or the seismographs. The correlation of these residuals with local gravity anomalies reinforces this interpretation (Figure 5). This suggests a method of exploration in which local structure is inferred from the residuals and gravity anomalies [Press and Biehler, 1964]. The improved travel times have led to better epicenter location, and the indication that deep mantle paths transmit short period *P*-waves with little change in pulse shape implies that information concerning the mechanism of the source is preserved. These last two observations have important applications to the problem of detection of nuclear explosions.

Revised travel times for paths through the core are beginning to yield new information concerning layering in the core [Bolt, 1964*a*, *b*]. The use of arrays to identify later core phases by precision apparent velocities will be an important new tool for core studies. One interesting new ray approach to obtain $Q$ in the mantle and rigidity in the core using multiply reflected rays is shown in Figure 6. Using several of the 125 standard stations recently distributed over the world, Anderson and Kovach [1964] found a record containing 5 multiple near-vertical reflections from the core-mantle interface. This enabled them to eliminate the source and place an absolute upper bound on core rigidity of $2 \times 10^{10}$ dynes/cm$^2$. Using earthquakes with varying depths, $Q$ data was obtained as a function of depth—a topic which will be discussed in a later section.

Where does ray theory end and wave theory begin? This question is not only of academic interest, because equalization of *P*-waves to remove the effects of crustal layering and the study of leaking modes to deduce crustal layering are topics of current research interest [Haskell, 1962; Oliver, 1964; Phinney, 1964]. Although the results are not yet competitive with pure ray or wave methods (mainly a matter of data availability rather than theory), there is no question that it is only a

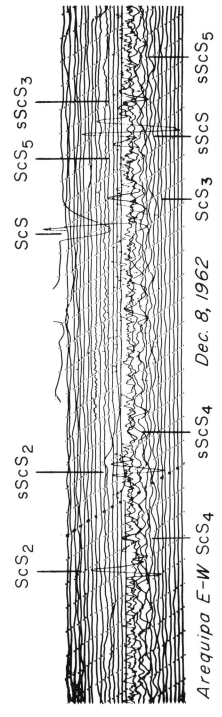

Fig. 6. Multiple reflected shear waves from the mantle-core interface. After Anderson and Kovach [1964].

matter of time when this method will become a significant tool for exploration.

## Propagating Long Waves and Free Oscillations

Because current interest in this subject is very high, several review papers have recently appeared [Press, 1964; Bolt, 1964a, b; Anderson, 1964a, b]. The following is an abbreviated and updated version of Press' discussion.

Long waves are involved with wave theory in the sense that the earth is treated as a wave guide, and the long waves are dispersed and also selectively excited according to the properties of the wave guide. The properties of the source and its location in the wave guide also enter into determining the excitation function. The effective penetration of long waves depends on wavelength and the velocity-depth function. Furthermore, long waves are particularly convenient for determining frequency-dependent absorption coefficients and hence the dissipation constant $1/Q$. Certain properties of the source not amenable to study by ray methods can be deduced from long-wave radiation patterns. Thus long waves supplement ray methods where the latter run into difficulties.

The principles of analysis of long waves can be understood from the following brief discussion. Consider the propagation of Rayleigh and Love waves in a wave guide formed by the variation of elastic velocities and densities with depth in the earth. As the wave propagation is dispersive, the phase velocity $c$ is dependent on circular frequency $\omega$ through a characteristic or frequency equation involving the velocity depth function.

The seismogram at a distance $r$ from a transient source can be written

$$f(t, r) = \int_{-\infty}^{\infty} g(\omega) \, d\omega \sum_{n} F_n(r, \omega) \exp \{i\omega[t - r/c_n(\omega)]\} \qquad (1)$$

Here $g(\omega)$ is the Fourier transform of the initial time variation of the source (assuming that the instrument response has been removed) and the summation over $n$ represents a summation over all the normal modes of the wave guide. The function $F$ includes the effect of the location and extent of the source as well as the response of the wave guide. It has been given explicitly by several recent papers [see for example, Harkrider, 1964]. Either by filtering or by excitation and absorption, the modes are usually separately identifiable on seismograms.

Integrals of the type in Equation 1 may be evaluated approximately by the method of stationary phase, which yields the predominant contribution for a narrow band of frequencies $\omega$ near $\omega_0$, where the exponent is stationary. Thus,

$$(d/d\omega)\left(\omega t - \frac{\omega r}{c}\right) = 0$$

or

$$t = r(d/d\omega)(\omega/c) = r/U$$

defines the arrival time of a predominant train of waves with frequency associated with the group velocity $U$. For large $r$, Equation 1 takes the approximate value

$$f(t, r) = \sqrt{2\pi}\, U \frac{g(\omega_0) F(r, \omega_0)}{\sqrt{r|dU/d\omega|_0}} \exp\left[i\omega_0(t - r/c_0) - i\epsilon \pm \pi/4\right] \quad (2)$$

where the subscript zero indicates that the quantities are those associated with group velocity $U$ given by Equation 4, and the upper or lower sign is taken in the exponent according as $dU/d\omega \gtrless 0$. The $r$ dependence in $F(r, \omega_0)$ is $(\sin r/a)^{-1/2}$ for a spherical earth of radius $a$; $\epsilon$ is prescribed by the source.

Equation 2 represents a train of waves which is both amplitude- and frequency-modulated. The amplitude modulation depends on the nature of the source and the properties of the wave guide. Since the source factor and the properties of the wave guide are variables in the problem, amplitude modulation is not easily used for deducing either one. Progress in the use of amplitude modulation is proceeding through the use of amplitude ratios of waves leaving the source in opposite directions and through the use of equalization to remove the propagation effect. The frequency modulation depends on the dispersion curves, that is, the curves of phase and group velocity versus frequency. Dispersion curves are prescribed only by the physical properties of the wave guide. Analysis of a seismogram for dispersion can, therefore, yield the properties of the wave guide—in this case, the properties of the earth's interior. The effective depth that can be probed by this technique is about $\frac{1}{3}$ wavelength, though velocity reversals can modify this. The analytical procedure is described next. In Figure 7 examples of phase velocity, group velocity, and amplitude modulation are indicated on a 2000-km array of matched seismographs. Figure 8 shows Love waves recorded at Pasadena before and after circling the earth; the phase shifts for the world-circling path are recoverable and yield phase velocity dispersion curves.

If the source distance $r$ and the propagation time $t$ are known, the

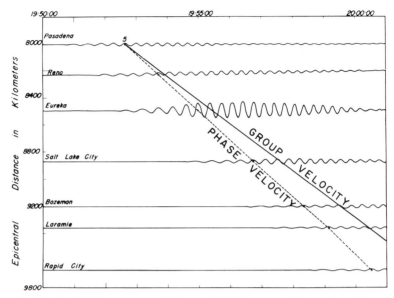

Fig. 7. Rayleigh waves recorded over a 2000-km array, illustrating difference between phase and group velocity.

dispersed train of surface waves yields an experimental group-velocity curve. Comparison with theoretical curves computed for several assumed structures permits selection of the most probable structure. Alternatively, theoretical phase- and group-velocity curves may be used with Equation 2 to synthesize theoretical seismograms, which may then be compared with actual seismograms. The most probable structure is that theoretical structure whose seismogram best fits the observed structure. Group-velocity analysis yields an average

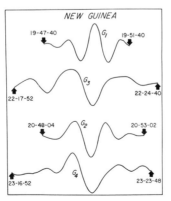

Fig. 8. Love waves: $G_1$ direct, $G_3$ direct plus one circuit of the earth; $G_2$ through antipodes, $G_4$ through antipodes plus one circuit of the earth.

structure representative of that region of the earth between the seismograph and the source.

An alternate procedure compares experimental and theoretical phase velocities and offers the advantage that it is sensitive to structure in the immediate vicinity of the phase velocity measurement [see for example, Press, 1960a,b]. Phase velocity can be measured in three ways. Fourier analysis of surface wave trains yields the phase factor of Equation 1. Cross-spectral analysis may be used to directly determine phase differences. An array of at least three stations can be used to separately determine from these phase factors the direction of approach and phase velocities.

Another approach is to use the direct train of surface waves which has traveled a distance $r$, and the train which has completely circled the earth and has traveled a distance $r_1 + 2\pi a$ [Sato, 1958]. The difference in arguments of the Fourier transform is $2\pi a \omega/c + 2n\pi$ where $n$ is an integer which is not too difficult to guess. For any $n$, differentiation with respect to $\omega$ gives $2\pi a/U$ from which the group curve can be constructed. Brune, Nafe, and Oliver [1960] showed how the phase factor of Equation 1 obtained from data of a single station can often be used to construct the phase velocity curve representative of the region between the epicenter and the seismograph.

We have passed over methods of computing theoretical dispersion curves—a procedure essential to the proper interpretation of the experimental data. Without going into details, the procedures may be briefly described as follows. The vertically inhomogeneous earth can be adequately represented by a large number of thin homogeneous layers. The boundary value problem for such a wave guide can be solved in particularly convenient fashion for computer calculation by matrix methods [Haskell, 1953; Dorman, Ewing, and Oliver, 1960]. Phase and group velocity curves are obtained as well as excitation functions for the wave guide. Curvature of the earth has been shown to be of importance. Curvature corrections may be applied by exactly computing phase velocities from free oscillations [Alterman, Jarosch, and Pekeris, 1961] or more conveniently and with sufficient precision by approximate methods based on mapping a sphere onto a half-space [Anderson and Toksöz, 1963]. Tables which give the variation of phase and group velocity for variations in elastic constants or densities at any depth have been published, which permit one to construct theoretical dispersion curves for almost any model of earth structure with the use of a hand calculator.

Observations of propagating Rayleigh and Love waves cover the period range from several seconds to several hundred seconds. A

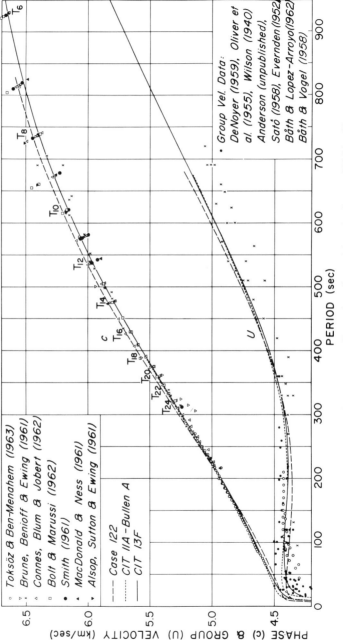

Fig. 9. Summary of Love wave dispersion data. After Anderson [1964a, b].

recent summary by Anderson [1964a, b] shows the experimental phase and group velocity curves together with several theoretical curves used to find the appropriate model (Figure 9). We selected Love wave data as an example, there being a complementary curve for Rayleigh waves.

Waves with periods below about 350 sec (wavelengths less than 1500 km) are sensitive to structure to a depth of about 600 km. Longer waves reflect the properties of the deep mantle and the data for these waves are derived primarily from free oscillation periods, using Equation 5 to make the transformation to phase velocity (Figure 9). We will interpret the data in terms of earth structure after a discussion of free oscillations.

Up to this point we have been dealing with propagating surface waves. The global standing-wave pattern formed by the interference of propagating waves is described by the free oscillations of the earth. Interfering Rayleigh waves form spheroidal oscillations and interfering Love waves lead to torsional oscillations. In principle, all of wave and ray seismology can be discussed in terms of a superposition of modes of free oscillations.

The periods of free oscillations depend only on properties of the earth. The relative excitation of different modes depends on earth structure and the position and nature of the source.

For sources with azimuthal symmetry the displacement $V$ corresponding to a free oscillation of the $n$th mode has the form

$$V_n = \sum_{m=0}^{\infty} F(r, {}_m\omega_n) P_n(\cos \theta) \exp(i {}_m\omega_n t) \qquad (3)$$

The summation over $m$ implies that for each mode there is a fundamental tone ($m = 0$) and an infinite number of overtones. The parameter $m$ determines the number of internal nodal surfaces (zeros of $F$). The standing-wave pattern on the surface is described by the surface zonal harmonic $P_n(\cos \theta)$. Thus for $n = 2$ (sometimes called the football mode) the spheroidal motions correspond to a sphere alternating between prolate and oblate forms. The relation between propagating waves and free oscillations is made plain when the asymptotic form of $P_n(\cos \theta)$ for large $n$ is used in Equation 3:

$$V_n \to \sum_m F(r, {}_m\omega_n)(2\pi n \sin \theta)^{-1/2} \{ \exp i[{}_m\omega_n t - (n + \tfrac{1}{2})\theta + \pi/4]$$
$$+ \exp i[{}_m\omega_n t + (n + \tfrac{1}{2})\theta - \pi/4] \} \qquad (4)$$

Here the first and second terms represent waves traveling in the positive and negative $\theta$ directions with phase velocity $c$ and wave number $k$ given by

$$n + \tfrac{1}{2} = a\omega/c = ka \tag{5}$$

Thus a knowledge of $n$ and $_m\omega_n$ from a free oscillation calculation can be used to obtain $c$ (hence $U$ from $U = (d/dk)kc = a\,d\omega/dn$) corresponding to propagating waves with exact sphericity correction included. Note also the factors $\pi/4$ in expression 4 which show the necessity of including a polar phase shift of $\pi/4$ when deriving phase velocity from propagating waves which have traversed paths through the origin and its antipodal point [Brune, Nafe, and Alsop, 1961].

Speculation concerning the possibility of observing free oscillations of the earth originated with conjectures made in the late nineteenth and early twentieth centuries by Lamb, Jeans, and Love that the period of the gravest mode would be about one hour. Benioff [1954] tentatively suggested that a 57-minute wave he observed from the Kamchatka earthquake of 1952 was indeed a spheroidal oscillation, although this was subsequently shown not to be the case (the duration of the oscillation was too short and the period was too far off). Benioff's suggestion stimulated a series of theoretical and experimental studies which culminated in the identification of the free oscillations. Pekeris and his colleagues undertook a series of theoretical studies [Alterman, Jarosch, and Pekeris, 1959] in which numerical methods suitable for electronic computers were outlined and applied to determining the eigenperiods for realistic earth models. Basically the method consists of numerically integrating the equation of motion assuming a given period. The process is repeated varying the period until a value is

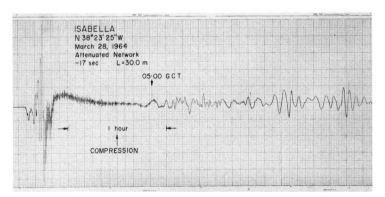

Fig. 10. Example of a seismogram which yields free-oscillation spectral peaks after power spectrum analysis.

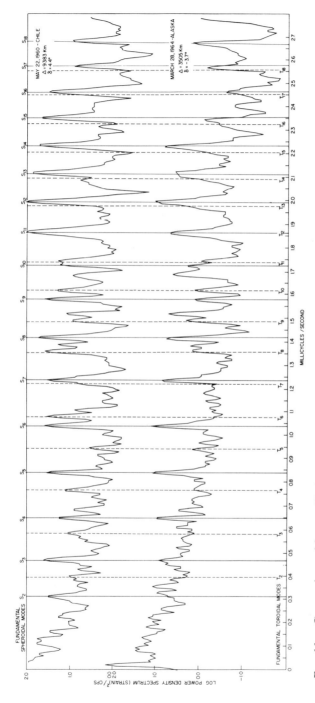

FIG. 11. Comparison of free-oscillation power spectra from earthquakes in Chile and Alaska recorded at Isabella, California. Courtesy S. Smith.

found for which the conditions of regularity at the origin and vanishing stresses on the earth's surface are satisfied.

An observation of free oscillations is made by performing a spectral analysis on records of long-period seismographs or gravimeters following a large earthquake. The records are in digital form originally, or are digitized with curve-following devices, and the analysis is performed on high-speed computers. Fourier analysis, power spectrum analysis, and cross-spectrum analysis using two distant stations have all been performed. The record length must be sufficient and the analysis methods fine enough to permit identification of eigenperiods with a precision of about 1 part in 500.

An example of a record which yields free oscillations after power spectrum analysis is shown in Figure 10. This is the Isabella strain seismograph recording for the great Alaskan earthquake of March 28, 1964. Examples of power spectra from two widely spaced stations for the same earthquake and two widely spaced earthquakes recorded at the same station are shown in Figures 11 and 12 [Benioff, Press, and Smith, 1961; Smith, in preparation]. It is evident that the periods and $Q$ factors are almost identical when energy is present, and that some modes are excited by one earthquake and not excited by the other.

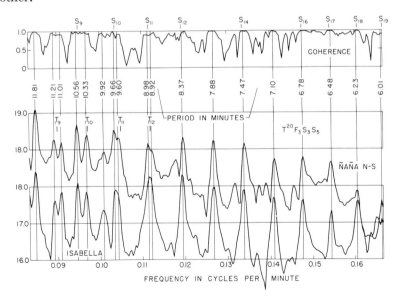

FIG. 12. Correlation of free-oscillation power spectra recorded at Isabella, California, and Ñaña, Peru for Chilean earthquake. After Benioff, Press, and Smith [1961].

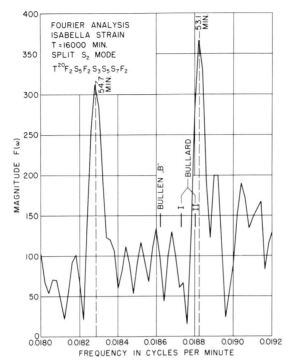

Fig. 13. Splitting of spectral peak of $_0S_2$ mode of free oscillation due to earth's rotation. After Benioff, Press, and Smith [1961].

The remarkable detection of split of spectral peaks is shown in Figure 13 for the gravest mode $_0S_2$. These and other split lines were shown to be a consequence of the earth's rotation [Pekeris, Alterman, and Jarosch, 1961; Backus and Gilbert, 1961].

The detection and identification of spectral peaks corresponding to the earth's free vibration is one of the most elegant experiments in geophysics. It occurred when it did because the necessary tools for observing free oscillations, namely, long-period instruments, computers, numerical spectral analysis methods and programs, and a sufficiently large seismic source were available in ensemble for the first time just prior to the great Chilean earthquake of May 22, 1960.* Not long after this event three teams [Benioff, Press, and Smith, 1961; Ness, Harrison, and Slichter, 1961; and Alsop, Sutton, and Ewing, 1961]

---

* Actually Press [1960b] reported his observation of several spheroidal modes a year earlier based on a Fourier analysis of the Palisades vertical pendulum recording of the Kamchatka earthquake of 1952. Without theoretical periods for these modes or corroborative data from other stations, he was only able to claim tentative identification which turned out to be correct for the higher modes and wrong for the lower modes.

## Table 1

### Theoretical and Observed Spheroidal Modes: Periods in Minutes

| Mode | Theoretical | | | | Experimental |
|---|---|---|---|---|---|
| | JB-A | JB-B | GB-A | GB-B | Mean |
| $_0S_2$ | 53.43 | 53.72 | 53.50 | 53.78 | 53.95 |
| $_0S_3$ | 35.25 | 35.52 | 35.32 | 35.58 | 35.62 |
| $_0S_4$ | 25.50 | 25.75 | 25.53 | 25.78 | 25.82 |
| $_0S_5$ | 19.62 | 19.85 | 19.65 | 19.88 | 19.86 |
| $_0S_6$ | 15.90 | 16.17 | 15.92 | 16.15 | 16.07 |
| $_0S_7$ | 13.42 | 13.65 | 13.43 | 13.67 | 13.52 |
| $_0S_8$ | 11.73 | 11.97 | 11.73 | 11.97 | 11.76 |
| $_0S_9$ | 10.53 | 10.77 | 10.53 | 10.77 | 10.57 |
| $_0S_{10}$ | 9.650 | 9.883 | 9.650 | 9.883 | 9.629 |
| $_0S_{11}$ | 8.950 | 9.183 | 8.950 | 9.183 | 8.917 |
| $_0S_{12}$ | 8.383 | 8.617 | 8.383 | 8.617 | 8.362 |
| $_0S_{13}$ | 7.887 | 8.105 | 7.887 | 8.108 | 7.875 |
| $_0S_{14}$ | 7.473 | 7.680 | 7.475 | 7.683 | 7.465 |
| $_0S_{15}$ | 7.107 | 7.300 | 7.108 | 7.305 | 7.102 |
| $_0S_{16}$ | 6.780 | 6.962 | 6.782 | 6.967 | 6.778 |
| $_0S_{17}$ | 6.488 | 6.658 | 6.492 | 6.663 | 6.492 |
| $_0S_{18}$ | 6.227 | 6.383 | 6.230 | 6.390 | 6.231 |
| $_0S_{19}$ | 5.988 | 6.133 | 5.993 | 6.142 | 5.998 |
| $_0S_{20}$ | 5.772 | 5.907 | 5.778 | 5.915 | 5.778 |
| $_0S_{21}$ | 5.575 | 5.698 | 5.582 | 5.708 | 5.595 |
| $_0S_{22}$ | 5.392 | 5.505 | 5.402 | 5.515 | 5.410 |
| $_0S_{23}$ | 5.225 | 5.327 | 5.235 | 5.338 | 5.256 |
| $_0S_{24}$ | 5.068 | 5.162 | 5.080 | 5.175 | 5.105 |
| $_0S_{25}$ | 4.923 | 5.007 | 4.937 | 5.022 | 4.959 |
| $_0S_{26}$ | 4.787 | 4.862 | 4.802 | 4.878 | 4.827 |
| $_0S_{27}$ | 4.660 | 4.727 | 4.675 | 4.743 | 4.694 |
| $_0S_{28}$ | 4.540 | 4.600 | 4.557 | 4.617 | 4.588 |
| $_0S_{29}$ | 4.427 | 4.478 | 4.445 | 4.498 | 4.469 |
| $_0S_{30}$ | 4.320 | 4.3650 | 4.340 | 4.385 | 4.368 |
| $_0S_{31}$ | 4.217 | 4.257 | 4.238 | 4.278 | 4.266 |
| $_0S_{32}$ | 4.120 | 4.155 | 4.143 | 4.178 | 4.167 |
| $_0S_{33}$ | 4.0283 | 4.058 | 4.052 | 4.082 | 4.085 |
| $_0S_{34}$ | 3.942 | 3.965 | 3.967 | 3.990 | 3.992 |
| $_0S_{35}$ | 3.857 | 3.877 | 3.883 | 3.902 | 3.920 |
| $_0S_{36}$ | 3.777 | 3.792 | 3.803 | 3.818 | 3.829 |
| $_0S_{37}$ | 3.700 | 3.712 | 3.727 | 3.738 | 3.743 |
| $_0S_{38}$ | 3.652 | 3.633 | 3.653 | 3.662 | 3.671 |
| $_0S_{39}$ | 3.553 | 3.560 | 3.583 | 3.588 | 3.612 |
| $_0S_{40}$ | 3.485 | 3.488 | 3.515 | 3.518 | 3.475 |
| $_0S_{41}$ | 3.420 | 3.420 | 3.452 | 3.450 | 3.405 |
| | 0.10* | 0.17* | 0.09* | 0.18* | |

* Mean absolute deviation between experimental and theoretical value for modes through $_0S_{20}$.

reported the observation of some 40 fundamental spheroidal modes, 25 torsional modes, and several spheroidal overtones. Early assignment of spectral peaks to appropriate modes was possible because of the remarkable theoretical predictions of Pekeris and his colleagues [Alterman, Jarosch, and Pekeris, 1961].

A summary of theoretical and experimental results for the eigenperiods is presented in Tables 1 and 2. The theoretical models used are based on the Jeffreys and Gutenberg velocity distributions (Figure

### Table 2

THEORETICAL AND OBSERVED TORSIONAL MODES: PERIODS IN MINUTES

| Mode | Theoretical | | | Experimental |
|---|---|---|---|---|
| | JB-A | JB-B | GB-A | Mean |
| $_0T_2$ | 43.46 | 44.16 | 43.61 | 43.78 |
| $_0T_3$ | 28.13 | 28.60 | 28.23 | 28.53 |
| $_0T_4$ | 21.54 | 21.91 | 21.62 | 21.76 |
| $_0T_5$ | 17.77 | 18.07 | 17.84 | 17.93 |
| $_0T_6$ | 15.28 | 15.54 | 15.36 | 15.36 |
| $_0T_7$ | 13.37 | 13.72 | 13.57 | 13.53 |
| $_0T_8$ | 12.13 | 12.33 | 12.22 | 12.25 |
| $_0T_9$ | 11.05 | 11.23 | 11.15 | 11.14 |
| $_0T_{10}$ | 10.20 | 10.34 | 10.28 | 10.32 |
| $_0T_{11}$ | 9.50 | 9.59 | 9.55 | 9.614 |
| $_0T_{12}$ | 8.86 | 8.95 | 8.93 | 8.975 |
| $_0T_{13}$ | 8.33 | 8.40 | 8.39 | 8.379 |
| $_0T_{14}$ | 7.86 | 7.92 | 7.93 | 7.942 |
| $_0T_{15}$ | 7.43 | 7.50 | 7.51 | 7.527 |
| $_0T_{16}$ | 7.07 | 7.12 | 7.15 | 7.161 |
| $_0T_{17}$ | 6.73 | 6.78 | 6.81 | 6.819 |
| $_0T_{18}$ | 6.45 | 6.48 | 6.52 | 6.502 |
| $_0T_{19}$ | 6.18 | 6.20 | 6.24 | 6.246 |
| $_0T_{20}$ | 5.93 | 5.95 | 5.99 | 6.006 |
| $_0T_{21}$ | 5.70 | 5.72 | 5.77 | 5.769 |
| $_0T_{22}$ | 5.50 | 5.50 | 5.55 | 5.545 |
| $_0T_{23}$ | 5.30 | 5.31 | 5.36 | 5.386 |
| $_0T_{24}$ | 5.12 | 5.12 | 5.18 | 5.200 |
| $_0T_{25}$ | 4.95 | 4.95 | 5.01 | 5.100 |
| | 0.13* | 0.09* | 0.06* | |

* Mean absolute deviation between experimental and theoretical value for modes through $_0T_{20}$.

16) each taken with a Bullen A and Bullen B density distribution. This gives the four models JB-A, JB-B, GB-A, GB-B. The theoretical values were computed by Alsop [1963], Pekeris, Alterman, and Jarosch [1961], and Sato, Landisman, and Ewing [1960]. The experimental values for the spheroidal modes were obtained by averaging the eigenperiods observed on the Isabella (California) strain seismograph, the Ogdensburg (New York) strain (or when lacking the Palisades, New York vertical pendulums), and the Los Angeles gravimeter. The torsional periods were obtained by averaging data from these stations as well as Ñaña (Chile) and Tiefenfort (Germany).

The mean absolute deviation between the several models and the observations are also given in Tables 1 and 2 for the first 20 modes. This test selects the GB-A model as the best-fitting one. Note that for the low-order modes the models with the same density distribution are closer together in period than the models with the same velocity. What is happening is that both velocity distributions satisfy to within a very few seconds the observed travel times of $P$- and $S$-waves for deeply penetrating waves (vertical shear wave reflections from the core differ by only one second between the two models). The lower-mode oscillations therefore select models on the basis of density. It is not until the modes beyond the thirtieth are reached that velocity differences between the models become significant, because these shorter-period oscillations are sensitive to upper-mantle velocities which differ markedly between the models.

We have mentioned $Q$ only as a source of difficulty in broadening spectral peaks. Actually it is potentially of great significance because methods have been developed to obtain $Q$ as a function of period $T$ and to invert these data to deduce $Q$ as a function of depth. This represents a new seismic parameter to add to elastic velocity and density from which composition and state in the interior are inferred.

There are several ways of obtaining $Q$ data. The most accurate data are derived from the decay in the power spectrum of propagating long waves with each passage around the earth. Data for periods longer than several hundred seconds must come from analysis of free oscillations. Here line splitting affects the broadening of the spectral peak so that power decay with time is probably the more precise method. Recent $Q(T)$ observations [Anderson and Archambeau, 1964] are shown in Figure 14. The inversion of $Q$ data by these investigators is based on the relation

$$\frac{1}{Q(T)} = \frac{\sum_i E_i(T) Q_i^{-1}}{\sum E_i(T)}$$

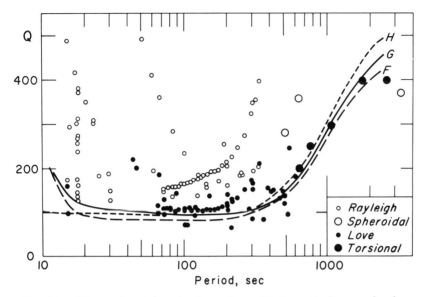

Fig. 14. Observational data for $Q$ together with theoretical curves for three models of $Q$ distribution. Theory predicts difference between $Q$ values for Love and Rayleigh waves. After Anderson and Archambeau [1964].

where the summation $i$ is over the layers of the model, $E_i$ is the relative energy in the layer, and $Q_i$ is the intrinsic $Q$ for the layer. Now the relative energy in each layer is known from normal mode theory. If there are $jQ(T)$ values such that $j \geq i$, then the equation can be inverted by well-known methods. An assumption implicit in this method of interpretation is that the intrinsic $Q$ is frequency independent, as implied by the limited data available [Knopoff and MacDonald, 1958].

Up to this point the source was a complication which had to be removed by one method or another to obtain dispersion curves or $Q$ data. New methods—experimental and theoretical—for deducing source properties have been used to recover such properties as source extent, rupture velocity, and mechanism. Prior to this new approach, the polarity of first motion of $P$-waves as recorded at a number of stations with varying azimuths and distances from the source was the only method for revealing source properties. Data from $S$-waves have been used to supplement $P$-data and give unique solutions of source mechanism as either single or double couple force systems.

Brune [1962] and Aki [1964] have been equalizing surface waves to remove effects of propagation and recover the initial amplitude and phase spectrum of long Love and Rayleigh waves to provide additional information about the source. A novel approach was proposed by Ben-Menahem [1961] when he defined a directivity function as the ratio of spectral amplitudes of surface waves leaving a source in opposite directions. A simple derivation shows that regardless of source mechanism, an earthquake can be viewed as a traveling, radiating disturbance, and the directivity function may be used to recover the horizontal extent of the fault and the velocity of propagation of the rupture.

The directivity function is defined as

$$D = \left| \frac{[(c/c_0) + \cos \theta_0] \sin (\pi b/cT)[(c/c_0) - \cos \theta_0]}{[(c/c_0) - \cos \theta_0] \sin (\pi b/cT)[(c/c_0) + \cos \theta_0]} \right|$$

where $c$ is the surface wave phase velocity, $T$ the period, $c_0$ the rupture velocity, $b$ is the horizontal fault extent, $\theta_0$ is the angle between the fault line and the great circle from the epicenter to the seismograph station. The method is used in the following way. Surface wave pairs such as $R_1$ and $R_2$ which have left the source in opposite directions and are recorded at the same station are Fourier analyzed. The directivity function, which is the ratio of the Fourier amplitudes, is plotted against frequency. The amplitudes are equalized for differences in absorption by deriving empirical absorption coefficients from $R_1$ and $R_3$ or $R_2$ and $R_4$ alone. No other equalization is needed since the instrument is the same, the geometric effect is the same, and Fourier analysis removes effects of dispersion. The preceding equation is used to compute theoretical directivities with fault length and rupture velocity as parameters ($\theta_0$ is typically known from local geology). Usually a fit can be found for unique values of $b$ and $c_0$. An illustration of the method is given in Figure 15 [Ben-Menahem and Toksöz, 1963] where it is shown that a fault length of 350 km and a rupture velocity of 3 to 3.5 km/sec may be associated with the Alaskan earthquake of July 10, 1958.

The analog of this technique using free-oscillation data was introduced by Benioff, Press, and Smith [1961]. In their method, the phase shift between vertical and horizontal displacements is measured for each mode. It varies between 0 and 180° in a regular pattern depending on fault length and rupture velocity. Reasonably good agreement was found between the two methods for the Chilean earthquake.

Before discussing the new information about the mantle which has emerged from long waves and free oscillations, a word is in order about the limitations of the experimental data.

Long propagating waves probe the upper mantle and offer the

possibility of detecting regional differences. Of particular importance is the elucidation of mantle structure under continents and oceans and the constraints the structure provides on theories of the thermal regimes and continental genesis. The experimental problem is one of obtaining long propagation paths over purely oceanic or continental structures for both Love and Rayleigh waves. Although this information will gradually become available as improved long-period seismo-

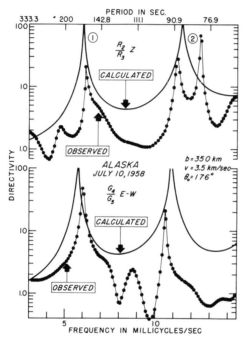

FIG. 15. Observed and theoretical directivity functions. After Ben-Menahem and Toksöz [1963].

graph systems are deployed around the world, the lack of such data is now the major bottleneck.

The free-oscillation data are limited in several ways. Earthquakes with magnitude $8\frac{1}{4}$ are still needed to recover a full suite of modes. The use of small tremors or of natural background has not as yet been successful and must probably await the introduction of arrays having continental dimensions. Another limitation is the uncertainty in spectral periods introduced by line splitting due to rotation of the earth and peak broadening due to finite $Q$. In principle, both of these effects can be removed, the first by an application of what

George Backus calls "geographic filters," the second by measuring $Q$ separately and then removing it.

The following results have emerged from long wave and free oscillation studies.

1. The presence of a low shear velocity zone is typical of the upper mantle under continents and oceans. It represents the mean condition for the world, although some local exceptions are possible [Dorman, Ewing, and Oliver, 1960; Takeuchi, Press, and Kobayashi, 1959]. Regional differences occur as indicated in Figure 16 where dispersion data for propagation paths containing increasing oceanic fractions show systematic increase in phase and group velocity [Anderson, 1964a].

Figure 17 shows three velocity models: The Jeffreys model without the low-velocity zone which was generally accepted before the long-wave data were available, the CIT 11 model which best fits oceanic Love waves, and the Gutenberg model which seems to be the limiting case for purely continental paths. CIT 11 implies a more pronounced

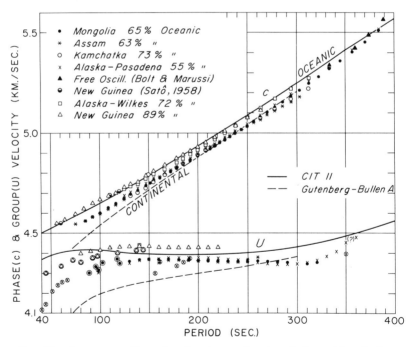

FIG. 16. Love wave dispersion showing regional variations. After Anderson and Toksöz [1963].

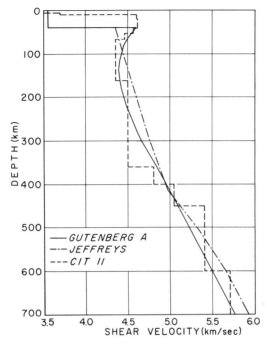

FIG. 17. Comparison of three models for the upper mantle shear velocity distribution.

and/or a shallower low-velocity zone beneath oceans. The rapid increase in velocity at depths between 350 and 450 km is striking. The most recent analysis shows this to be a feature common also to continental mantle [Anderson, 1964a].

The velocity reversal could be an effect of the temperature coefficient of velocity exceeding the pressure effect [MacDonald and Ness, 1961] or it could imply temperatures near the melting point and partial melting of the basaltic fraction of mantle rock [Press, 1961]. The rapid velocity increase in the range 350 to 450 km may indicate phase changes to the denser silicate forms of spinel or corundum, as suggested by Anderson [1964a].

2. By requiring consistency between earth models derived from free oscillation data and short-period body-wave travel-time curves, free-oscillation periods then depend primarily on the density distribution. This implies a powerful new technique for arriving at the earth's density distribution. Landisman, Sato, and Nafe [1964], and Anderson [1964b] used total mass and new moment of inertial results as further constraints on the first application of the method. No assumption of

chemical homogeneity, adiabaticity, absence of phase changes need be made. The early results from free oscillation data selected between existing models, e.g., Bullen A densities over Bullen B, Gutenberg velocities over Jeffreys. The most recent results imply abrupt density changes in the upper mantle coinciding with the rapid velocity changes [Anderson, 1964b]. The startling result which emerged was a

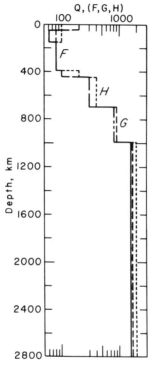

FIG. 18. Three theoretical models for $Q$ distribution which fit Love wave data. After Anderson and Archambeau [1964].

revision of mantle densities below 1600 km. The densities in this zone are either constant with depth [Landisman et al., 1964] or increase slightly with depth [Anderson, 1964b].

3. The inversion of $Q$ data by Anderson and Archambeau [1964] gives $Q$-depth functions shown in Figure 18. These three models are consistent with Love wave and torsional mode observations. Although they differ in detail, their common features are striking. A low $Q$ approximately coincident with the low-velocity zone and a $Q$ higher by an order of magnitude for the deep mantle below 1000 km. This is a new seismic parameter—the first new one introduced in several decades. Detailed $Q$-depth curves will certainly be forthcoming as the subject develops, but it is difficult to see how this preliminary result

can change. The high $Q$ deep mantle implies that rocks at these depths are probably not near the melting point as some theories require. The correlation of low $Q$ values with the low-velocity zone is suggestive of a common cause. Whether this is simply a temperature effect ($Q$ and velocity decrease with temperature) or an indication of partial melting is a key question which will certainly receive attention. On most thermal models the temperature at these depths is not far from the melting point for the basaltic fraction of the mantle material.

4. The source mechanism for the large earthquakes studied with long surface waves is double couple without moment, as mechanical

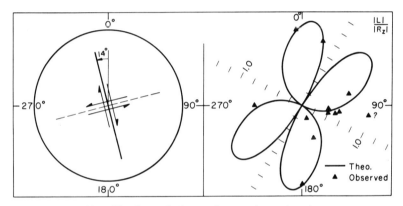

FIG. 19.  Shoal: explosion and tectonic strain release.

equilibrium considerations require [Brune, 1962, Ben-Menahem and Toksöz, 1963]. Earlier methods using body waves and short-period surface waves had difficulty distinguishing between these two mechanisms. (However, Stauder and Bollinger [1964] showed that the polarization of $S$ also indicates double couple sources for all cases studied by them.) A surprising result is the double couple mechanism implied by the radiation patterns of certain underground nuclear explosions (Figure 19). Brune and Pomeroy [1963], Toksöz, Harkrider, and Ben-Menahem [in press], and Press and Archambeau [1962] discuss these and other data with the suggestion that the explosion could actually have triggered the release of preexisting tectonic strain.

The directivity method (Figure 15) has produced evidence for fault lengths ranging from 300 to 1000 km, depending on the magnitude with rupture velocities of 3 to 4 km/sec. This represents the first time that source dimensions and faulting rate could be recovered instrumentally [Ben-Menahem and Toksöz, 1963].

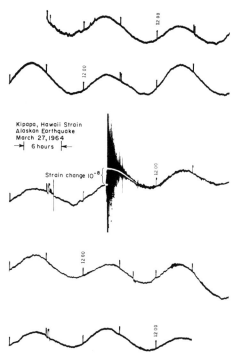

Fig. 20. Strain seismograph in Hawaii showing permanent offset following great Alaskan earthquake of 1964. One week of recording is presented to show that offsets did not occur before or after quake.

## Deformation Field at Teleseismic Distances

Up to this point we have been dealing with the seismic spectrum from 1 cycle per second to 1 cycle per hour. Some recent results imply that much useful information will become available, particularly about the source mechanism from observations of permanent deformation produced by earthquakes to distances of thousands of kilometers. In a sense we are concerned with zero frequency seismology.

An example of the kind of data that generated interest in this subject is shown in Figure 20 where the record from a strain seismograph in Hawaii shows the disturbance due to the great Alaskan earthquake. Clearly indicated is a permanent offset corresponding to a strain change of about $10^{-8}$. No such offset occurred for at least a week centered at the time of the earthquake, as the figure shows. Such permanent changes in strains and tilts have been observed before by Benioff, Tomascheck, and Nishimura, but were always in doubt due to the possibility that they were instrumental and not real.

A calculation just completed shows that the distant strain field indicated in Figure 20 is probably real. The calculation represents a fault by a sheet of dislocations having the fault dimensions and dislocation deduced from field observations in Alaska. The strain field at a point corresponding to Hawaii is seen in Figure 21 to be close to the observed value.

This result suggests the experiment of deploying strain meters, tiltmeters, and displacement meters around an active seismic belt in order to recover the deformation field. Since data would be forthcoming even if the seismic event were thousands of kilometers away (assuming a large magnitude event), the chances of accumulating data in a reasonable time are increased.

The permanent deformation field is probably a better indicator of source mechanism than methods using propagating waves. One would hope to recover the vertical extent of major faults, the elastic

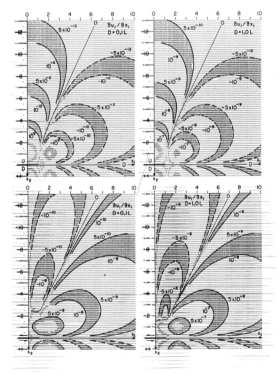

FIG. 21. Theoretical fields for one component of strain. $2L$ is fault length, $D$ is depth to bottom of vertical fault plane. Strike slips and dip slips are $0.33 \times 10^{-4}$ $L$. Coordinates of Hawaii strain seismograph are $x_1 \sim -9L$, $x_2 \sim 4L$. $2L$ is 600 to 800 km for Alaskan earthquake.

energy release, and mechanism (dislocation faulting, phase change, etc.).

Strain and tilt measurements in a tectonically active region have been widely suggested for testing the possibility of earthquake prediction.

### Detection and Identification of Underground Explosions

A chapter with the title of this one would be incomplete without a discussion of this question which attracts much of the efforts of the seismological community. I should like to describe progress in the field by comparing capability of detection and identification as we know it today with what we thought it to be at the Geneva conference of experts of 1958.

In the years since 1958 much experience has been accumulated in: array monitoring and processing of small teleseisms, magnitude-yield relationship for nuclear explosions in media of different types, epicenter location of small teleseisms and explosions, statistics of small earthquakes.

You may recall that the original Geneva system called for rather simple arrays, spaced at intervals of 1000 to 2000 km. Almost exclusive emphasis was placed on near zone monitoring, i.e., monitoring at

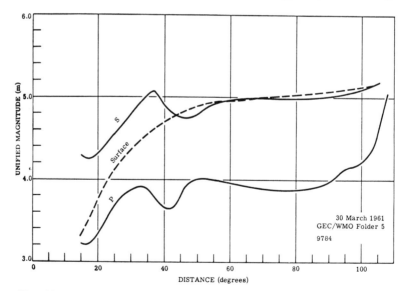

FIG. 22. Detection capability of array WMO in interior of United States. After Ruina [1963].

distances less than 2000 km. Thus some 20 monitoring posts were called for on the territory of the Soviet Union. The detection capability for such a system was thought to be about 1 kt in tuff. Figure 22 [Ruina, 1963] shows the detection limits at WMO, an array at a quiet site in the interior of the United States. Nothing fancier than simple superposition of the array elements was used in achieving this detection capability. The impressive feature is the ability of the station to detect events of magnitude 4 (1 to 2 kt in tuff) at appreciable distances beyond 20°. This implies that a smaller network of arrays, entirely outside a country can detect events with magnitudes near 4 in that country. The concept of third zone or teleseismic monitoring is a major advance for it offers the same capability as the administratively and politically more complex Geneva system. That signal levels in the third zone are comparable to those in the 1000 to 2000 km region is probably due to the shadow cast by the low-velocity zone in the upper mantle, discussed in an earlier section.

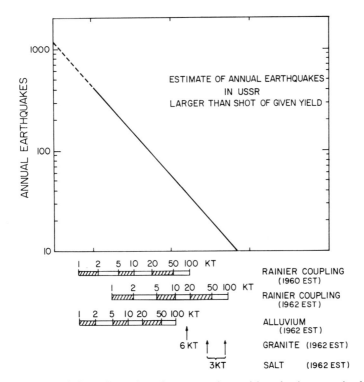

FIG. 23. Statistics of earthquakes occurring with seismic magnitude of explosions of given yield in several different media. After Romney [1963].

The relationship between magnitude and yield and magnitude and numbers of earthquakes is of major importance because it leads to the number of natural events that produce seismic signals comparable to an explosion of given yield. The problem here is the effect of medium on seismic signal excited by an explosion. A further difficulty arises

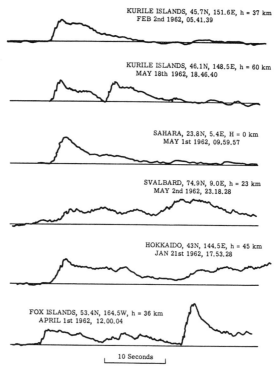

FIG. 24. Array correlator outputs for five earthquakes and one explosion in the Sahara. After Carpenter.

from the fact that the magnitude scale for teleseisms was for large earthquakes and the extrapolation to small distant events was open to question. The multiplicity of underground explosions in which yields, media, and locations were varied and the accumulation of statistics on the occurrence of small earthquakes all lead to revisions. Romney [1963] summarized the results in Figure 23. It is seen that from 1960 to 1962 a downward revision could be justified by a factor 2 to 3 in the number of earthquakes associated with an explosion of given yield. Whereas we associated 500 events with a 1-kt explosion (in tuff) in 1960, a monitoring system must now contend with the smaller number of about 250 earthquakes. The effect of medium is also indicated on the horizonal scale.

This is still a large number of events, and methods must be developed to identify earthquakes as such, in order to eliminate them from the suspicious category. Rather simple concepts such as depth of focus, first motion polarity, epicenter location in unlikely places for testing (near borders or under water) have been shown to be capable of reducing the number of suspicious events to about 40. Some calibration explosions and deep-sea seismographs may be needed to achieve this result. In 1958 the identification capability in the magnitude 4 range was essentially nonexistent, and 450 of the 500 detected events would have remained in the suspicious category.

There is a growing body of evidence that a teleseismic identification scheme may be avilable which would reduce the remaining unidentified events to a number significantly less than 40. It was proposed by Thirlaway for the nuclear detection problem [see Carpenter, 1965], and is based on a suggestion of Gutenberg in a paper written 20 years earlier. It consists of an examination of $P$-waves which have been array processed to remove clutter induced by the signal in the vicinity of the station. The clutter is recognized and eliminated using the property that its phase velocity is much less than the signal velocity. Explosion waves so treated would yield simple wave forms of relatively brief duration in all azimuths. An earthquake signal would be complex, if not in one azimuth then in another. The complexity arises from mode conversion of shear to compression energy in the vicinity of the source—an effect which is reduced for explosions because of poor shear wave excitation. Figure 24 taken from Carpenter's report shows correlation outputs from two segments of a crossed array. The simple form of the Sahara explosion and the complex form of four of the five earthquakes is evident.

### References

Aki, K., Study of Love and Rayleigh waves from earthquakes with fault plane solutions or with known faulting, *Bull. Seismol. Soc. Am.*, *54*, 529–558, 1964.

Alsop, L. E., Free spheroidal vibrations of the earth at very long periods, *Bull. Seismol. Soc. Am.*, *53*, 483–502, 1963.

Alsop, L. E., G. H. Sutton, and M. Ewing, Free oscillations of the earth observed on strain and pendulum seismographs, *J. Geophys. Res.*, *66*, 631–641, 1961.

Alterman, Z., H. Jarosch, and C. L. Pekeris, Oscillations of the earth, *Proc. Roy. Soc. London*, *A*, *252*, 80, 1959.

Alterman, Z., H. Jarosch, and C. L. Pekeris, Propagation of Rayleigh waves in the earth, *Geophys. J.*, *4*, 219–241, 1961.

Anderson, D. L., Recent evidence concerning the structures of the upper mantle from the dispersion of long-period surface waves, in *Proc. Vesiac Conference on Variations of the Earth's Crust and Upper Mantle*, Institute of Science and Technology, University of Michigan, Ann Arbor, 1964a.

Anderson, D. L., Densities of the mantle and core, *Trans. Am. Geophys. Union*, *45*, 101, 1964b.

Anderson, D. L., and C. Archambeau, The anelasticity of the earth, *J. Geophys. Res.*, *69*, 2071–2084, May 15, 1964.

Anderson, D. L., and R. Kovach, Attenuation in the mantle and rigidity of the core from multiply reflected core phases, *Proc. Natl. Acad. Sci.*, *51*, 168–172, 1964.

Anderson, D., and M. N. Toksöz, Surface waves on a spherical earth, 1. Upper mantle structure from Love waves, *J. Geophys. Res.*, *68*, 3483–3500, 1963.

Backus, G., and F. Gilbert, The rotational splitting of the free oscillations of the earth, *Proc. Natl. Acad. Sci., U.S.*, *47*, 362–371, 1961.

Benioff, H., in Progress Report, Seismological Laboratory, *Trans. Am. Geophys. Union*, *35*, 979–987, 1954.

Benioff, H., F. Press, and S. Smith, Excitation of the free oscillations of the earth by earthquakes, *J. Geophys. Res.*, *66*, 605–620, 1961.

Ben-Menahem, A., Radiation of seismic surface waves from finite moving sources, *Bull. Seismol. Soc. Am.*, *51*, 401–435, 1961.

Ben-Menahem, A., and M. N. Toksöz, Source mechanism from spectra of long-period seismic surface waves, *Bull. Seismol. Soc. Am.*, *53*, 905–919, 1963.

Bolt, B. A., The velocity of seismic waves near the earth's center, *Bull. Seismol. Soc. Am.*, *54*, 191–208, 1964a.

Bolt, B. A., Recent information of the earth's interior from studies of mantle waves and eigenvibrations, *Phys. Chem. Earth*, *5*, 55–120, 1964b.

Brune, J. N., Correction of initial phase measurements for the southeast Alaska earthquake of July 10, 1958, *J. Geophys. Res.*, *67*, 3643, 1962.

Brune, J. N., J. E. Nafe, and J. Oliver, A simplified method for the analysis of dispersed wave trains, *J. Geophys. Res.*, *65*, 287–304, 1960.

Brune, J. N., J. E. Nafe, and L. E. Alsop, The polar phase shift of surface waves on a sphere, *Bull. Seismol. Soc. Am.*, *51*, 247–258, 1961.

Brune, J. N., and P. W. Pomeroy, Surface wave radiation patterns for underground nuclear explosions and small magnitude earthquakes, *J. Geophys. Res.*, *68*, 5005–5028, 1963.

Carpenter, E. W., Explosion seismology, *Science*, *147*, 363–373, 1965.

Dorman, J., M. Ewing, and J. Oliver, Study of shear velocity distribution by Mantle Rayleigh waves, *Bull. Seismol. Soc. Am.*, *50*, 87–115, 1960.

Gamburtsev, G. A., Deep seismic sounding of the earth's crust, *Dokl. Akad. Nauk SSSR*, *87*, 943–945, 1952.

Harkrider, D. G., Surface waves in multilayered elastic media 1. Rayleigh and Love waves from buried sources in a multilayered elastic half space, *Bull. Seismol. Soc. Am.*, *54*, 627–679, 1964.

Haskell, N., The dispersion of surface waves on multilayered media, *Bull. Seismol. Soc. Am.*, *43*, 17–34, 1953.

Haskell, N., Crustal reflection of plane $P$ and $SV$ waves, *J. Geophys. Res.*, *67*, 4751–4767, 1962.

Knopoff, L., and G. J. F. MacDonald, Attenuation of small amplitude stress waves in solids, *Rev. Mod. Phys.*, *30*, 1178–1192, 1958.

Landisman, M., Y. Sato, and J. Nafe, Free vibrations of the earth and the properties of its deep interior regions, *Geophys. J.*, in press, 1964.

MacDonald, G. J. F., and N. F. Ness, A study in the free oscillations of the earth, *J. Geophys. Res.*, *66*, 1865–1912, 1962.

Ness, N. F., J. C. Harrison, and L. B. Slichter, Observations of the free oscillations of the earth, *J. Geophys. Res.*, *66*, 621–629, 1961.

Oliver, J., Propagation of PL waves across the U.S., *Bull. Seismol. Soc. Am.*, *54*, 151–160, 1964.

Pakiser, L., Continental crust, *Encyclopedia of Earth Sciences*, in press, Reinhold Publishing Co.

Pekeris, C. L., Z. Alterman, and H. Jarosch, Rotational multiplets in the spectrum of the earth, *Phys. Rev.*, *122*, 1692–1700, 1961.

Phinney, R., Structure of the earth's crust from spectral behavior of long period body waves, *Trans. Am. Geophys. Union*, *45*, 92, 1964.

Press, F., Crustal structure in the California-Nevada region, *J. Geophys. Res.*, *65*, 1039–1051, 1960a.

Press, F., Observations of the free vibrations of the earth, *Program Am. Geophys. Union Meetings*, *30*, Abstract, 1960b.

Press, F., The earth's crust and upper mantle, *Science*, *133*, 1455–1463, 1961.

Press, F., Long-period waves and free oscillations of the earth, *Research in Geophysics*, vol. 2, Chap. 1, pp. 1–26, The M.I.T. Press, Cambridge, Mass., 1964.

Press, F., and C. Archambeau, Release of tectonic strain by underground nuclear explosions, *J. Geophys. Res.*, *67*, 337–343, 1962.

Press, F., and S. Biehler, Inferences on crustal velocities and densities from $P$ wave delays and gravity anomalies, *J. Geophys. Res.*, *69*, 2979–2995, 1964.

Romney, C., Hearings before Joint Committee on Atomic Energy, U.S. Congress, March 5–12, p. 91, 1963.

Ruina, J., Hearings before Joint Committee on Atomic Energy, U.S. Congress, March 5–12, p. 55, 1963.

Sato, Y., Attenuation, dispersion and the wave guide of the G wave, *Bull. Seismol. Soc. Am.*, *48*, 231–251, 1958.

Sato, Y., M. Landisman, and M. Ewing, Love waves in a heterogeneous, spherical earth: Part 2, *J. Geophys. Res.*, *65*, 2399–2404, 1960.

Shimshoni, M., and S. Smith, Seismic signal enhancement with 3-component detectors, *Geophysics*, in press.

Smith, S., Free vibrations of the earth, *International Dictionary of Geophysics*, Pergamon Press, in press.

Stauder, W., S.J., and G. A. Ballinger, The S wave project for focal mechanism studies of earthquakes of 1962, *Report under AFOSR grant 62-458*, St. Louis University, 1964.

Takeuchi, H., F. Press, and N. Kobayashi, Rayleigh wave evidence for the low velocity zone in the mantle, *Bull. Seismol. Soc., Am.*, 355–364, 1959.

Texas Instruments, Inc. Seismometer array and data processing system, *Final report phase 1, AFTAC Proj. VT/077, Contract No. AF 33(600)-41840*, 1962.

Toksöz, M. N., D. G. Harkrider, and A. Ben-Menahem, Determination of source parameters by amplitude equalization of surface waves. 2. Release of tectonic strain by underground nuclear explosions and mechanism of earthquakes, *J. Geophys. Res.*, in press.

# THE CHEMICAL COMPOSITION AND ORIGIN OF THE EARTH

*A. E. Ringwood*

*Department of Geophysics and Geochemistry, Australian National University,
Canberra, Australia*

## 1. Introduction

The author was originally invited to present a paper to this International Conference under the title "Composition and Phases of the Mantle." The mantle, however, accounts for nearly 70 per cent of the mass of the earth. Consequently, a broad discussion of the composition of the mantle did not appear profitable without considering evidence bearing upon the composition of the entire earth. This topic in turn was intimately connected with the problem of the earth's origin, and some discussion of this problem, therefore, appeared desirable. Thus the scope of the paper soon became broader than was originally intended, and it was decided to cover the topic "Mineralogy of the Mantle" separately.

A principal objective of this chapter will be to compare estimates of the present chemical composition of the earth with estimates of the primordial abundances of the elements in the solar system, and, from this comparison, to inquire into the chemical fractionations which occurred when the earth formed from primordial material and the physical conditions under which these fractionations were established. Despite the accumulation of a vast amount of data on the chemical compositions of crustal and possible mantle rocks, meteorites, and the sun, the subject remains beset with uncertainties. Inferences based upon anomalies in abundances of single elements are of doubtful value. At the present stage, it is desirable to search for systematic fractionation patterns involving groups of elements possessing analogous properties and to inquire into the significance of the major over-all abundance trends. Although many of the conclusions drawn are necessarily highly tentative, they may be of value since they direct attention to fields where new data are most urgently required and suggest hypotheses to be tested. One finds when attempting a general survey that during the last twenty years much of the geochemical work aimed at establishing concentrations of elements in terrestrial and meteoritic material has lacked direction and purpose, and, as a result, our knowledge of the composition of the earth and of meteorites is not nearly so advanced as might have been hoped from the effort which has been expended and the techniques which have been available.

## 2. Selective Volatility as a Factor Controlling the Composition of the Earth

It is generally agreed that the solar system was formed by the gravitational collapse of an initially cold interstellar cloud of dust and gas, and that the earth was formed by accretion of solids comprised of the less volatile components of the primordial gas-dust cloud. It follows that the over-all composition of the earth is related in some way to that of the primordial dust. Knowledge of this primordial composition would accordingly place general limits on the composition of the earth.

In recent years, information on this subject from three different directions has shown encouraging signs of convergence. First of all, knowledge of the abundance of elements in the sun has steadily improved. Goldberg *et al.* [1960] have published relative abundances in the solar atmosphere for forty-three elements. It would be widely agreed that the solar abundances reflect closely the composition of the primordial solar nebula. Second, the study of chondritic meteorites has revealed the presence of a class—the Type I carbonaceous chondrites—which are more primitive in their chemical composition than any other meteorites. The constitution of these objects which are highly oxidized and volatile-rich resembles that which would be expected if primordial dust had accreted into a small parent body and had been subjected to a very mild degree of metamorphism [Ringwood, 1965]. It is clear that they have never been subjected to temperatures above about 200°C since their formation [DuFresne and Anders, 1961] and, consequently, it appears unlikely that significant over-all fractionation of nonvolatile elements has occurred. The view that these objects were of primitive origin was originally stated by Urey [1953]. Subsequently discovered irregularities in the compositions of other groups of carbonaceous chondrites caused him to discard this view, which, however, was restated by Mason [1960] and Ringwood [1961a]. Since then a large amount of new analytical information has become available, and it is now accepted by most students of meteorites that Type I carbonaceous chondrites have retained to a considerable degree the primordial abundances of the elements, apart from those

Table 1

RELATIVE ATOMIC ABUNDANCES OF METALS IN TYPE I CARBONACEOUS CHONDRITES AND IN THE SUN NORMALIZED ACCORDING TO THE RELATIONSHIP

$$\frac{M \text{ chondrite}}{Si \text{ chondrite}} \div 1.6 \frac{M \text{ solar}}{Si \text{ solar}}$$

(Data from Urey, 1964, and Goldberg et al., 1960)

| Element | Relative Abundance | Element | Relative Abundance |
|---|---|---|---|
| Na | 0.6 | Fe | (0.2)† |
| Mg | 0.8 | Co | 1.1 |
| Al | 1.1 | Ni | 1.1 |
| Si | 0.6 | Cu | 0.1 |
| P | 1.2 | Zn | 0.7 |
|   |   | Ge | 1.4 |
| S | 0.5 | Rb | 0.5 |
| K | 1.4 | Sr | 3.0 |
| Ca | 1.1 | Sr | (1.5)‡ |
| Sc | 1.0 | Y | 0.5 |
| Ti | 1.0 | Cd | 1.6 |
| Cr | 1.1 | Ba | 0.7 |
| Mn | 2.0 | Yb | 0.1 |
| Fe | 4.8 | Pb | 1.5 |
| Fe | (1.2)* |   |   |

\* Claas [1951], corrected for latest $f$ values [Goldberg et al., 1960].
† Pottasch [1963].
‡ Claas [1951].
  Note: The normalization factor of 1.6 was chosen to provide the best match for the abundances of the more common elements between Na and Ni, excluding Fe and S.

of high volatility [Ringwood, 1962a; Lovering, 1962; Mason, 1962; Greenland, 1963; Urey, 1964; and Anders, 1964].

Comparison of abundances in Type I carbonaceous chondrites and the sun is given in Table 1. It is seen that out of the 25 elements for which satisfactory data exist, the relative abundances of 21 of these elements in the sun and within Type I carbonaceous chondrites differ by less than a factor of 2, which is within the limits of combined determinative error. In two out of four exceptional elements, there are discrepancies in the solar abundances by different workers of such a magnitude as to suggest a rather large uncertainty.

A third approach to primordial abundance determination arises from arguments based upon nucleosynthesis models. Considerable progress has been made in this field during recent years [Burbidge et al., 1957; Cameron, 1959; Clayton and Fowler, 1961; Hoyle and Fowler, 1964]. The arguments are not entirely independent of chondritic abundances, since the latter are used to fix the abundance curves in critical mass regions. Nevertheless, they form an invaluable complement to the chondritic abundances. Indeed, one of the principal reasons for the wide degree of current agreement regarding the significance of the Type I carbonaceous chondrites has been the discovery that their compositions agree very well with primordial abundances estimated on nucleosynthetic grounds. In contrast, the abundances of certain elements in other groups of meteorites diverge widely from estimated primordial values. Elements which have been shown to be depleted in ordinary chondrites but present in their primordial abundances in Types I (and II) carbonaceous chondrites are Pb, Bi, Tl, and perhaps Hg [Reed et al., 1960]; Te and I [Goles and Anders, 1962]; Cd [Schmitt et al., 1963a, b]; Zn and Ge [Greenland, 1963]; and U and Th [Lovering and Morgan, 1964].

Thus we find a wide measure of agreement from three different sources concerning the composition of the material from which the solar system formed. The agreement between solar abundances, abundances estimated from nucleosynthetic arguments, and abundances determined in Type I carbonaceous chondrites is of great significance. The results of future investigations on other elements, and on the presently discrepant elements, will be awaited with great interest. Since the abundances have been determined with greatest precision in the Type I carbonaceous chondrites, it is appropriate to use these as a basis for subsequent discussion. The composition of this class of meteorites is given in Table 2.

We now consider the manner in which the present net chemical composition of the earth might be related to the primordial dust

## Table 2

MEAN CHEMICAL COMPOSITION OF TYPE I CARBONACEOUS CHONDRITES
[ORGUEIL AND IVUNA]

| Major Elements [Wiik, 1956] (Weight per cent) | | Rare Elements (Relative to Si = $10^6$) Compiled by Urey [1964] | | | |
|---|---|---|---|---|---|
| $H_2O$ | 19.29 | F* | 2200 | La | 0.36 |
| C | 3.97 | Na | $6.40 \times 10^4$ | Ce | 1.17 |
| Organic Matter | 5.53 | P | $1.27 \times 10^4$ | Pr | 0.17 |
| $Na_2O$ | 0.74 | S | $5.05 \times 10^5$ | Nd | 0.77 |
| MgO | 15.96 | Cl | 2200 | Sm | 0.23 |
| $Al_2O_3$ | 1.64 | K | 3540 | Eu | 0.091 |
| $SiO_2$ | 22.63 | Sc | 33 | Gd | 0.55 |
| $P_2O_5$ | 0.35 | Ti | 2300 | Tb | 0.037 |
| $K_2O$ | 0.07 | Cr | $1.24 \times 10^4$ | Dy | 0.36 |
| CaO | 1.56 | Co | 2300 | Ho | 0.09 |
| $TiO_2$ | 0.07 | Zn | 930 | Er | 0.22 |
| $Cr_2O_3$ | 0.34 | Ge | 135 | Tm | 0.035 |
| MnO | 0.22 | Rb | 7.1 | Yb | 0.21 |
| FeO | 10.42 | Sr | 58.4 | Lu | 0.035 |
| FeS | 16.73 | Y | 4.6 | Hg | 11.5 |
| CoO | 0.06 | Cd | 2.37 | Tl | 0.18 |
| NiO | 1.29 | Cs | 0.37 | Pb | 1.6 |
| | | Ba | 4.7 | Bi | 0.17 |
| | | | | Th† | 0.076 |
| | | | | U† | 0.028 |

* Reed, G. [1964].
† Lovering and Morgan [1964].

composition. Latimer [1950] and Urey [1952] showed that at low temperature ($<0°C$) the dust in the parental solar nebula would be highly oxidized (as is the case with the Type I carbonaceous chondrites). Furthermore, the dust would probably contain water, methane, and ammonia hydrates as ices. These components have since been largely lost by the carbonaceous chondrites. The formation of the earth and other classes of meteorites from this primitive material is a highly controversial subject. However, some aspects are clear. Much or most of the primitive material was subjected to chemical reduction at high temperatures ($>1000°C$), resulting in the formation of a metal phase and the loss of volatiles. A major difference between the earth and the primitive material is that the earth became relatively depleted in many elements which are volatile at high temperatures—particularly under reducing conditions. It follows that the similarities in chemical composition between the earth and primitive material are likely to be greatest for those elements which possess the lowest volatilities under high-temperature reducing conditions [Ringwood, 1962a]. These elements include most of the electropositive (oxyphile) elements and transition elements. Virtually all theories of the origin of the solar system hold or imply that selective volatility (or condensation) has been an important factor during the formation of the terrestrial planets from primitive material. Some workers, e.g., Urey [1952, 1962, 1963], believe that additional mechanisms involving physical fractionation of metal from silicates have also been operative. Others [Ringwood, 1959, 1960, 1965] disagree with this view. The difficulties of obtaining silicate/metal fractionation in the solar nebula by physical means are severe and, in the author's opinion, no plausible models for such fractionation have been proposed.

For the present, we will accept the view that selective volatility under high-temperature reducing conditions is the principal property which has influenced the composition differences between the earth and primitive material. It is suggested that the abundances of nonvolatile oxyphile elements and transition metals are similar in carbonaceous chondrites and in the earth. The relative abundances of elements which are volatile under high-temperature reducing conditions, such as many $B$ subgroup elements of the periodic table, electronegative elements, and the alkali metals, may, however, differ considerably. The way in which the earth may have formed from primitive material and the nature of the fractionations to be expected were discussed by Ringwood [1959, 1960, 1962a].

We have not yet defined adequately the usage of the term "volatile" or "nonvolatile." This is difficult since the terms are used in a relative sense. Whether or not a particular element is volatile depends upon a combination of many factors—chiefly, of course, temperature and chemical environment. It is apparent that any distinction must be arbitrary. Rather than formulate a rigid definition, an arbitrary relative classification for a selected group of elements has been given in Table 3. It is seen that the elements classed as "volatile" can be volatilized from silicate melts under moderately reducing conditions around 1300 to 1500°C. Those in the "nonvolatile" group (Column I) are not readily volatile under these conditions. According to the model under consideration, the abundances of nonvolatile elements (Table 3, Column I) in the earth should be similar to their primordial abundances as given by Type I carbonaceous chondrites (Table 2).

Comparison of the chemical compositions of the different groups of

Table 3

CLASSIFICATION OF SOME ELEMENTS ACCORDING TO THEIR PROBABLE RELATIVE VOLATILITIES FROM BASIC SILICATE MELTS UNDER HIGH-TEMPERATURE REDUCING CONDITIONS

| I | II | |
|---|---|---|
| Nonvolatile Group | Volatile Group | |
| A. Oxyphile Elements | | Probable volatile species |
| Be, B, Mg, Al, Si, P, Ca, Sc, Ti, Sr, Y, Zr, Nb, Ba, Rare earths Hf, Ta, Th, U | H, C, N | $H_2O$, CO, $N_2$ |
| | F, Cl, Br, I | Silicon and metal halides |
| B. Siderophile Elements | S, Se, Te | Hydrides |
| | (Li?), Na, K, Rb, Cs | Elements |
| Fe, Co, Ni | | |
| V, Cr, Mn | | |
| Cu, Ag, Au (Section 6B) | Zn, Cd, Hg, Tl, Pb, As, Sb, Bi, Se, Te | Elements |
| Mo, (Sn?), W | | |
| (Ru?), Rh, Pd, Re (Os?) | Ga, Ge, Sn, In | Suboxides, sulfides |
| Ir, Pt | | |

chondrites reveals that in general the nonvolatile group, particularly the siderophile elements, are only slightly fractionated. In cases where the abundance of such an element in a Type I carbonaceous chondrite has not been determined, it appears that its abundance in an ordinary chondrite may be tentatively used in its place. Abundances of these elements have been tabulated by Urey [1964].

Finally, it should be emphasized that the presence of an element in the "volatile" column does not imply its *absence* from the earth, but merely a probability of relative depletion compared to primordial abundances. As Brown [1952], Chamberlin [1952], and Urey [1952] pointed out, the earth contains some volatile components, and their presence implies that at least some of the material from which the earth was formed was trapped at low temperatures. The reconciliation of this consequence with the evidence that much of the material of the earth has been subjected also to very high temperature before or during its formation is one of the principal difficulties faced by theories of the origin of the solar system.

The limitations on the composition of the earth have so far been proposed on the basis of rather general evidence, not connected directly with the earth. However, in Sections 3 and 4 we shall describe direct evidence on the composition of the earth which shows that a self-consistent chemical model can be obtained from the primordial abundance of nonvolatile elements. In contrast, there is now strong evidence that many of the volatile elements are not present in the earth in their primordial abundances.

## 3. Mantle-Core Relationships

The two major divisions of the earth are the mantle and core. According to the Bullen Model $A$, the mantle (+ crust) contains 69 per cent of the earth's mass and the core contains 31 per cent. It is generally believed that the core consists dominantly of nickel-iron. However, an alternative hypothesis proposed by Ramsay [1948, 1949] maintains that the core-mantle boundary is an isochemical phase transition caused by pressure degeneracy in the silicates of the mantle. This hypothesis has been widely discussed and is still supported by some workers. A realistic appraisal of the Ramsay hypothesis has been given by Birch [1952, pp. 275–280] on the basis of a broad range of physical, geophysical, and geochemical evidence. He showed clearly that the evidence, when considered as a whole, was highly unfavorable to the Ramsay hypothesis. At the same time, all known physical properties of the core were consistent with it being composed dominantly of a nickel-iron alloy containing some light elements in solution.

Recent shock-wave investigations of the densities of metals and silicates up to pressures exceeding those in the center of the earth have confirmed Birch's appraisal. The evidence has been summarized by Birch [1961, 1963, 1964]. From the systematic relationships between density, pressure, seismic velocity, and mean atomic number which have been disclosed by the experimental work, it is extremely improbable that mantle material consisting dominantly of Mg, Si, and O in the metallic state could provide the densities and seismic velocities observed in the core.

Recent experimental and theoretical investigations have also provided a deeper understanding of transitions from nonmetals to metals of the type considered by Ramsay [Drickhamer, 1963; Alder, 1961, 1963]. Such transitions are accompanied by large (>20 per cent) density changes only when the nonmetal possesses an open structure, usually characterized by directed covalent bonds. The increase in density is caused dominantly by the closer atomic packing of the metallic state, rather than by the accompanying change in electronic structure. In the lower mantle, the silicates are already close-packed and presumably ionic. The Ramsay hypothesis requires that this material, a multicomponent system, should undergo a first-order transformation into a metal accompanied by a density increase of 65 per cent. These requirements do not appear reasonable in the light of current experimental and theoretical investigations. For these reasons, the established view that the core is composed dominantly of nickel-iron is maintained in this paper.

The next question to discuss is whether the primordial abundances of "nonvolatile" elements, as obtained from the composition of Type I carbonaceous chondrites (Table 2), are capable of yielding an earth model with the correct core-to-mantle ratio and acceptable compositions for both these major phases. This is examined in Table 4, which is a modification of the argument by Ringwood [1959]. Only

### Table 4

COMPOSITION OF EARTH AS DERIVED BY REDUCTION FROM COMPOSITION OF TYPE I CARBONACEOUS CHONDRITES

|  | 1 | 2 | 3 | 4 |
|---|---|---|---|---|
| $SiO_2$ | 33.32 | 35.85 | 29.84 | 43.25 |
| $MgO$ | 23.50 | 25.19 | 26.29 | 38.10 |
| $FeO$ | 35.47 | 6.14 | 6.38 | 9.25 |
| $Al_2O_3$ | 2.41 | 2.59 | 2.69 | 3.90 |
| $CaO$ | 2.30 | 2.47 | 2.57 | 3.72 |
| $Na_2O$ | 1.10 | 1.18 | 1.23 | 1.78 |
| $NiO$ | 1.90 | — | — | — |
|  | 100.00 | 73.52 | 69.00 | 100.00 |
| Fe |  | 24.88 | 25.87 |  |
| Ni |  | 1.60 | 1.66 |  |
| Si |  | — | 3.47 |  |
|  |  | 26.48 | 31.00 |  |

Column 1. Average composition of principal components of Type I carbonaceous chondrites [Orgueil and Ivuna] on a C-, S-, and $H_2O$-free basis [analyses by Wiik, 1956].

Column 2. Analysis from Column 1 with [FeO/(FeO + MgO)] reduced to be consistent with probable value for earth's mantle (0.12).

Column 3. Analysis from Column 2 with sufficient $SiO_2$ reduced to elemental silicon to yield a total silicate to metal ratio 69/31 as in the earth.

Column 4. Model mantle composition: silicate phase from Column 3 recalculated to 100 per cent.

the principal components of the primordial oxidized composition are considered. Sodium has been included, although it now appears likely that it should be classed with the volatile elements (Section 6a).

The primitive oxidized composition for the earth is given in Column 1. Geochemical and geophysical evidence indicates that the FeO/(FeO + MgO) (molecular) ratio of the mantle is between 0.1 and 0.2 [Clark and Ringwood, 1964; this chapter]. We will assume a value of 0.12 for the mantle. To obtain this value it is necessary to reduce all the nickel and an appropriate amount of iron, giving Column 2, which has a mantle/core ratio of 73.5/26.5. To obtain the correct mantle/core ratio of 69/31, it is necessary to remove another component from the oxidized mantle, reduce it to metal, and transfer it to the core. After the oxides of iron and nickel, $SiO_2$ is the common oxide most readily reduced to metal. Accordingly, to obtain an earth model from the primitive composition, it is necessary to transfer some $SiO_2$ to the core as elemental silicon. This is done in Column 3. The core would then contain about 11 weight per cent or 20 atomic per cent of silicon.

The suggestion that the earth's core might contain silicon as a major component was made by Ringwood [1958, 1959, 1961b], MacDonald and Knopoff [1958], and MacDonald [1959a, b]. As developed above it may appear that the suggestion is rather arbitrary, depending critically upon the composition adopted for the chosen model. However, there is independent evidence supporting the presence of silicon in the earth's core.

Comparisons of the density of the core and the probable density of iron at high pressure [Birch, 1952] indicated that the material of the core is about 10 to 20 per cent less dense than would be expected for iron under similar pressure-temperature conditions. Support for this view has been provided by Knopoff and MacDonald [1960] and Birch [1961] using shock-wave data on the density of iron.

Birch [1952] has pointed out that the elastic ratio $K/\rho$ ($K$ = incompressibility, $\rho$ = density) of the core appears much higher than that of iron under similar $P$-$T$ conditions. Shock-wave evidence indicates that the corresponding seismic velocities in the core are substantially higher than those of pure iron [Knopoff and MacDonald, 1960; Birch, 1961].

It therefore appears that the earth's core contains a substantial amount of an element with a low density which can also increase the elastic ratio and seismic velocity of iron. Limitations upon possible choices are that this element must be reasonably abundant, miscible with liquid iron, and possess chemical properties which would allow it to enter the core. Elements which have been considered in this respect by various authors are H, He, C, O, N, Mg, Si, and S. We may reject H, He, C, O, and N since they are known to form interstitial solid solutions with iron. Additions of these elements do not significantly decrease the density of iron since they occupy holes already present in the lattice. Magnesium is unlikely to be present in substantial amounts since it has a much greater affinity for oxygen than has silicon. Accordingly, any chemical conditions which may have led toward the incorporation of magnesium in the core would inevitably have caused the incorporation of much larger amounts of silicon.

This leaves us with silicon and sulfur as possibilities. It would require about 15 weight per cent of sulfur in the core to decrease the density by 10 per cent. This would imply that the earth captured almost all of the sulfur originally present in the primitive material. The question is further discussed in Section 6*b*, where it is concluded on geochemical grounds that the core does not contain more than a small fraction of the sulfur required to explain the discrepancies in density and seismic velocity.

A process of elimination thus points toward silicon as the most likely extra component of the earth's core. It forms extensive substitutional solid solutions with iron, causing a marked reduction of density. Furthermore, the elastic ratio of silicon is twice that of iron. Silicon also satisfies the criteria of abundance, miscibility with liquid iron, and chemical compatibility. Finally, it has been found to occur as a significant constituent of the metal phase in enstatite chondrites [Ringwood, 1961*b*].

The chemical conditions under which silicon may have become incorporated in the earth's core have been extensively discussed by the author [1959, 1960, 1961*b*]. Urey [1960] has objected that they imply that the earth's core and mantle are out of equilibrium and that this consequence is intrinsically unlikely. The author agrees with the first part of Urey's argument but not with the second. The question was extensively discussed in the foregoing papers and is further considered later in this paper.

We have now reached the stage where we have derived a model composition both for the whole earth (nonvolatile elements of Table 2) and for the mantle (Table 4, Column 4). The next step is to discuss other more direct sources of information regarding the composition of the mantle and their compatibility with the model compositions.

The subject matter of the next section is treated in somewhat greater detail than is needed for the purposes of this chapter. This is required in order to provide adequate background for the accompanying chapter on "Phases of the Mantle."

## 4. Composition of the Upper Mantle

A synthesis of geophysical, geological, and geochemical information leads to some useful inferences concerning the composition of the upper mantle. The velocity of $P$ seismic waves in the mantle immediately beneath the Mohorovicic Discontinuity under stable continental areas and deep oceanic basins averages about 8.2 km/sec. For most practical purposes, this property limits the essential mineralogy of the upper mantle to some combination of olivine, pyroxene, garnet, and perhaps, in restricted regions, amphibole. The two principal rock types carrying these minerals are peridotite (olivine-pyroxene) and eclogite (pyroxene-garnet). Both types may carry some amphibole. Complete mineralogical transitions between the two major rock types are rare, and usually of local significance only when they occur. Eclogite may sometimes carry a little olivine, and peridotite often contains some garnet, but the relative lack of true intermediate members seems to indicate a fundamental dichotomy between the two rock types. This dichotomy has given rise to two hypotheses regarding the composition of the upper mantle. The first holds that the upper mantle is ultramafic in composition and that the Moho represents a chemical change. The second hypothesis holds that the upper mantle is eclogitic in composition and that the Moho represents an isochemical phase transition from a basaltic lower crust into eclogite. The latter hypothesis is derived from the work of Fermor [1913, 1914] and Goldschmidt [1922]. More recently it has been revived by Sumner [1954], Robertson *et al.* [1957], Lovering [1958], and Kennedy [1959]. The revival has created a great deal of interest and stimulated much new work. A critical discussion follows.

### a. *The Moho as a Phase Change from Basalt to Eclogite*

Advocates of this hypothesis point to several interesting properties. If the Moho is a phase change, temperature perturbations could cause basalt to change into eclogite or vice versa, resulting in substantial crustal uplift or depression. This would provide a possible explanation for one of the most difficult geotectonic problems—namely, the ultimate cause of vertical movements of the earth's crust. Second, complete melting of eclogite in the mantle would provide a copious and readily available source of basaltic magma. Furthermore, it was suggested that an eclogitic upper mantle would explain the heat flow from oceanic areas which, according to some workers, is too high to be readily reconciled with an ultramafic upper mantle. Finally, experimental investigations on synthetic systems [Robertson *et al.*, 1957; Birch and Le Compte, 1960; Boyd and England, 1959; Yoder and

Chinner, 1960; Yoder and Tilley, 1962] indicated that the $P$-$T$ conditions required for the transition may well be close to those occurring at the Moho.

Despite these interesting features, it has become clear during recent years that the hypothesis is subject to a number of serious objections. Some of these are summarized in the following.

i. If the Moho under oceans and continents is caused by a basalt-eclogite transition, a strong correlation between surface heat flow (which is related to temperature at the Moho) and crustal thickness should be found. Such a correlation is not found either within continents, within ocean basins, or between ocean basins and continents [Bullard and Griggs, 1961; MacDonald and Ness, 1960; Ringwood and Green, 1964].

ii. Harris and Rowell [1960] and Bullard and Griggs [1961] showed that the Moho cannot be explained by the same phase transition under both oceans and continents as long as the increase of temperature with depth is more rapid beneath oceans than beneath continents. This latter feature is common to almost all possible thermal models for the mantle.

iii. The transition interval over which basalt changes to eclogite has been experimentally determined by Ringwood and Green [1964] and was found to correspond to a depth interval of 20 km or more. Later work by the same authors [unpublished] has shown that the density (and hence also the seismic velocity) increase is spread uniformly across this interval. It is extremely difficult to reconcile these properties of the basalt-eclogite transition with seismic evidence on depth-velocity profiles in the lower crust and upper mantle.

iv. Formation of a basalt magma in the mantle would require complete fusion if the upper mantle were eclogitic. It is difficult to understand how complete fusion would arise in the mantle. Plausible mechanisms of magma formation lead to fractional melting, which must be much more common than complete melting [Hess, 1960]. Yet the fractional melting of an eclogite would not yield a basaltic magma.

v. The density of true eclogites of basaltic composition is close to 3.55 gm/cc [Clark and Ringwood, 1964; Ringwood and Green, unpublished]. The preferred density for the upper mantle used in gravimetric studies is between 3.3 and 3.4 gm/cc [Talwani and Worzel, 1959], which corresponds well with that of ultramafic rocks but is well below that of true eclogite.

vi. Eclogites are extremely rare rocks compared to ultramafic rocks. This is sometimes explained by their tendency to transform to amphibolite under crustal conditions. This explanation is not entirely convincing. The tendency of eclogite to transform to amphibolite is comparable with that of peridotite and dunite to transform to serpentine. Yet there is still vastly more recognizable peridotite and dunite in the crust than eclogite [see also next section dealing with diamond pipes].

Other contrary arguments could be proposed, but those mentioned should suffice. It is possible to avoid individual arguments by introducing special assumptions. However, the combined weight of the preceding arguments is, in the author's opinion, sufficient to relegate the phase-change hypothesis as a general explanation of the Moho to a state of extremely low probability.

*b. An Ultramafic Upper Mantle*

The hypothesis that the upper mantle is composed dominantly of rocks of the dunite-peridotite family is probably supported by most geologists. Indeed much of the evidence is of a geological nature and can only be fully appreciated by adequate references to the voluminous literature on ultramafic rocks and, ideally, some familiarity with their occurrence in the field.

*i. The Occurrence and Significance of Alpine Peridotites.* Benson [1962], Hess [1939, 1955a, b], and Thayer [1960] have extensively discussed the relationships of this class of rocks. Alpine peridotites are characteristically intruded close to the axes of maximum deformation along mountain building belts and island arcs. Intrusion has frequently been controlled by major faults which may extend for hundreds of miles and which almost certainly extend into the mantle. Ultramafic bodies may occur in the form of innumerable separated bodies along these fault zones, as in the Appalachians and the Great Serpentine Belt of New South Wales [Turner and Verhoogen, 1960]. Elsewhere, as in New Caledonia, Cuba, the Philippines, New Guinea, Newfoundland, and British Columbia, ultramafics occur in the form of large individual intrusions, covering hundreds of square miles and closely connected with major tectonic features. An example of the latter class is the Nahalin Peridotite in British Columbia. This is an elongate body 100 miles long and about 5 miles wide, closely associated with a major fault along its principal axis [MacGregor, personal communication].

It is believed by many geologists that alpine peridotites of this type are derived directly from the upper mantle and are representative of the rocks occurring in that part of the earth. This hypothesis is supported by their tectonic setting and by the evidence of extreme solid-state deformation which they often display. It is also supported by their physical properties. They yield, when fresh, the required mantle seismic velocities and possess a density close to 3.32 gm/cc.

*ii. Inclusions in Kimberlite Pipes.* These have been described by Wagner [1914], Williams [1932], Dawson [1962], and Nixon *et al.* [1963] and their geologic significance discussed. A classic paper dealing with the interpretation and significance of kimberlite inclusions was published by Wagner [1928]. Kimberlite pipes carrying diamonds are of frequent occurrence over 1,000,000 square miles of northern Africa, and are also known in India, Brazil, the United States, Siberia, and Australia. In Africa, these pipes carry numerous xenoliths of crustal rocks which they are known to have intruded on their journey upward. They contain, also, large numbers of xenoliths of rocks which are not known to occur in the vicinity, particularly peridotites, pyroxenites, and eclogites. The presumption is that these inclusions have been derived from deeper levels in the earth and represent a random sample of deep-lying rock types cut by the pipes. The occurrence of diamonds both in the pipes and in the inclusions implies that the pipes are derived from depths of 120 km and more. Accordingly it appears that the kimberlites have presented to us a random sample of mantle rocks over a vast area and extending down to a depth of at least 120 km.

It is of great significance that in all cases where adequate sampling has been carried out, *peridotitic inclusions are found to be much more common than eclogitic inclusions* [Wagner, 1928; Williams, 1932; Dawson, 1962; Nixon *et al.*, 1963]. If the sampling is representative, then the upper mantle must be of peridotitic composition, with eclogite a minor, but widely distributed, constituent.

The kimberlite forming the host to these inclusions is a rock of variable composition and degree of alteration. It is noteworthy that the dominant primary mineral is magnesian olivine which originally comprised between 50 and 75 per cent of kimberlites. This suggests that the region in the mantle where kimberlite originates is one where magnesian olivine is also abundant.

*iii. Further Development of an Ultramafic Model for the Upper Mantle.* The discovery [Maxwell and Revelle, 1951; Bullard, 1952, 1954] that the mean heat flow from the deep oceanic crust was approximately equal to the mean continental heat flow proved to be a milestone in our understanding of the constitution of the upper mantle. The result was surprising to those who had previously advocated an upper mantle of alpine-type peridotite, since it was known that the radioactivity of these rocks was far too low to yield the observed oceanic heat flow. Objections on these grounds to the hypothesis of an ultramafic upper mantle were made by Lovering [1958], Kennedy [1959], and Tilton and Reed [1963]. It was also pointed out that alpine peridotites were characteristically deficient in many other elements, e.g., Na, Ca, Al, and it was difficult to see how they might yield a basaltic magma when fractionally melted.

A way out of this dilemma was suggested by two parallel developments. Rubey [1951, 1955] produced a series of powerful arguments showing that the earth's atmosphere, hydrosphere, and crust had been formed gradually, over geological time, by degassing and fractional melting of the upper mantle. This implied that beneath continents there must exist a zone which has been deprived of its low-melting and volatile components. Might not the alpine peridotites be representative of this barren zone? If so, then beneath the residual, refractory ultramafic zone there should exist a more primitive material, which could yield crustal material on partial fusion, leaving behind refractory dunite and peridotite. Bullard's [1952] interpretation of the oceanic heat flow data complemented this hypothesis rather satisfactorily. He proposed that, to a first approximation, the mean chemical composition of the mantle was the same both over ocean and continent. The principal difference was that continental regions had become more strongly differentiated, and the low melting point components and radioactive elements were concentrated near the surface. This general model of continental evolution is of ancient lineage. Nevertheless, the reasons for accepting it have only become compelling within the last decade or so. An excellent account of geological aspects of the model has been given by Wilson [1954] and Engel [1963]. The model has also received strong support from studies of the development of strontium 87 in the crust and mantle—particularly by Hurley and coworkers at the Massachusetts Institute of Technology [Hurley *et al.*, 1962]. Further geochemical-geophysical consequences have been explored by Ringwood [1962*b*, *c*], MacDonald [1963], and Clark and Ringwood [1964].

The second development affecting the status of ultramafic models for the upper mantle has arisen from petrology and can be traced back to Bowen [1928], who maintained that the primary composition of the upper mantle was that of a felspathic peridotite and that basalt magmas were formed by fractional melting of this primary material, leaving behind a "barren" residuum. (According to previous discussion we would now identify this "barren" residuum with alpine peridotites.)

Bowen's lead was not generally pursued by succeeding geologists who were more interested in the fractional crystallization of basalt magma than in its ultimate origin. Many geologists believed that alpine peridotites were early crystallizing products of basalt magmas. This view was challenged by Hess [1955b], who pointed out significant differences in the mineralogy of alpine peridotites compared to peridotites which had clearly formed from basaltic magma in layered complexes. An important paper was published by Ross, Foster, and Myers [1954], who demonstrated an extremely close relationship between the mineralogies of alpine peridotites and ultramafic inclusions in basalts. The relationship argued rather convincingly for a common origin. The composition of olivines, pyroxenes, and spinels in these assemblages also supported Hess's claim that they were not derivatives of basalt magma, but were, on the contrary, more primitive. They concluded, with Hess, that alpine peridotites and ultramafic inclusions were samples of the mantle.

In the 1950's, petrologists displayed more interest in the origin of basalt, and many papers appeared, reviving Bowen's hypothesis that basalts were formed by fractional melting of peridotite in the mantle, e.g., Verhoogen [1954, 1956], Powers [1955], Kuno [1957], Wager [1958], Wilshire and Binns [1961], Yoder and Tilley [1962], and Kushiro and Kuno [1963]. These authors were somewhat vague about the nature of the parental peridotite, usually identifying it with types of alpine peridotites and selected basalt nodules richer than usual in $Na_2O$, $Al_2O_3$, and $CaO$. Although a step in the right direction, this assumption is not entirely satisfactory, since such rocks do not appear to have the correct trace element chemistry to yield basalt magmas on fractional melting, even if the major element partitions are satisfactory.

Ringwood [1958, 1962a, b], Green and Ringwood [1963], and Clark and Ringwood [1964] sought to follow more closely the logical consequences of the earlier hypothesis of Bowen [1928], Rubey [1951], and Bullard [1952]. It was found necessary to postulate a primitive, parental mantle material, which was arbitrarily defined by the property that on fractional melting it would yield a typical basaltic magma and leave behind a residual refractory dunite-peridotite of alpine type. The composition of this primitive material must lie between those of basalt and dunite-peridotite. Such a primitive material would be unlikely to contain less $Al_2O_3$ and $CaO$ than are present (relatively to $MgO$ and $SiO_2$) in chondrites (Section 1). This would indicate a limit of about 1 basalt to 3 of dunite-peridotite. On the other hand, the ratio is hardly likely to be less than 1 basalt to 1 of dunite-peridotite if basalt is to be produced by *partial* melting in the mantle [Hess, 1960]. Such a hypothetical primitive material was called "pyrolite"-pyroxene-olivine rock. While there are often valid objections to inventing new names for hypothetical materials, in this case the author believes a new term was a necessity. The composition and properties of pyrolite do *not* match those of natural peridotites, particularly in trace element chemistry. Furthermore, the assumption that a rock with the properties of pyrolite exists is necessitated by the now widely recognized complementary differentiation relationships of peridotite and basalt and also by the need for a primitive material which can explain the oceanic heat flow. As has been mentioned, alpine peridotites cannot provide this.

Green and Ringwood [1963, and in progress] have studied primary ultramafic rocks in order to find how closely the hypothetical pyrolite composition is approached. Starting with a "typical" alpine ultramafic containing virtually no Ca, Al, and Na and composed of olivine and orthopyroxene, one can find complete continuum of compositions over to rocks containing about 4 per cent $Al_2O_3$, 3 per cent CaO, and 0.4 per cent $Na_2O$. Such rocks are found as garnet peridotites in diamond pipes, as inclusions in basalt (rather rarer than the published analyses based upon selected samples suggest), and as high-temperature peridotites [Green, 1964]. Rocks with this composition lie very close to a 3:1 peridotite/basalt composition, as given by chondrites (Table 4). Nevertheless, even in such rocks, the content of minor elements, particularly $Na_2O$, $K_2O$, $TiO_2$, $P_2O_5$, and probably Ba, Th, and U, are too low for them to be regarded as true pyrolite. Another interesting fact is that representative ultramafic rocks of the type under consideration, possessing compositions corresponding to peridotite/basalt ratios smaller than 3:1 (based on major elements) are very rare. There seems to be some sort of discontinuity in natural occurrences at the 3:1 ratio. It appears that nearly, if not all, ultramafic rocks exposed at the surface of the earth have been subjected to some degree of fractionation. It is important to note, however, that the degree of fractionation varies over a very wide range. It is most interesting, and it may be significant, that high-temperature peridotites, such as the Lizard [Green, 1964] and Tinaquillo [Mackenzie, 1960; Green, 1963] are among those closest to the ideal pyrolite composition and are accordingly least fractionated.

Is the apparent absence of ideal pyrolite at the earth's surface an objection to the concept? The "deficient" elements (K, Ti, P, Ba, Sr, Th, U, and others) are unable to enter the major phases of the mantle because of their ionic charges and sizes (Section 5) and accordingly are fractionated rather readily [Harris, 1957]. It appears that they may be strongly concentrated in the liquid phase, which forms at a comparatively low temperature and at a very early stage of fractional melting in the mantle when the amount of liquid formed is small. This would account for the high mobility of these elements under mantle conditions which seem necessary to account for the formation of kimberlites.

The ultramafic rocks accessible for sampling by the geologist at the earth's surface represent a very specialized class. To be raised from the mantle so that they outcrop at the surface of the earth, exceptional conditions including intense tectonic deformation and, often, elevated temperatures are required. Under such conditions, fractionation may be expected. Also, according to previous discussion, the mantle beneath continents is believed to have been strongly fractionated, and this characteristic is likely to be reflected in ultramafics derived from this region. In oceanic areas, ultramafics are found as inclusions in basalts and as intrusions in island arcs, which are dominantly composed of volcanic rocks. It is highly probable that the mantle underlying such centers of volcanic activity has been fractionated. It is only in the deep oceanic basins, away from islands, and hence presently inaccessible to the geologist, that ideal pyrolite may be expected to occur close to the Moho. Thus, in the author's opinion, the differentiated character of the ultramafics which are observed in the crust is to be regarded as a necessary consequence of their presence in the crust.

Thayer [1960] has recently examined some of the larger terrestrial ultramafic intrusions. He found in all of these that they do not consist of pure peridotite, but rather of mixed peridotite and gabbro. The two rocks clearly have a common origin. The most straightforward and in many ways the most satisfactory explanation of this association is to regard them as complementary differentiates of primary mantle material. If their original net composition were known, it would yield the pyrolite composition. From the descriptions, an average ratio of 3 peridotite to 1 gabbro might not be far out. The ratio could perhaps be 2:1, but it does not appear likely to be as low as 1:1. Henceforth we will assume that the composition of pyrolite is given by 1 basalt to 3 dunite-peridotite. The composition so obtained will be adopted as a model so that its properties can be explored.

The logical development of the hypotheses of Rubey [1951] and Bullard [1952], as discussed in this section, leads to a chemically zoned upper mantle [Ringwood, 1958], as depicted in Figure 1. The depth of the refractory dunite-peridotite zone underneath continents

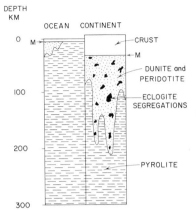

Fig. 1. Chemically zoned model for the upper mantle.

would vary, the maximum thickness, perhaps around 150 to 200 km, being reached beneath Precambrian shields. Beneath deep oceanic basins, on the other hand, the primitive pyrolite may extend to the Moho. Alternatively a thin layer of dunite-peridotite a few tens of kilometers thick may be present.

### c. *Derivation of a Model Pyrolite Composition*

The following procedures differ slightly from that adopted by Green and Ringwood [1963]. These authors used the composition of dunite for the refractory component of pyrolite. The author now believes it preferable to use typical alpine peridotite composition rather than dunite. Because of the mineralogical heterogeneity of alpine peridotites, individual chemical analyses of samples are unlikely to yield compositions corresponding to the bulk composition of the intrusion. Accordingly, a "synthetic" alpine peridotite was constructed, consisting of olivine 80 per cent, orthopyroxene 19.3 per cent [Thayer, 1960], and chrome-spinel 0.7 per cent. Individual compositions of these minerals are taken from analyses by Ross, Foster, and Myer [1954] of typical olivine (Table 4, No. 14), enstatite (Table 5, No. 16), and spinel (Table 7, No. 12) from alpine peridotites. References are to analyses tabulated in Ross, Foster, and Myers' paper. The composition of the basalt used was that of a typical primitive Hawaiian olivine tholeiite described by MacDonald and Katsura [1961, Table 1, Column 1].

It is seen from Table 5 that the model pyrolite composition so derived is a close match to the over-all composition of the mantle obtained from the chondrite model. Of course it should be remembered that the particular ratio of basalt to peridotite of 3:1 was adopted with

## Table 5
### Derivation of Pyrolite Composition

|  | I<br>Alpine Peridotite | II<br>Hawaiian Olivine Tholeiite | III<br>Pyrolite (Peridotite, 3 Basalt, I) | IV<br>Mantle Composition Derived from Chondrite Table 4 Column 4 |
|---|---|---|---|---|
| $SiO_2$ | 43.98 | 48.72 | 45.16 | 43.25 |
| $Al_2O_3$ | 0.25 | 13.40 | 3.54 | 3.90 |
| $Fe_2O_3$ | 0.04 | 1.70 | 0.46 | } 9.25 |
| $FeO$ | 7.42 | 9.88 | 8.04 |  |
| $TiO_2$ | 0.03 | 2.77 | 0.71 |  |
| $Cr_2O_3$ | 0.57 | — | 0.43 |  |
| $CaO$ | 0.33 | 11.30 | 3.08 | 3.72 |
| $MgO$ | 46.96 | 8.98 | 37.47 | 38.10 |
| $Na_2O$ | 0.01 | 2.23 | 0.57 | 1.78 |
| $K_2O$ | — | 0.58 | 0.13 |  |
| $MnO$ | 0.14 | 0.18 | 0.14 |  |
| $CoO$ | 0.01 | — | 0.01 |  |
| $NiO$ | 0.26 | — | 0.20 |  |
| $P_2O_5$ | — | 0.24 | 0.06 |  |
|  | 100.00 | 100.00 | 100.00 | 100.00 |

the chondrite composition in mind. Nevertheless, there were independent reasons for this choice. Had ratios of 2:1 or 4:1 been chosen, the correspondence between the models would still have been strong.

## 5. Differentiation of the Mantle

### a. Lack of Strong Fractionation of Major Components

The close correspondence between the model pyrolite composition for the *upper mantle* with the composition for the *entire mantle* obtained from chondrites (Table 5) suggests that fractionation of major rock-forming elements within the mantle has been surprisingly small. When large basic intrusions cool very slowly, they usually display strong differentiation due to crystal fractionation. If a body the size of the mantle had melted and subsequently cooled under conditions permitting crystal fractionation, it would be expected that strong differentiation would have occurred resulting in an upper mantle composition which was very different from that of the lower mantle.

As previously noted, this does not appear to have occurred. Indeed, petrologists have previously commented that there is much less sialic rock material near the surface than might be expected if the whole earth had melted and differentiated. The absence of strong crystal fractionation for this group of elements is demonstrated by some examples.

*Iron.* A characteristic feature of the crystallization of magmas and artificial melts is the increase in FeO/MgO ratio as fractionation progresses. If fractional crystallization had affected the whole mantle, a strong enrichment of iron toward the surface would be expected. However, the FeO/(FeO + MgO) ratio of pyrolite is only about 0.12, and Fe/Mg fractionation has apparently been insignificant. It might be suggested that there may be practically no FeO in the lower mantle and that, therefore, the low FeO/(FeO + MgO) ratio near the surface could represent a relative enrichment. However, the occurrence of oxidized nickel and other siderophile elements (see Section 5e) in appreciable quantities in the mantle shows that such complete reduction of FeO is unlikely. Nickel is more noble than iron. Furthermore, complete absence of FeO in the deep mantle would make it difficult to explain the elastic ratio and density of this region [Clark and Ringwood, 1964].

*Nickel and Chromium.* During the crystallization of basic magmas these elements become highly concentrated in the earliest ferromagnesians to separate, leading to impoverishment and virtual elimination from residual magmas [Wager and Mitchell, 1951]. It is not known whether these tendencies would be retained throughout the whole mantle—they may well be changed by pressure. However, in the upper mantle, where olivine remains a stable phase, it appears likely that this behavior would persist. If the upper mantle had been subjected to normal crystal fractionation, very little Ni or Cr should be found in the higher levels. Yet we find that alpine peridotites, believed to be derived from these regions, carry substantial quantities of nickel and chromium. Considering the amounts which have entered the core, (Section 5e), the upper mantle does not appear to be significantly depleted in these elements.

The significance of the preceding observations will be discussed in the next section, after consideration of the behavior of another group of elements which, to the contrary, display strong fractionation.

*b. Fractionation of Some Nonvolatile, Oxyphile Trace Elements*

In Table 3 the elements of which the earth is composed were divided into two groups—a nonvolatile group and a volatile group.

The former can be divided into two chief groups—oxyphile elements and transition (siderophile) elements. We will now consider the behavior of some oxyphile trace elements—U, Th, Ta, Ba, Sr, Zr, Hf, and the rare earths.

For many years it has been known from geothermal arguments that the radioactive elements have been strongly concentrated in the earth's crust. Assuming that the earth had the same uranium concentration as chondrites, Birch [1958] showed that a large proportion of the total uranium in the earth is present in the crust. Since the crust amounts to less than 1 per cent of the mass of the mantle, it appeared that the latter had been subjected to a remarkably efficient differentiation process. Gast [1960] extended Birch's calculations and estimated that 62 per cent of the earth's uranium, 49 per cent of the barium, and 16 per cent of the strontium were in the crust. More recently, Taylor [1964a, b] has repeated and extended the analysis to many other elements and showed that Ta, Zr, Hf, and many of the rare earths were similarly concentrated in the crust.

It is possible that the estimates of crustal abundances used by these workers were too high, and in some cases the abundances in chondrites have been revised upwards. For example, it appears from the results of Reed et al., [1960], Goles and Anders [1962], and Lovering and Morgan [1964] that the mean abundance of uranium in Types II and III carbonaceous chondrites is about twice as high as the normal chondritic abundance used by Birch, Gast, and Taylor. Nevertheless, allowing for such uncertainties, their major conclusions are not likely to be fundamentally changed—namely, that all or most of the mantle has been subjected to a remarkably efficient differentiation process, resulting in the intense upward differentiation and concentration in the crust of the elements under discussion.

We have previously adopted the view that the crust was derived from the upper mantle by fractional melting. Accordingly the differentiation has probably proceeded in two steps. First, the trace elements were concentrated from the whole mantle into the upper mantle, forming a protocontinental layer [Patterson and Tatsumoto, 1964]; subsequently, continents developed by differentiation of the protocontinental layer. In the present paper, this protocontinental layer is identified with pyrolite. Strong upward concentration of heat-producing elements is required in the pyrolite of the suboceanic upper mantle on geothermal grounds. It is also required if the common types of basalt are to be produced in this region by plausible fractional melting processes [Clark and Ringwood, 1964].

The strong fractionation of U, Th, Ba, Sr, Ta, Zr, Hf, and the rare earths is clearly connected with their crystal chemical properties, as was recognized many years ago by Goldschmidt. The radii and/or charges on these ions are larger than those of the major elements (Figure 2) which determine the principal phases in the mantle ($Si^4$, $Mg^2$, $Fe^2$, $Al^3$, $Ca^2$, $Na^1$, $Cr^3$), and it appears that they are unable to enter the principal minerals of the mantle [Ringwood, 1960]. It is probable that the difficulty of entry into major phases would be increased in the lower mantle, where the silicates are in a state of close packing. Taylor [1964b] has given a convincing discussion and interpretation of the data (Figure 2), which demonstrate well the effect of ionic size and charge in determining the degree of crustal enrichment.

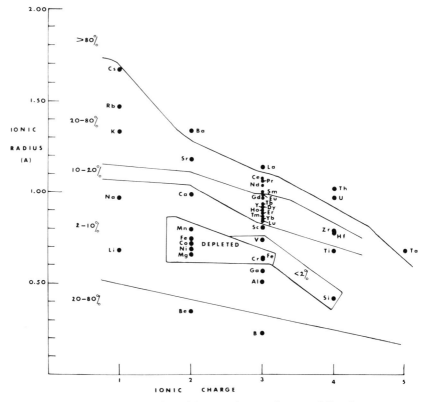

FIG. 2. Estimated crustal enrichment factors for oxyphile elements as a function of ionic radius and charge. Percentages refer to

$$\frac{\text{Estimated mass of element in crust}}{\text{Estimated total mass of element in earth}}$$

Diagram after Taylor [1964a].

The rare-earth elements demonstrate the dependence of fractionation upon ionic radius to a remarkable degree. Figure 3 shows the relative rare-earth abundances in 7 sediments [Hasking and Gehl, 1962] which are probably representative of the relative crustal abundances compared to chondrites [Schmitt et al., 1963a, b, 1964]. There is a net enrichment of the entire rare-earth group in the crust. In addition there is a further fractionation which is a regular function of ionic

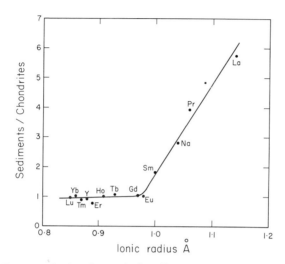

FIG. 3. $\frac{\text{Rare-earth abundances in 7 sediments}}{\text{Rare-earth abundances in chondrites}}$ as a function of rare-earth ionic radius. Data from Haskin and Gehl [1962].

radius. It is seen from Figure 3 that there is no relative fractionation of rare earths possessing radii smaller than that of $Ca^2(0.99 Å)$. However, the relative fractionation increases regularly with radius from 1.0 to 1.14 Å. It seems clear that fractionation has been controlled by substitution for the $Ca^2$ ion, which is the largest formed by the common rock-forming elements.

Schmitt has shown that the fractionation patterns displayed by Kilauea basalt and by a diamond-pipe peridotite generally resemble those of the crustal sediments, showing that the fractionation is characteristic of some regions of the upper mantle. It is interesting that Frey and Haskin [1964] did not find strong fractionation of rare earths in some oceanic basalts. They suggest that these basalts may have been derived from rather deep in the mantle. The discovery is clearly of considerable interest and significance.

*c. Causes and Significance of Fractionation Pattern of Nonvolatile Oxyphile Elements*

The only physically reasonable explanation known to the author which is capable of explaining the strong upward differentiation of U, Th, Ba, Sr, Ta, Zr, Hf, and the rare earths is a process involving crystal-liquid equilibria. This explanation implies that all or most of the mantle has been subjected to complete or partial melting and that the trace elements which could not enter the major crystalline phases (incompatible elements) became strongly concentrated in residual melts, which, in turn, migrated to the upper levels of the mantle. Masuda and Matsui [1963] analyzed the fractionation of the rare earths quantitatively and showed that their behavior is explained very completely by crystal-liquid equilibria.

We must now inquire how it was that the incompatible trace elements were so strongly fractionated whereas the major rock-forming elements and ions with charges and sizes close to them display a negligible degree of fractionation (Section 5*a*). Two possible explanations have been suggested.

*1.* Wager [1958] and Ringwood [1960] have suggested that convection occurred in the crystal mush during primary crystallization of the mantle from a molten state. This would mix up and homogenize the major phases while the small quantity of residual liquid containing ions unable to enter the major phases would be successively squeezed upward, leading to strong concentration of incompatible trace elements near the surface. Convection in the mush might be driven by the normal tendency for FeO/MgO fractionation. The earliest phases to crystallize near the base of the mantle are enriched in magnesium, and these become overlaid by denser crystals increasingly rich in iron. The unstable density distribution thus generated may well lead to repeated convective overturn and mixing of solid phases. The small amount of residual liquid greatly decreases the effective viscosity and enhances convective overturn in the mush. A second tendency encouraging convection would be the initial melting point gradient which is much higher than the adiabatic. A few convective overturns in the crystal mush during crystallization would cause rapid cooling so that the initial temperature distribution in a mantle formed in this manner would not depart far from the adiabatic temperature gradient.

*2.* A second method of explaining the contrasting differentiation of major elements and incompatible elements is zone melting, as originally suggested by Harris [1957] and further developed by Vinogradov [1961]. As described by its advocates, zone melting is capable of causing extreme fractionation of incompatible elements, while not affecting the composition of the major phases. Another property is that although the entire mantle must pass through a fractional melting stage, only a relatively small portion need be molten at any one time. The zone melting process commences when the temperature at the core-mantle boundary becomes high enough to cause partial melting. The liquid thus formed moves upward, solid phases dissolve at the top and precipitate at the bottom, while incompatible ions are strongly concentrated in the liquid. It is seen that the zone melting process requires an initial temperature distribution very close to the minimum melting point curve for the entire mantle and that after completion of the process, the mantle will retain this temperature distribution.

*d. Lateral Segregation in the Mantle*

The model for the upper mantle previously proposed was based upon the assumption that, to a first approximation, the mean chemical composition of continental and oceanic sectors was the same when averaged to depths of a few hundred kilometers. If this assumption were strictly correct, it becomes difficult to understand why continents have evolved in some regions and not in others. The possibility should be entertained that the assumption is not strictly correct and that there exist small but significant differences in mean chemical composition between oceanic and continental sectors. Such differences are more likely to involve the concentrations of fractionated trace elements than of the major elements.

Clark and Ringwood [1964] compared the average heat flows from Precambrian shields (26 observations, average 0.96 μcal/cm²sec), off-shield continental areas (42 observations, average 1.56 μcal/cm² sec), and ocean basins with water depths between 4 and 7 km (207 observations, average 1.22 μcal/cm²sec). The deficiency of heat flow in the shields was extensively discussed and taken to imply a net deficiency of radioactivity in the crust-mantle sector caused by the operation of the major geological cycle. The results just quoted also imply a higher mean content of radioactive elements in off-shield continental areas, compared to deep oceanic basins. Because of the much larger area of the off-shield regions compared to shields, it does not appear likely that the higher off-shield values can be quantitatively explained by incorporation into reworked sediment of radioactive elements derived by erosion from the shields. The results, therefore, suggest that normal continental regions possess significantly higher net contents of radioactive elements than deep oceanic basins, when averaged down to 100 km or so.

In discussing data on the contents of radioactive elements in basalts, Clark and Ringwood [1964, p. 47] noticed the possibility that oceanic tholeiites may contain less uranium than continental tholeiites. Engel and Engel [1964] have since recognized tholeiites as the most abundant type of volcanic in oceanic regions. They also pointed out the close chemical resemblances of a large group of widely separated samples. They were all characterized by unusually low contents of $K_2O$, $TiO_2$, and $P_2O_5$ compared to continental tholeiites. If U and Th follow K, as is well established [Heier and Rogers, 1963], the suspicion that oceanic areas are depleted in radioactivity compared to off-shield continental areas is strengthened.

In the context of previous discussion, it might be suggested that rather small but significant differences in mean radioactivity of the protocontinental pyrolite have determined the locations of continents and oceans. These differences might have arisen from the proposed convective overturns in the crystal mush during primary crystallization of the mantle, leading to a relative concentration of readily fractionated trace elements in regions determined by the convective pattern. Vening Meinesz [1947] and Wager [1958] have previously suggested that the location of primitive continental nuclei may have been governed by such an early system of currents.

*e. Fractionation of Nonvolatile Siderophile Elements*

The siderophile elements are those which become preferentially partitioned in the metal phase when molten iron and a complex silicate melt of chondritic composition are equilibrated. Generally speaking, the free energies of formation of the oxides of siderophile elements are comparable or smaller in magnitude than that of ferrous oxide. Such elements are spoken of as being more "noble" than iron. Most possess very low volatility, either as metals or oxides, so that they are likely to be present in the earth in their primordial abundances, according to the discussion in Section 1. The siderophile elements are listed in Table 2. They include Cu and Ag. These are usually classed as chalcophile, but in the absence of sulfur (Section 5*b*) they are siderophile.

In Table 6 we have compared the abundances of the siderophile elements estimated to be present in pyrolite with their primordial abundances (all values normalized to $Si = 10^6$ atom). This comparison was possible only in a few cases, since analytical data on this group are scanty.

## Table 6

COMPARISON OF ABUNDANCES OF SOME SIDEROPHILE ELEMENTS IN "PYROLITE" WITH THEIR CHONDRITIC ABUNDANCES, ASSUMED TO BE PRIMORDIAL

| Element | Chondritic Abundance (per $10^6$ Si atoms) | Pyrolite Abundance (per $10^6$ Si atoms) | Refs. | Ratio $\frac{\text{Pyrolite Abundance}}{\text{Chondrite Abundance}}$ |
|---|---|---|---|---|
| Co | 2300 | 280 | 1 | 0.12 |
| Cu | 490 | 64 | 1, 11 | 0.13 |
| Ni | 46,000 | 3573 | Table 5 | 0.08 |
| Pt | 0.9 | 0.014 | 2, 6, 7, 8, 9, 10 | 0.015 |
| Au | 0.13 | 0.004 | 3, 4, 5 | 0.03 |

Chondritic abundances were obtained from Urey [1964].
Pyrolite abundances calculated on basis of 1 basalt:3 peridotite.
1. Turekian, K., and K. Wedepohl, *Geol. Soc. Am. Bull*, 72, 175–192, 1961.
2. Wright, T. L., and M. Fleischer, *Geochemistry of the Platinum Metals*, Unpublished Report, U.S. Geol. Survey.
3. Vincent, E. A., and A. A. Smales, *Geochim. Cosmochim. Acta*, 9, 154, 1956.
4. Vincent, E. A., and J. H. Crocket, *Geochim. Cosmochim. Acta*, 18, 130, 1960.
5. Vincent, E. A., and J. H. Crocket, *Geochim. Cosmochim. Acta*, 18, 143, 1960.
6. Hagen, J. C., Unpublished Ph.D. thesis, M.I.T., 1964.
7. Hahn-Weinheimer, P., and F. Rost, *Geochim. Cosmochim. Acta*, 21, 165–181, 1961.
8. Stockley, G. M., Tanganyika Territory, *Geol. Div. Bull.*, 18, 60 pp., 1953.
9. Neiva, J. M., Cotelo, *Estud. Notas Trabal. Serv. Formento Mineiro Port.*, vol. 2, 1–19, 1946.
10. Ivanov, A., and N. V. Lizunov, *Bull. Acad. Sci. USSR, Ser. Geol.*, No. 5, 78–86, 1944.
11. Schmitt, R. A., and R. H. Smith, *Quarterly progress report for period ending No. 30, 1963*, General Atomic N.A.S.A. Contract N.A.S.W.-843.

It is seen from Table 6 that all of these elements are strongly depleted in the upper mantle (pyrolite) compared to their primordial abundances. It may confidently be assumed that they entered the metal phase and have been carried into the core. It is, however, somewhat surprising that the depletions of these elements in the upper mantle are not stronger than observed. The ionic radii and charges of Ni, Co, and Cu fall within the range between $Mg^2$, $Fe^2$, and $Ca^2$. Accordingly it is possible for these elements to enter into solid solution in the principal phases of the mantle. From the discussion of Sections 5a and 5b, it would not be expected that they should be strongly fractionated upward in the mantle in the same way as the trace elements which are unable to be accepted by the major phases. Indeed, it is well known that nickel tends to become concentrated in the earliest crystals to separate from the magma. Its abundance is, therefore, more likely to *increase* with depth. There are grounds, therefore, for suspecting that the pyrolite abundance may represent a lower limit to the abundance for the entire mantle.

When the core separated from the mantle, it appears that about 20 per cent of the total nickel was left in the mantle. This is inexplicable if the formation of the core was an equilibrium process. Fredriksson [personal communication] has found that the nickel content of olivines coexisting with the metal phase ($\sim$ 12 per cent Ni) of ordinary chondrites is about 0.01 per cent. Lovering [1953] found an average of 0.025 per cent of nickel in the olivines from pallasites. Considering the smaller proportion of nickel probably present in the earth's core, it is evident that the nickel abundance in the upper mantle is at least an order of magnitude higher than would be expected if the mantle had maintained equilibrium with the metal phase during segregation of the earth's core. There is a possibility that the discrepancy is due to a pressure effect [Ringwood, 1959], but the fact that it is displayed by the other members of Table 6 makes this explanation somewhat contrived.

The behavior of cobalt is closely analogous to nickel. Lovering *et al.* [1957] and Lovering [1957] found that cobalt was strongly siderophile in pallasites. The ratio of cobalt in the metal phase to cobalt in the olivine phase was approximately 200. On this basis the abundance in the upper mantle is more than an order of magnitude higher than would be obtained for an equilibrium metal-silicate distribution. These references also reveal that copper behaves similarly to cobalt and nickel in the absence of sulfur.

The geochemistry of gold and platinum is also interesting in this respect. The oxides of these elements are unstable and decompose to the native element at low temperatures. Vincent and Smales [1956] found that gold did not display significant fractionation between the various silicate and oxide phases of the Skaergaard and concluded that it occurred as free metallic atoms in the magma. Accordingly, it is unlikely that gold has been significantly enriched in the upper mantle by the major fractionation discussed previously. The geochemistry of platinum is not well known. However, it is known to be most strongly concentrated in ultramafic rocks. It is not likely that its abundance will decrease with increasing depth in the mantle. Referring to Table 6 and taking a mantle-to-core ratio 7 to 3, we see that on the average about 4 to 7 per cent of the total platinum and gold have remained in the mantle.

The partition coefficients for these elements between iron and silicate melts have not been determined. It appears certain, however, that under equilibrium conditions they will be concentrated in the metal phase by an enormous factor. The partitition coefficients in favor of the metal phase would greatly exceed that of nickel. Even if the estimates of mantle abundances of these elements are incorrect by a factor of 10, it appears that the earth's core was not in chemical equilibrium with the mantle when it segregated.

This conclusion has a bearing on the manner by which the core formed. Two rather different hypotheses have been proposed. One holds that the primitive material from which the earth formed was subjected to strong heating, reduction, and loss of volatiles *before* the earth accreted. According to this widely held view, the material from which the earth formed may have consisted of a rather intimate mixture of silicate and metal phases, rather analogous, but not necessarily chemically identical to the chondrites. If this view were correct, we would expect that the partition of siderophile elements between the silicate and metal phase should have approached equilibrium because of the close association of the two phases and their previous high-temperature history. If the material had melted after accretion, the mantle should contain virtually no Ni, Co, Cu, Au, and Pt. The observations previously discussed show that these expectations are grossly violated and the earth is therefore unlikely to have formed from any material of high-temperature origin consisting of intimately mixed silicate and metal.

An alternative view of the formation of the core was suggested by Ringwood [1959, 1960, 1961$b$]. A fundamental aspect of this theory was that the core did *not* form in equilibrium with the mantle. The occurrence of siderophile elements in the mantle appears to support this view.

The atomic abundances of Pd, Pt, Os, and Ru in chondrites are rather similar (1, 0.9, 0.6, and 1.6 atoms/$10^6$ Si, respectively). It is intriguing that the abundances of osmium and ruthenium compared to platinum and palladium in the crust and upper mantle appear to be depleted by a substantial factor [Wright and Fleischer, unpublished summary of results of several authors]. For example, the palladium/osmium ratio of basalt W1 is about 50 times higher than the chondrite value [Bate and Huizenga, 1963; Vincent and Smales, 1956]. Osmium and ruthenium both form oxides which are stable at elevated temperatures and are also readily volatile. Bate and Huizenga noticed that some of the osmium and ruthenium in chondrites was definitely associated with the silicate phase. This is to be attributed to their oxyphile properties. The fact that these elements enter the silicate phase makes it surprising that they have been relatively depleted in the upper mantle more than Pt and Pd, which do not enter the silicates. There is a possibility that osmium and ruthenium have been fractionated by volatilization during or before the earth's formation and should be classed with the volatile elements.

A final note on manganese, chromium, and vanadium. These elements are somewhat deficient in pyrolite compared to chondritic abundances. When normalized to silicon on an atomic abundance basis, pyrolite appears to contain about one-quarter as much chromium and vanadium and one-half as much manganese as chondrites. The missing complement may well be in the core. We have suggested that the core was formed under reducing conditions strong enough to cause entry of some elemental silicon. Since the oxides of Mn, Cr, and V are more readily reduced to metals than $SiO_2$, it would be expected that substantial proportions of these elements would enter the core.

## 6. Fractionation of Volatile Elements in the Earth

In Section 2 the elements were somewhat arbitrarily divided into two groups—nonvolatile and volatile (Table 3). It was suggested that the earth probably retained the primordial abundances of the former, but not necessarily of the latter, which may be depleted. The concentrations of volatile elements in the upper mantle and crust will, therefore, be mainly controlled by two factors: (*1*) the proportion which was retained by the earth; (*2*) the behavior of the elements in the earth during the inferred major differentiation. Of those volatile elements which are mainly present in the mantle, we may expect that their fractionation will be governed by ionic radii and charges in a similar manner to the nonvolatile elements (Section 5). Strong upward enrichment would be predicted for all ions with radii greater than 1.0 Å. On the other hand, monovalent and divalent ions possessing radii smaller than 1.0 Å could not be expected to fractionate strongly upward, since they are capable of substituting for the common elements in the principal phases of the mantle.

The relative atomic abundances of some volatile elements in pyrolite and in Type I carbonaceous chondrites are compared in Table 7. The table is incomplete, owing to lack of abundance data for many elements.

### a. The Alkali Metals

The ionic radii of these elements are as follows: ($Li^1$ 0.68 Å; $Na^1$

0.97 Å; K$^1$ 1.33 Å; Rb 1.47 Å; Cs 1.67 Å). The radii of potassium, rubidium, and cesium are too large to permit them to enter the principal phases in the mantle; hence we would expect them to be concentrated in the crust and upper mantle to a relative extent

### Table 7

ESTIMATED ATOMIC ABUNDANCES OF SOME VOLATILE ELEMENTS IN PYROLITE COMPARED TO THE PRIMORDIAL ABUNDANCES AS GIVEN BY THE TYPE I CARBONACEOUS CHONDRITES

| Element | Ionic Radius Å$^1$ | Atomic Abundance in Type I Carbonaceous Chondrites (per 10$^6$ Si) | Ref. | Atomic Abundance in Pyrolite (per 10$^6$ Si) | Ref. | Relative Abundance* |
|---|---|---|---|---|---|---|
| S$^{--}$ | 1.84$^2$ | 505,000 | 4 | 1250 | 7 | 0.0025 |
| Zn$^{++}$ | 0.74 | 932 | 5 | 130 | 8, 9, | 0.14 |
| Cd$^{++}$ | 0.97 | 2.4 | 4 | 0.1† | 10, 11, 15 | 0.04 |
| Hg$^{++}$ | 1.10 | 11.5 | 4 | 0.03† | 8 | 0.003 |
| Ge$^{++++}$ | 0.48$^3$ | 135 | 5 | 1.8 | 12, 13, 14 | 0.013 |
| Tl$^+$ | 1.47 | 0.18 | 4 | 0.07† | 8 | 0.4 |
| Pb$^{++}$ | 1.20 | 1.6 | 4 | 1.3 | 8 | 0.8 |
| F$^-$ | 1.36$^3$ | 3940 (av) | 4, 6 | 1230 | 8 | 0.3 |
| Cl$^-$ | 1.81$^2$ | 2200 | 5 | 300 | 8 | 0.14 |

\* Relative abundance = $\frac{\text{Pyrolite abundance}}{\text{C. Chondrite abundance}}$.

† Probable upper limit.

1. Ahrens, L. H., *Geochim. Cosmochim. Acta*, 2, 168–169, 1952.
2. Stilwell, C. W., *Crystal Chemistry*, pp. 417–419, McGraw-Hill Book Company, Inc., New York, 1938.
3. Ahrens, L. H., to S. R. Taylor, personal communication, revised value.
4. Urey, H. C., *Rev. Geophys.*, 2, 1, 1964.
5. Greenland, L. J., *Geophys. Res.*, 68, 6507, 1963.
6. Reed, G. W., *Trans. Am. Geophys. Union.*, 45, 87, 1964.
7. Ricke, W., *Geochim. Cosmochim. Acta*, 21, 35, 1960.
8. Turekian, K. K., and K. H. Wedepohl, *Bull. Geol. Soc. Am.*, 72, 175, 1961.
9. Rader, J., *Geochim. Cosmochim. Acta*, 27, 695, 1963.
10. Brooke, R. R., and L. H. Ahrens, *Geochim. Cosmochim. Acta*, 23, 100, 1961.
11. Brooke, R. R., and L. H. Ahrens, *Geochim. Cosmochim. Acta*, 23, 145, 1961.
12. Wardani El, S. A., *Geochim. Cosmochim. Acta*, 13, 5, 1957.
13. Onishi, H., *Bull. Chem. Soc. Japan*, 29, 686, 1956.
14. Horman, P. K., *Geochim. Cosmochim. Acta*, 27, 861, 1963.
15. Vincent, E. A., and Bilefield, L., *Geochim. Cosmochim. Acta*, 19, 63, 1960.

comparable to barium and uranium. On the other hand, lithium and sodium are able to occupy $Mg^2$ and $Ca^2$ sites in minerals. Following previous discussion, strong upward concentration would not be expected.

Gast [1960] has made a major contribution toward the geochemistry of alkalis in the upper mantle. He demonstrated from a study of the evolution of radiogenic strontium in the upper mantle that if initial chondritic abundances were assumed, rubidium had not been concentrated upward in the mantle nearly as strongly as strontium. This was surprising, since in crustal differentiation processes the Rb/Sr ratio was known to rise rapidly with increasing fractionation. From further abundance studies, independent of the strontium isotope evidence, Gast showed that the anomalous behavior of rubidium was shared also by potassium and cesium. On the assumption of initial chondritic abundances, these elements were not nearly as strongly enriched in the upper mantle as U, Ba, and Sr. Gast also demonstrated that rubidium and potassium had become slightly fractionated relative to one another, but in a direction opposite to that expected for normal crystallization differentiation. An analogous inference can be drawn in the case of cesium, using the latest abundance data.

Two alternative interpretations of these observations were offered.

1. A large proportion of the heavy alkali metals in primordial material was not retained by the earth. Their loss may have been caused by the relatively high volatility of alkali metals and compounds compared to those of uranium, barium, and strontium.

2. During the major differentiation of the earth when U, Ba, and Sr were enriched in the upper mantle, the heavy alkalis did not fractionate and hence either became uniformly distributed in the mantle or concentrated in the deep mantle. Since such behavior is completely contrary to the known crystal chemical properties of K, Rb, and Cs, it was hypothesized that the properties of the alkali ions had been radically altered by high pressure.

Gast also drew attention to two other important consequences. Whichever of the two previous explanations was applicable, rubidium was separated from strontium more than 4.0 billion years ago and probably close to 4.5 billion years ago. Hence either a melting and differentiation process or a high-temperature volatilization process occurred at or before this time. The second important consequence was connected with the heat balance of the earth. Since potassium produces three-fifths of the heat generated in a chondrite, the loss of potassium by the earth as in explanation 1 or in the retention of most of the primordial potassium in the deep mantle would profoundly affect the subsequent thermal evolution of the earth.

Ringwood [1960, p. 225; 1962a] discussed the significance of Gast's results. He argued that explanation 2 (deep burial of potassium) was improbable on several grounds and that explanation 1 must, therefore, be invoked. He suggested a mechanism by which most of the heavy alkali metals may have become volatilized and removed from the earth during its formation.

Results of subsequent geochemical investigations on the abundances of potassium and uranium in the crust and mantle support these views [Wasserburg et al., 1964; Clark and Ringwood, 1964]. Hoyle and Fowler [1964] have advocated a mechanism identical to that suggested by Ringwood [1960, 1962] for volatilization and removal of the heavy alkalis from the earth during its formation.

Finally, the abundance of sodium in the earth is considered. It is not likely to be strongly fractionated because the ionic radius of sodium permits it to enter the $Ca^2$ lattice sites in minerals. Boyd [1964] has drawn attention to the fact that the chondritic abundance of sodium is too high to yield a satisfactory earth model. The feldspar in chondrites is a sodic oligoclase. Fractional melting of material closely related to chondrites could not, therefore, yield a basaltic magma, which is characterized by the crystallization of a feldspar comparatively poor in soda. The problem is seen in Tables 4 and 5, in which the chondritic abundance of sodium was retained when deriving the model mantle composition. Pyrolite contains only about a third of the sodium which is present in the chondrite. Accordingly, it appears that sodium has been lost from the earth in a similar degree to the heavy alkalis.

### b. Sulfur

From Table 7 it is seen that the atomic abundance of sulfur in carbonaceous chondrites is 505,000 compared to only 1300 in pyrolite. If the pyrolite value is characteristic for the whole mantle, this indicates an enormous relative depletion of sulfur in this region of the earth. The ionic radius of sulfur is 1.84 Å compared to 1.40 Å for oxygen. The sulfur ion is thus too large to readily replace oxygen in silicates. Accordingly if $S^{--}$ ions were present originally in the mantle, we would expect them to be strongly concentrated upward during the major differentiation. The value quoted for pyrolite is, therefore, likely to overestimate the sulfur abundance for the entire mantle.

It appears that the initial sulfur endowment of the earth has either entered the core or it has been lost from the earth. The available evidence points strongly towards the second alternative.

It was inferred in Section 5c that the earth has passed through a completely or partially molten stage during which strong upward differentiation of incompatible trace elements occurred. It is probable that this melting was accompanied by segregation of the earth's core (Section 7). Since sulfur is substantially soluble in silicate melts closely related in composition to the mantle [Bishop et al., 1956; Grant and Chipman, 1946; Hatch and Chipman, 1949; Ol'shanski, 1950], we must consider the partition of sulfur between the silicate melt (as $S^{--}$ ions) and the metal (as S atoms). The partition coefficients in closely related systems have been measured and are of the order of unity [Bishop et al., 1960; Grant and Chipman, 1946]. Therefore, we should expect the sulfur content of the mantle during the melting process to be approximately the same as that of the core. However, because of the inability of sulfide ions to enter crystalline silicates, the concentration of sulfur will build up during crystallization of mantle liquid until it is saturated and troilite is precipitated simultaneously with silicates. This condition of sulfur saturation would be maintained through the entire crystallization of the mantle, and pyrolite should, therefore, contain a considerable amount of sulfur (from 1 to 9 per cent, according to whether the mantle was completely or partially molten). In fact, pyrolite contains only 0.03 per cent of sulfur, which represents an enormous depletion compared to expectations based upon partitions of the primordial sulfur abundance between core and mantle. It therefore appears that the core contains only a small fraction of the primordial sulfur, which must have been lost from the earth before or during accretion.

In Section 5e the significance of the presence of substantial quantities of siderophile elements (Ni, Co, Cu, Au, Pt) in the earth's mantle was discussed. It was clear that partition of these elements between core and mantle had not reached equilibrium during segregation of the core. Yet the proportion of sulfur remaining in the upper mantle (in relation to primordial abundances) is an order of magnitude smaller than the proportion of Au and Pt and 40 times smaller than the corresponding proportions of Co and Ni in the upper mantle. (Compare Tables 6 and 7.) The siderophile tendencies of these metals are far stronger than those of sulfur. It would be inexplicable if sulfur had been more efficiently concentrated in the core than these elements. The low concentration of sulfur in pyrolite, therefore, indicates a corresponding deficiency in the core.

The behavior of sulfur in meteorites may also have some relevance to the issue. Iron meteorites contain on the average about 0.5 weight per cent sulfur. On an atomic basis the S/Fe ratio in meteoritic irons is less than 0.01 compared to 0.6 in Type I carbonaceous chondrites. It is evident that the processes by which metal separated from silicates in the meteoritic parent body did not lead to concentration of sulfur in the metal.

The conclusion that most of the cosmic complement of sulfur was not retained by the earth is of importance in considering the behavior of the so-called chalcophile elements. In the absence of a sulfur-rich core, or of a separate immiscible layer of sulfide (Goldschmidt's chalcosphere) for which there is no geophysical support and which is also precluded by the argument relating to sulfur in the core, the behavior of the chalcophile elements will be determined by their oxyphile or siderophile tendencies. For this reason copper and silver were grouped with the siderophile elements in Table 3.

### c. *Zinc, Cadmium, Germanium, and Mercury*

The radii of the ions of the above elements are $Zn^{++}$ 0.74 Å, $Cd^{++}$ 0.97 Å, $Ge^{++++}$ 0.48 Å, and $Hg^{++}$ 1.10 Å. Zinc and cadmium are able to enter the lattice sites of $Fe^{++}$ (0.74) and $Ca^{++}$ (0.99 Å) in mantle minerals, and hence strong upward fractionation in the mantle is not expected (Section 5). $Ge^{++++}$ readily substitutes for $Si^{++++}$ (0.42 Å) in silicates, and again strong upward fractionation is not expected. The lack of Ge-Si fractionation in igneous silicates is well demonstrated by the data of Onishi [1956] and Hormann [1963].

Referring to Table 7, we see that pyrolite is strongly depleted in these elements compared to the primordial abundances. Nishimura and Sandell [1964] established that zinc is oxyphile in meteorites, and this is also probably true of cadmium. It is clear that the deficiencies of zinc and cadmium cannot be attributed to their presence in the core. The most probable interpretation is that they were not retained by the earth in their primordial abundances. Schmitt *et al.* [1963*a*, *b*] have also drawn attention to the small proportions of zinc and cadmium in the earth's crust, compared to other lithophile elements, on the basis of primordial abundances.

Germanium is a siderophile element [Wardani, 1957], and the deficiency of germanium in pyrolite (Table 7) may be partly connected with this property. It is doubtful whether this can account for the entire deficiency, however. Germanium is not so strongly siderophile as cobalt or nickel, yet on an atom fraction basis it is depleted in pyrolite by a factor of about 6 compared to these elements. It is unlikely that germanium has been concentrated in the core more strongly than Co and Ni. Accordingly it appears that the earth retained only a small proportion of primordial germanium.

Table 7 also shows an enormous depletion of mercury in pyrolite compared to the primordial abundances. It is possible that the latter abundance has been overestimated by a factor of 10. Even if we make an allowance for this, mercury would be strongly depleted in pyrolite. The ionic radius of the $Hg^2$ ion is 1.10 Å, which is too large to permit its ready entry into close-packed silicates. Accordingly a marked upward enrichment of mercury in the mantle, comparable to that displayed by strontium (1.12 Å), would be expected. The pyrolite abundance quoted is, therefore, a maximum. There are no grounds to expect mercury to become strongly concentrated in the core. Accordingly it appears that the earth has retained only a small proportion of the primordial mercury.

*d. Thallium and Lead*

The data in Table 7 indicate that the lead and thallium abundances in pyrolite are only slightly depleted compared with the primordial abundances. This might suggest complete or essentially complete retention of these elements by the earth. However, it is more probable that the agreement is coincidental, being caused by the self-cancellation of two opposing factors. The radius of $Tl^+$ is 1.47 Å and that of $Pb^{++}$ is 1.20 Å. These ions are far too large to be accommodated in the principal phases of the mantle and accordingly it is to be expected that they would become fractionated strongly upward in the same general manner but not necessarily to the same extent, as U, Th, and Ba, Ta, etc. If this is correct, then the average mantle abundance of lead and thallium will be much smaller than the pyrolite value, and it would be inferred that they have not been retained by the earth in their primordial abundances. According to this interpretation, the behavior of Pb and Tl has been similar to that of Zn, Cd, Ge, and Hg, which would be expected in view of the resemblances in many chemical properties.

Loss of lead by volatilization from primordial terrestrial material must have occurred about $4.55 \times 10^9$ years ago if average modern crustal lead is to lie on the meteoritic $(Pb^{206}/Pb^{204}) - (Pb^{207}/Pb^{204})$ isochron. This points again to a high-temperature process acting during or immediately before the formation of the earth.

*e. Fluorine and Chlorine*

The relative abundances of fluorine and chlorine in pyrolite are 0.3 and 0.14. Ionic radii of $F^-$ and $Cl^-$ are 1.36 Å and 1.81 Å respectively. As is well known, $F^-$ ions readily substitute for $O^{--}$ (1.40 Å) in silicates and hence are not expected to be strongly fractionated in the mantle. The data in Table 7, therefore, suggest only a modest depletion of fluorine in the earth compared to primordial abundances.

On the other hand, the chlorine ion is far too large to enter the lattice of the dominant silicates replacing oxygen. Accordingly strong upward concentration in the mantle would be predicted so that the mean concentration of $Cl^-$ in the mantle is probably much smaller than indicated in Table 7. It follows that the earth has retained a rather small amount of the primordial chlorine.

*f. Hydrogen, Carbon, and Nitrogen*

A lower limit to the abundances of these elements in the earth is given by the amount of $H_2O$, $N_2$, and $CO_2$ in the atmosphere, hydrosphere, and in sedimentary rocks. Rubey [1955] has estimated the amount of these "excess volatiles" relative to the mass of the earth as $H_2O$—280 ppm, $CO_2$—15 ppm, and $N_2$—0.7 ppm. According to his model, the excess volatiles are liberated by degassing of the earth's interior during continent formation. Since this has occurred over about one-third of the earth, the above minimum figure just given should perhaps be multiplied by three.

Vinogradov *et al.* [1963] have determined the nitrogen contents of many igneous rocks with the following average results: dunites 14 ppm, basalt 49 ppm, and granite 27 ppm. Turekian and Wedepohl [1961] give corresponding figures of 6, 20, and 20 ppm. Taking an average, the nitrogen content of pyrolite would be 16 ppm. If this applied to the entire earth (the core would probably contain some nitrogen), there would be about 20 times as much nitrogen remaining within the earth as has been lost to the atmosphere. This is probably an upper limit, since nitrogen may be somewhat concentrated in the upper mantle. Dunite contains only about one-third as much nitrogen as basalt. On the other hand, granite and basalt have similar abundances. Evidently nitrogen is not strongly fractionated. Apparently it occurs in the form of nitrogen molecules in the magma, which combine at submagmatic temperatures with hydrogen to form ammonia, which is the form in which nitrogen presently occurs in rocks [Vinogradov *et al.*, 1963]. It does not appear that nitrogen occurs as $NH_4^+$ at magmatic temperatures. The ionic radius of $NH_4^+$ is similar to $K^+$, and strong fractionation of nitrogen between basalts and granites would be expected if nitrogen were present in this form.

Thus we have estimated a lower limit of 3 and an upper limit of 20 for the ratio of the amounts of volatiles still remaining within the earth to those in the atmosphere, hydrosphere, and sediments. We will take a value of 5 for the ratio as a rough estimate. The atomic abundances of hydrogen, carbon, and nitrogen in the earth in Type I carbonaceous chondrites [Mason, 1963] and in the sun [Goldberg et al., 1960] are given in Table 8.

Table 8

ESTIMATED ATOMIC ABUNDANCES OF HYDROGEN, CARBON, AND NITROGEN IN THE SUN, TYPE I CARBONACEOUS CHONDRITES, AND IN THE EARTH

|    | Sun | Type I Carbonaceous Chondrites | Earth |
|----|-----|-------------------------------|-------|
| H  | $3.2 \times 10^4$ | 5.6 | $2.5 \times 10^{-2}$ |
| C  | 17  | 0.81 | $2.8 \times 10^{-4}$ |
| N  | 3   | 0.05 | $4.1 \times 10^{-5}$ |
| Si | 1   | 1    | 1 |

Table 8 demonstrates the well-known extreme impoverishment of the earth in these very volatile elements compared to the sun and to Type I carbonaceous chondrites. It is clear also that the carbonaceous chondrites have been unable to retain more than a small fraction of the primordial abundance of these elements. The ratio H/C in the earth is approximately 13 times higher than in carbonaceous chondrites, suggesting that hydrogen may have played a more important role in the chemical evolution of the earth than it is believed by Ringwood [1965] to have played in the evolution of meteorites. The significance of the terrestrial H/C ratio as obtained from excess volatiles is, however, not entirely unambiguous. It may be that C is preferentially retained in the deep interior of the earth as graphite or diamond and that the formation of $CO_2$ and $CO_3^{--}$ is inhibited by high pressure.

## 7. Recapitulation and Discussion

In Section 2 a basic hypothesis was stated. Elements were classified according to their volatility from basic silicate melts at temperatures in the vicinity of 1500°C and under reducing conditions. It was suggested that the earth retained the primordial abundances of elements which were not volatile under these conditions, whereas it may well have lost varying proportions of the volatile elements. Subsequent discussion of the abundances of many elements supported this hypothesis. It was found that a self-consistent earth model could be constructed from the primordial abundances of nonvolatile elements (as given by the Type I carbonaceous chondrites). On the other hand, the earth appears not to have retained a large proportion of the primordial abundances of many volatile elements of widely varying chemical properties—the alkali metals, Zn, Cd, Hg, Ge, Pb, Tl, Cl, F, S, N, C, and H. It appears probable that this list may be extended when sufficient data become available for many other elements, e.g., Bi, As, Sb, Br, I, Se, Te, Ga, and In. From the strontium and lead isotope data, it is clear that the loss of these elements from terrestrial matter occurred at or before approximately 4.5 billion years ago. It is important to notice that loss has not been complete and that a substantial quantity of these volatile elements have nevertheless been retained.

It has previously been inferred that the original state of the solar nebula was cold and the solid phase (dust particles) was highly oxidized. In order to form the metal phase now in the earth's core, heating and partial reduction of the primitive material occurred. It seems reasonable to assume that loss of the volatile elements already discussed occurred at this stage and that the heating and reduction must have been strong enough to cause this loss.

The construction of a self-consistent earth model from the abundances of nonvolatile elements requires a very strong enrichment of U, Th, Ba, Sr, Ta, Zr, Hf, and the rare earths in the upper mantle and crust. The sizes and charges of the ions of these elements are such that they are not readily accepted into the major phases of the mantle. It was concluded that their upward concentration implied that all or most of the mantle had been subjected to a partial or complete melting process at some stage.

The formation of the earth's core also involves a major differentiation process. Formation of the core is most readily understood if the mantle passed through a molten or partially molten stage as suggested above. In this case, differentiation of the mantle and formation of the core were part of a single decisive high-temperature event in the earth's history. It is tempting to identify this decisive event with the high-temperature process which caused reduction of primordial material to metal and loss of volatile elements as previously discussed. In this case, all of these processes occurred during the earth's formation, about 4.5 billion years ago.

Construction of a self-consistent earth model from primordial abundances of nonvolatile elements required the presence of some elemental silicon in the earth's core. This requirement was supported by other geophysical evidence. The presence of silicon in the core, whereas the mantle contains substantial quantities of oxidized iron, implies that the mantle is not in chemical equilibrium with the core. This implication is supported by studies of the abundances of several siderophile elements—Ni, Co, Cu, Au, Pt—in the mantle which are present in far greater concentration than would be expected for an equilibrium partition between metal and silicate phase. We will find that these observations and inferences have an important bearing on the mode of formation of the core.

An alternative and currently more popular view regarding formation of the core is that the earth accreted as a mixture of silicate and metal and that its initial temperature was low. According to this model, therefore, formation of metal by reduction of primordial material and loss of volatile elements occurred during an earlier preterrestrial stage of evolution of the solar system. Subsequent radioactive heating caused temperatures to rise so that partial melting, differentiation of the mantle, and formation of the core occurred long after the earth's formation. This stage must have preceded the formation of the earliest continental crust, which is about 3.5 billion years old. Hence widespread melting and differentiation occurred within the first billion years according to this model.

There is little direct evidence bearing upon the time of melting and differentiation. However, consideration of the "age of the earth" as determined by the lead-uranium method tends to support the view that melting and differentiation occurred during or very soon after the earth's formation.

The "age of the earth" as determined by the lead-uranium method [Patterson, 1956] is established by the observation that average modern crustal lead falls on or close to the meteoritic isochron for $4.55 \times 10^9$ years. This means that, to a first approximation and neglecting second-order effects, the upper mantle and crust have behaved as a closed system for lead and uranium over the past 4.55 billion years. However, it has been argued that uranium has been very strongly concentrated in the upper mantle-crust system by a melting process. Since lead has rather different crystal properties than uranium, it would not normally be expected that a differentiation process which strongly enriched uranium near the surface would similarly enrich lead in precisely the same ratio. The studies of Marshall [1959], Patterson and Tatsumoto [1964], and Patterson [1964] support this expectation. Although lead and uranium show the same general tendencies to become concentrated in residual magmas, these are not equally strong, and hence lead and uranium become significantly separated during differentiation processes occurring in the upper mantle and crust. If this observed behavior had been followed during the major upward differentiation and concentration of uranium, then the Pb/U ratio of the lower mantle would differ significantly to that of the upper mantle. The major upward differentiation of uranium and other incompatible trace elements would, therefore, have occurred at or close to 4.55 billion years ago [Patterson and Tatsumoto, 1964]. According to the previous discussion this would imply that the earth passed through the completely or partially molten stage during or very soon after its formation.

Urey [1952] argued that the earth formed at a low temperature as an intimate mixture of iron and silicates. Upon subsequent heating, the iron melted but not the silicates. Urey then suggested that liquid iron might segregate and form a core by a convection process while the mantle remained solid.

If the effect of pressure on melting point is allowed for, the mean temperature throughout the earth required to cause melting of metal but not silicate would be about 2000°C. Urey [1952] showed that formation of the core from an initially uniform state would liberate about 800 cal/gm for the whole earth. Making a correction for strain in the interior, Birch [personal communication] suggests a smaller value of 600 cal/gm. The liberation of this additional energy after the mean temperature of the earth had reached 2000°C would cause complete melting. Because the effective "viscosity" of the earth's solid mantle decreases exponentially with rising temperature, segregation of the core according to Urey's model is an inherently unstable process, and the rate of segregation would be expected to increase rapidly with time, culminating in complete melting.

## 8. The Origin of the Earth

In the context of the previous discussions, theories of the origin of the earth may be divided into two classes which may be called single-stage and multistage, according to the number of separate major steps proposed. Both classes start with the postulate that the ultimate parent of the earth was a cold dust-gas cloud of solar composition.

### a. Multistage Theories

The first class of theories maintains that the primitive oxidized dust was subjected to a high-temperature stage—either in a dispersed state, or in an earlier generation of parent bodies, perhaps of lunar size [Urey, 1956], before the earth was formed. During this preterrestrial stage an extensive and complex chemical and physical processing occurred, during which a metallic phase was produced by reduction, and volatiles were lost. After the high-temperature stage, the mixed silicate and metal phases were cooled. The earth then formed by accretion of this mixed silicate-iron material. Accretion took place sufficiently slowly so that the temperature of the earth remained low. The mean temperature after accretion was perhaps less than 1000°C.

Theories of this type are by far the most widely held at present. Professor Urey has been the chief advocate [Urey, 1952, 1956, 1957, 1958, 1962, 1963], but theories advocated by Kuiper [1952], Wood [1962], and many others belong in this class. The assumption that the earth formed by accretion of partially degassed silicate-iron material generally similar to chondrites is widespread—e.g., Runcorn [1962, 1964]; Elsasser [1963]; Munk and Davies [1964], and Birch [1964].

Urey has made by far the most comprehensive and thorough attempt to develop workable hypotheses of this kind. As I interpret them, Urey's theories have been governed to a large extent by the following four boundary conditions or assumptions:

1. Escape of gases with a molecular weight greater than about 10 from the earth's gravitational field is not possible. Accordingly, reduction of metal and loss of volatiles occurred *before* the earth accreted to its present size. These processes probably occurred on an earlier generation of parent bodies of lunar size.

2. The varying zero-pressure densities of the terrestrial planets imply that a variable physical fractionation of metal from silicates has occurred in the solar system. Again, this requires formation of metal and silicate by high-temperature processes before the terrestrial planets accumulated.

3. Although an earlier high-temperature reduction and fractionation process occurred, the presence of volatile components such as $H_2O$ in the earth implies that accretion of the earth took place at comparatively low temperatures.

4. Theories for the development of the earth must be consistent with evidence yielded by chondritic meteorites. Urey believes that formation of these also demands a complex multistage process.

Within this framework, Urey has produced several ingenious theories of the origin of the solar system [Urey, 1952, 1956, 1957, 1958, 1962, 1963]. However, they are all extremely complex and, in the author's opinion, lack plausibility because of their complexity. There are also some specific difficulties arising from the present discussion of the earth's composition. The first concerns the inferred non-equilibrium between core and mantle. It is difficult to understand how this was established if the earth formed from an intimate mixture of silicate and metal similar to chondrites, which had adequate opportunity to reach equilibrium. Second, the material from which the earth was formed possessed a chemical composition very different from ordinary chondrites. It was depleted in alkali metals and sulfur and at the same time required a degree of reduction much higher than exhibited by chondrites to explain the inferred high silicon content of the core.

### b. A Single-Stage Hypothesis

Because of the difficulties, mentioned and unmentioned, in the multistage theories, Ringwood [1960] proposed that the earth formed directly by accretion from the primitive oxidized dust in the solar nebula and that reduction to metal, loss of volatiles, melting, and differentiation occurred simultaneously with, and as a direct result of, the primary accretion process. The assumption of intermediate stages of reduction and fractionation was avoided, so that the proposal amounts to a single-stage process. Advocacy of this model implied rejection of the boundary conditions and assumptions accepted by Urey and previously mentioned. Thus Ringwood [1959, 1961a, 1965] argued that chondrites and other terrestrial planets also formed by a single-stage process and did not demand an earlier silicate-metal fractionation. The incorporation and retention of some volatiles within the earth was found not to be inconsistent with an accretion process occurring dominantly at high temperature. However, the most fundamental difference between the author's approach and Urey's was the former's assumption that virtually complete escape of a dense primitive atmosphere from the earth had occurred. This assumption, which was vital to the hypothesis, has been criticized by Urey [1960, 1962] and others as being highly improbable. We will discuss this aspect further in Section 8 (5), after outlining the single-stage process as proposed by the author [1960].

*1. Chemical Composition of Accreting Solids.* The earth is assumed to have formed by direct accretion from the solid phases of the solar nebula which had developed as a disk surrounding the sun. The nebula is assumed to be cold ($\ll 0°C$), because of its opacity to radiation [Öpik, 1962a], and composed of the solar abundance of elements. Under these conditions, Latimer [1950] and Urey [1952] have shown that the common metals (particularly iron) in the nonvolatile solid phase would be completely oxidized. The solid phase would consist principally of a mixture of the nonvolatile silicate-iron oxide dust, ice, and the hydrates of $CH_4$ and $NH_3$. The suggested over-all composition of the solids, therefore, resembles current models for comet nuclei [Donn, 1963].

The relationship between this primitive material and the Type I carbonaceous chondrites deserves some comment. The author [1963, 1965] has previously concluded that Type I carbonaceous chondrites were derived from the primitive solid phase of the nebula which accreted into a small parent body and was heated to a temperature around 100°C. Under these conditions, loss of the more volatile ices occurred, leaving behind a reconstituted and metamorphosed mixture of hydrated silicates, iron oxides, and complex less volatile carbonaceous compounds. Ringwood [1961a, 1963] argued that the other groups of chondrites were formed when parent bodies of Type I carbonaceous chondrite composition were heated internally by short-lived radio-activities. Under these conditions the carbonaceous compounds reacted with the oxidized iron to form a metal phase *in situ*. The chemical evidence given by the chondrites strongly indicated that carbon was the dominant reducing agent.

In the present model it is suggested that the earth formed directly from the primitive dust in the solar nebula without passing generally through a Type I carbonaceous chondrite stage. This is a consequence of the different energy sources involved. As we shall see, the formation of the earth was controlled by the gravitational energy of accretion which was liberated practically instantaneously during the collision of any particular planetesimal with the accreting earth. The stage of slow heating in a parent body by radioactive elements, leading to gradual loss of volatiles and a low grade of metamorphism of the residual nonvolatiles, accompanied by an increase in the C/H ratio, which was passed through by the carbonaceous chondrites, did not generally occur on the earth.

Accordingly, the solid material from which the earth formed possessed a somewhat higher H/C ratio than the Type I carbonaceous chondrites. This makes it easier to understand the H/C ratio of the excess volatiles of the earth [Rubey, 1951] and the high depletion of the earth in sulfur (lost as $H_2S$) compared to chondrites. Also we shall see that the assumption of a higher H/C ratio during formation of the earth makes it easier to understand the subsequent loss of the primitive atmosphere.

*2. The Accretion Process.* We now consider the development of the earth from the parent gas-dust cloud. For some reason, as yet unknown, the cloud becomes unstable. Dust particles when they collide begin to stick to one another [Urey, 1952] instead of flying apart or volatilizing, and thus a series of condensations or planetesimals are formed. Current estimates suggest that these condensations may have had varying sizes up to perhaps 100 km diameter. However, they are continually colliding and disintegrating so there will be a wide range of size distributions within the dust cloud. Eventually a fluctuation arises, such that a large condensation forms. This is sufficiently large to exert substantial gravitational attraction upon other condensations

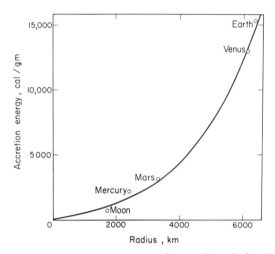

FIG. 4. Relationship between energy of accretion (cal/gm) and radius during the growth of a terrestrial planet.

and not to be broken up by the resulting collisions. When this stage is reached, the primary accretion process commences, and dust and condensations begin to fall on the central nucleus.

The mean gravitational energy liberated during formation of a body the size of the earth from a dispersed dust cloud is about 10,000 cal/gm. The accretion energy increases roughly as the square of the radius as is shown in Figure 4. During the early stages of accretion, the gravitational energy is low and the condensations are not strongly heated when they fall upon the nucleus. As the latter grows and the energy of accretion increases, the heating on impact becomes sufficient to vaporize the more volatile components of the dust—principally $CH_4$, $NH_3$, and water in excess of that held by hydrous silicates. However, the temperature of the less volatile components of the condensate remains low. Thus we see that the first stage of accretion results in the development of a cool, oxidized nucleus, rich in volatile components, particularly water. Perhaps the size of this nucleus would be about that of Mars.

A critical stage of evolution is reached when the nucleus is large enough to retain (even temporarily) some of the escaping volatiles in an atmosphere. The situation is then radically changed—instead of dust and planetesimals falling on to a cold solid surface, they fall into a blanket of gas, and the interactions are fundamentally different. As the central body grows and its gravitational potential increases, smaller condensations will fall into the atmosphere and become completely evaporated. Ionization and radiation occur, followed by chemical recombination of the less volatile components. However, the recombination occurs under reducing conditions in the presence of an excess of hydrogen and carbon, so that the interaction of dust and atmosphere causes reduction of oxidized iron to metal. Larger condensations may possess sufficient coherence to completely penetrate the atmosphere, although they would suffer considerable ablation in so doing. They would explode on hitting the surface of the growing earth and, in the presence of reducing agents at transient high temperatures, would suffer reduction of iron and rather complete degassing.

Ultimately the atmosphere would reach a size after which practically all of the infalling material was reduced and degassed before reaching the surface of the earth. As the accretion energy increased, the temperatures and intensity of reduction would also increase, so that not only iron but also silicates were reduced to metal. The condensates reaching the surface of the growing earth during the later stages of accretion would thus probably consist entirely of iron silicide and the nonvolatile oxides referred to in Section 2. Virtually all of the elements which were volatile at high temperature in the presence of excess hydrogen and carbon would enter the atmosphere (Table 3). These include the alkali metals, Zn, Cd, Hg, Ge, Pb, Tl, S, Cl, F, and probably many others—e.g., Br, I, Se, Te, Bi, Sb, As, In, Ga. In this manner, the growing earth may have become depleted in the volatile elements.

*3. Melting of the Earth.* For the volatile elements mentioned earlier to remain in the atmosphere, it is necessary for the temperature of the atmosphere and earth's surface to be high—at least 1500°C during the later stages of accretion. We have previously discussed evidence which implied that the earth passed through a completely or partially molten stage very early in its history, and most probably during its formation. The only plausible energy source is partial conservation of the gravitational energy of accretion. This amounts to 10,000 cal/gm for the earth, whereas the energy required to cause melting (from low temperature) is probably in the vicinity of 800 cal/gm, with due allowance for the effect of pressure on melting temperatures.

The surface temperature on the accreting planet is determined by the balance between the rate of accretion and rate at which energy can be radiated away. The two most important parameters are the time scale of the accretion process and the opacity of the medium surrounding the earth. The former is not well known. It appears that accretion may proceed very slowly at first, giving rise to a cool central body, as discussed previously. However, as the mass of the nucleus grows, the accretion rate will probably increase rapidly [Hoyle, 1946; Ter Haar and Wergelund, 1948]. The latter authors suggest that the rate of accretion may be proportional to the square of the mass of the central nucleus. It is apparent that the final stages of accretion may proceed rapidly. Hoyle [1946] found that even assuming a transparent atmosphere, the outer parts of the earth may have melted during the final stages of accretion.

However, it is improbable that the atmosphere was transparent to radiation during accretion. It has been suggested here that a large absorbing atmosphere developed. This would increase the effective temperatures developed during accretion. Öpik [1960; and personal communication] has drawn attention to a factor which is probably more important. During accretion the earth's environment will be effectively opaque to radiation because of screening by the surrounding dust particles. The energy of accretion is liberated near the earth's surface and is blanketed by the surrounding dust. Öpik has calculated that the time of accretion could be extended over $10^6$ years and still lead to melting. This time interval need apply only to the later, rapid stages of accretion, so that the total period of accretion may have been substantially longer than $10^6$ years. In view of these calculations, the assumption which is made in this paper, that perhaps two-thirds of the material in the earth accreted under sustained high-temperature conditions, does not appear unreasonable.

At the conclusion of accretion the outer regions of the earth would be close to or above the melting point. These outer regions, being highly reduced, will be metal-rich and dense, whereas the deep interior of the earth consists of a cool nucleus of oxidized silicate and iron oxides and is rich in volatiles. Such a state is gravitationally unstable. Convective overturn must follow, leading to a sinking of the metal-rich outer region into the center and segregation of a metallic core. This causes further evolution of heat due to the energy of gravitational rearrangement. Previously it was pointed out that the formation of a core from an initially homogeneous earth would evolve

600 to 800 cal/gm for the whole earth. Since the initial state is already unstable in the present case (being denser near the surface than in the center), the amount of gravitational energy evolved may substantially exceed the foregoing estimate. This would be sufficient to melt the entire earth. Since the rate of metal segregation is likely to be accelerated as internal temperature rises, the whole process is likely to be catastrophic.

Alternatively, it is possible that segregation of the core occurred gradually during the accretion and not after. Once the melting point at the surface of the earth was exceeded during accretion, the metal may have collected into masses large enough to flow directly to the center of the earth. This would take place as a continuous process during the later stages of accretion, and the gravitational energy would be liberated more slowly, again leading to complete melting. The gradual segregation of the core in this manner appears in some ways more probable than the catastrophic version suggested above.

*4. Mantle-Core Equilibrium.* According to the previous model, the earth develops in a state which is grossly out of chemical equilibrium. The deep interior is initially highly oxidized and rich in volatiles, whereas the outer regions are progressively more reduced and poor in volatile components. After melting near the surface, the metal phase, consisting of iron-nickel-silicon alloy, collects into bodies which are large enough to sink into the core. Equilibrium between metal and silicates can only be attained by diffusion across the interface of the sinking bodies of metal. If the rate of sinking of the metal is high compared to the rate of attainment of equilibrium by diffusion, the core which separates will not be in equilibrium with the mantle. In the author's opinion this situation is likely to arise, particularly in the later stages after melting when metal segregation is comparatively rapid. After the core has separated and the mantle is molten, both regions will become homogenized by convection, but they will remain out of equilibrium with one another. The situation has been more fully discussed by Ringwood [1959, 1961$b$]. The presence of silicon as an important component of the earth's core while the mantle contains oxidized iron can be understood on these grounds. Also, the presence in the mantle of a significant proportion of the primordial nickel, cobalt, copper, gold, and platinum is explicable. These elements are derived from the cool oxidized central nucleus, much of which entered the mantle in bulk during core segregation and was not intimately associated with a metal phase under such conditions that equilibrium could be reached.

The retention of volatiles,* particularly water, nitrogen, carbon, and inert gases, within the earth is also ascribed to the nonattainment of equilibrium. The volatiles are incorporated in the earth at a very early stage. After melting they become uniformly distributed throughout the mantle. The only way they can escape is by diffusion processes at the earth's surface, which, considering the vast volume concerned, are prohibitively slow. The difficulty of degassing industrial glass melts in large furnaces is well known. Escape of volatiles by formation of bubbles in the upper few kilometers of the molten earth is unlikely since the pressure of hydrogen and carbon monoxide above the earth's surface would prevent cavitation. During crystallization of the mantle, some of the volatiles would probably be expelled but it is probable that a large proportion would be trapped within crystals and as separate minor phases—e.g., amphiboles in the upper mantle and various high-pressure equivalents in the deeper mantle.

5. *Escape of Primitive Atmosphere.* The single-stage model for the formation of the earth involves the production of an enormous atmosphere consisting chiefly of hydrogen and carbon monoxide, with smaller quantities of $H_2O$. In order to achieve the high degree of reduction believed to have occurred during the later stages of accretion, a high $H_2/H_2O$ ratio is needed. At 1500°C, this is about 200 and it decreases with increasing temperature [Mueller, 1964; Ringwood, 1965]. Because of this, hydrogen does not play an important part in the reduction process during the latter stages of accretion, although it is believed to be present in great abundance. The effective reducing agent is carbon, derived mainly from methane. During formation of the earth's core by reduction, the carbon monoxide produced would amount to about 20 per cent of the earth's mass. This, however, is a minimum figure, since a considerable amount of carbon is consumed by reduction of $H_2O$ in the accreting ices, so that the required high equilibrium $H_2/H_2O$ value is maintained. Altogether it seems possible that the carbon monoxide produced during the earth's formation may have amounted to half the earth's mass and that a roughly comparable mass of free hydrogen derived from the decomposition of methane and ammonia, by reduction of $H_2O$ and also from excess free hydrogen trapped within the accreting ices, may have been present.

* The composition of the earth's excess volatiles (dominantly $H_2O$ and $CO_2$) also supports the previous arguments that the core and mantle did not develop in equilibrium. If the material of the mantle had been equilibrated with metallic iron at temperatures above 100°C early in the earth's history, the excess volatiles remaining in the mantle should consist dominantly of $H_2$ and CO rather than $H_2O$ and $CO_2$ [Mueller, 1964; Ringwood, 1965].

There is no evidence from crustal rocks of the presence of such an enormous former atmosphere. There is strong evidence that the present atmosphere has developed from degassing of the earth's interior. Obviously, therefore, a critical requirement of the hypothesis is that complete escape of the primitive atmosphere together with all its minor components, such as volatile metals, occurred at a very early stage of the earth's history.

Ringwood [1959, 1960] explicitly assumed that this was possible and qualitatively discussed several possible factors which may have contributed toward escape. On the other hand, Urey [1960, 1962] and several others, on the basis of the Jeans-Spitzer theory of selective escape of gases from a gravitational field, maintained that the required escape was impossible and that the theory must, therefore, be discarded.

Recently, Öpik [1962a, 1963a,b,c] has extensively reinvestigated the escape of planetary atmospheres and has concluded that the Jeans-Spitzer formula is not applicable to the escape of a hot dense atmosphere. Öpik has shown that the escape of such an atmosphere will not be selective with respect to molecular weight. If the temperature is sufficiently high, the atmosphere will simply "blow-off" indiscriminately. The critical parameter for determining escape then becomes the *mean* molecular weight of the atmosphere. If this is sufficiently low, the escape of a primitive terrestrial atmosphere must be seriously considered. In the model under discussion, the mean molecular weight of the earth's atmosphere is greatly lowered by the presence of large quantities of hydrogen. If there is about half as much hydrogen present by mass, as carbon monoxide, the mean molecular weight would be reduced to 4.

Öpik [1963b; and personal communication*] believes it possible that escape of the earth's primitive atmosphere may have occurred by blowing off simultaneously with the primary accretion process. This requires blanketing of the earth by dust, so that a high accretion temperature is maintained, as previously discussed. If the mean molecular weight of the gases is low enough, a substantial proportion of the gravitational energy of the accreting solids may be conserved to blow off the volatile components. Clearly then, Öpik's investigations place the gas escape problem in an entirely different light.

There are other factors also which may have been involved in the escape of an atmosphere. Ringwood [1960] suggested that if the sun had passed through a T-Tauri stage, characterized by large-scale

---

\* The author is greatly indebted to Dr. Öpik for a stimulating conversation on this subject and for drawing his attention to the significance of the *mean* molecular weight of the primitive atmosphere.

ejection of mass and strong particle radiation, the impact on the earth may have blasted off a primitive atmosphere. Herbig [1962] has pointed out that some T-Tauri stars are losing mass at the rate of a solar mass per million years. The interaction of such particle fluxes with the earth's atmosphere deserves study. It is possible that in addition to heating, extensive dissociation and ionization of the primitive terrestrial atmosphere might be caused. This further reduces the mean molecular weight of the terrestrial atmosphere, thus lowering the temperature at which it might blow off. Recently Hoyle and Fowler [1964] have also suggested the possibility that a primitive terrestrial atmosphere might be scoured away by solar corpuscular streams.

Suess [1949] suggested that a high rate of rotation of the earth early in its history might have facilitated escape of the atmosphere. If the earth formed in a turbulent dust cloud, one might expect a higher rate of rotation than it now possesses. If the primitive atmosphere was appreciably ionized either thermally or by solar radiation, it might have become magnetohydrodynamically coupled to the earth in the later stages of accretion when the core was separating and a terrestrial magnetic field was generated. Is it possible that at this stage the interaction between the rapidly accreting earth and the coupled atmosphere was such that the latter was driven outward from the earth in order to conserve angular momentum and maintain stability? In such a manner the rate of rotation of the earth might have been slowed down, while the primitive atmosphere when pushed far enough outward would be readily dissipated by intense solar particle bombardment if the earth's magnetic field temporarily weakened, as probably occurs during reversals.

The analogy, of course, is with current theories of the evolution of the sun and the transfer of its angular momentum to a surrounding nebula by magnetohydrodynamic coupling. Recent observations on T-Tauri stars [Herbig, 1958, 1962] which are undergoing Kelvin contractions show that they are ejecting large amounts of matter, apparently by processes analogous to those discussed earlier. Whatever the details may be, it is tempting to compare the Kelvin contractions of such stars with that of the earth.

Finally there is, in the author's opinion, a serious objection to the widely accepted belief that the earth did *not* possess a large atmosphere immediately after its formation. The advocates of this view, together with the author, agree that the present atmosphere and hydrosphere are of secondary origin and have been formed by degassing the earth's interior. In Section 6f it was argued that this process has not gone to completion and that the amount of excess volatiles remaining within the earth exceeds that which has been outgassed by a substantial factor, perhaps about 5.

Referring to Figure 4, we see that when the mass of the earth has grown to about one-tenth the present value (Mars), any solid matter accreting at the earth's surface arrives with such a high velocity that it is subjected to intense transient heating during impact, leading to melting and degassing. As the earth grows and the energy of accretion increases, the transient heating during explosive impact with the solid surface causes almost complete vaporization and degassing. This stage occurs before the earth has grown to more than one-quarter of its present mass. After this stage, nearly all the incoming material will be thoroughly degassed. At the conclusion of accretion, the amount of degassed volatiles must exceed, on any reasonable assumptions, the amount of volatiles trapped in the interior. This implies that the earth formed with an atmosphere and hydrosphere much larger than it now possesses. Since it is agreed that the present atmosphere is of secondary origin, it follows that some mechanism for escape of a primitive atmosphere must have existed.

6. *Origin of the Moon.* If, toward the end of the accretion process, the rate of rotation of the earth was close to the instability limit, there is the possibility that segregation of the core may have decreased the earth's moment of inertia and accordingly increased its angular velocity sufficiently to cause profound effects, including fission. Ringwood [1960] suggested the possibility that this had occurred, that the moon was the result of fission, and that the excess angular momentum of the earth-moon system had been carried away by the primitive atmosphere, which was also disrupted and escaped during the cataclysm. This suggestion was essentially a variation of the classic fission hypothesis of Darwin.

Wise [1963] also suggested that the moon formed by fission when segregation of the earth's core caused rotational instability. He discussed this hypothesis in some detail without being aware that it had been previously proposed by Ringwood. More recently, Cameron [1964] has supported this hypothesis and considered it at some length. Interest in the fission hypothesis has been revived partly because of the possibility that tektites come from the moon. The chemical and isotopic compositions of these objects are so similar to terrestrial surface material that if a lunar origin is ever conclusively demonstrated, it will constitute strong evidence for the ultimate derivation of the material from which the moon accumulated from the earth's outer regions, after the core had segregated [O'Keefe,

1963]. At the present stage the fission hypothesis must be regarded as rather extreme. Nevertheless, the origin of the moon is at present a matter of such mystery that it would be unwise not to afford it due consideration.

In the event of fission, a primitive terrestrial atmosphere would be lost. This adds one more mechanism to the possibilities for atmosphere escape previously discussed.

There is also another way in which the origin of the moon might be connected with loss of a primitive atmosphere. During the final stages of accretion of the earth, the atmospheric temperature may have risen to well over 1500°C as discussed previously. At temperatures between 1500 and 2000°C, and depending upon the redox conditions, there is the possibility that major chemical fractionation occurred and a large proportion of components previously regarded as nonvolatile—e.g., $SiO_2$, $MgO$, $Al_2O_3$, and $CaO$—were in fact volatilized and entered the atmosphere. Urey [1952] has shown that in the presence of gases possessing the solar hydrogen/oxygen ratio, there is a substantial $P$-$T$ field over which silicates are volatile but iron is not. It is suggested that analogous conditions occurred during the final stages of formation of the earth and that fractionation of silicates from metallic iron was thereby caused, with the iron continuing to accrete on the earth and silicates entering the atmosphere as gaseous components.

Consider now the consequences of escape of the primitive atmosphere according to one of the mechanisms previously suggested. As the atmosphere was driven outward, it expanded and cooled. This would have caused a large proportion of the least volatile components, principally silicates depleted in iron, to condense from the gas phase in the form of dispersed solid particles or liquid drops. It is conceivable that this material may have collected into a ring of planetesimals which surrounded the earth. The moon might then have formed by coagulaion of this "sediment-ring" as suggested by Öpik [1955, 1962a, b].

> 'If the primeval earth was more massive than the present one, and decreased in size through loss of light gases to space, the orbit of the particles of the ring would have increased (radius of orbit inversely proportional to mass of earth) until Roches Limit was reached. At this stage, the particles would coagulate under the influence of mutual gravitation, the moon being formed precisely in Roches Limit at "zero hour"' [Öpik, 1955, p. 247]

Öpik's model for the formation of the moon possesses many attractive aspects. However it does not explain the origin of the sediment ring or why the sediment was depleted in iron compared to terrestrial material. The present hypothesis attempts to answer these questions and is thus complementary to Öpik's model.

A detailed investigation of the chemical equilibria involved in the selective volatilization of silicates from iron during the final stages of accretion of the earth is nearly complete and will be published separately [Ringwood, 1965]. A summary of the preliminary results follows. Fractionation of silicates into the primitive terrestrial atmosphere can be considered in two stages. First, reduction according to the basic equation

$$MO + H_2 = M + H_2O$$

Second, the reduced metal M may or may not be volatilized, according to its vapor pressure and abundance. It is evident from the foregoing equation that the $H_2/H_2O$ ratio in the primitive atmosphere exercises an important control over the equilibrium. Previous investigators, e.g., Urey [1952], Wood [1963], have used an H/O ratio calculated from the total abundances of hydrogen and oxygen in the solar atmosphere in discussing such equilibria. This procedure cannot be used in the present model according to which the earth accreted from material which was strongly depleted in hydrogen compared to solar abundances. The use of the solar H/O ratio can also be criticized on general grounds, since it does not necessarily define the $H_2/H_2O$ ratio in the solar atmosphere. This latter ratio is critically dependent upon the carbon to oxygen ratio, as is seen from the equilibrium

$$CH_4 + H_2O = CO + 3H_2, \quad Kp = \frac{P_{CO} \cdot P^3_{H_2}}{P_{CH_4} \cdot P_{H_2O}}$$

The equilibrium constant for this reaction at 2000°K is $2.2 \times 10^7$. It can be seen that the equilibrium is driven very strongly to the right at high temperature because of the much higher affinity of oxygen for carbon than for hydrogen under these conditions. In the case of the primitive terrestrial atmosphere, it can be shown that if the C/O ratio in the accreting material exceeds unity, it causes the establishment of an $H_2/H_2O$ ratio in the atmosphere of between $10^3$ and $10^7$ in the temperature range 1500 to 2000°K. The establishment of such high $H_2/H_2O$ ratios is essentially independent of the ratio of total hydrogen to total oxygen in the system, which may be smaller than 10.

We will explicitly assume that the C/O ratio during the final stages of accretion was greater than or approximately equal to unity. This assumption can be justified in terms of the nonequilibrium processes by which carbonaceous material is incorporated in the nonvolatile oxide dust in the solar nebula. It is also important to notice that if a conservative possible error of a factor of 2 is applied to the solar carbon and oxygen abundances, the solar C/O ratio may lie anywhere between 0.15 and 2.6. In view of the profound effects on the $H_2/H_2O$ ratio of a solar C/O ratio exceeding unity, it is clear that there is little justification for assuming a fixed solar $H_2/H_2O$ ratio of 1000 as a boundary condition in discussion of equilibria in the solar nebula, as is commonly done.

In view of the above considerations, the volatilities of common

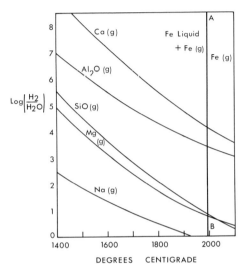

FIG. 5. Calculated volatilities of major elements and components from primitive condensed material possessing chondritic elemental abundances as a function of $H_2/H_2O$ ratio and temperature, and at a pressure of 1 atmosphere. For any given element, selective volatilization relative to iron will occur in the area above the curve pertaining to the particular element and to the left of the line $AB$. Iron is completely volatilized to the right of the line $AB$. After Ringwood [1965].

elements in a primitive terrestrial atmosphere were calculated as a function of $H_2/H_2O$ ratio and temperature. Separate allowance can be made for pressure. Preliminary results are set out in Figure 5. We see that whereas iron is not appreciably volatilized at temperatures below 2000°C, the common silicates will be volatilized at temperatures much lower than this if the $H_2/H_2O$ ratios are sufficiently high. From a separate study, it was concluded that the $H_2/H_2O$ ratios in the terrestrial atmosphere might vary between $10^3$ and $10^7$ according to the local $P$-$T$ conditions (for C/O ⩾ 1). Accordingly if the temperature attained during the final stages of accretion of the earth were in the vicinity of 1500 to 2000°C as previously argued, a large proportion of the accreting silicate components would be reduced and volatilized into the primitive atmosphere. On the other hand, iron would not be volatilized under these conditions and would hence continue to accrete upon the earth.

Figure 5 shows also that large differences exist in the relative volatilities of major silicate components as previously shown by Urey [1952]. Thus we might expect selective fractionation among silicates to occur during the proposed volatilization process. Although a considerable range of $H_2/H_2O$ ratios are possible, it appears that values between $10^4$ and $10^6$ would be attained most widely in the primitive terrestrial atmosphere. This would lead to selective fractionation of magnesium, silicon, and alkali metals from calcium, aluminum, and iron. However, under more restricted conditions, we can also expect aluminum and calcium to become fractionated from iron. A second fractionation in the reverse direction occurs during condensation caused by expansion and cooling of the primitive atmosphere. This leads to further separation of components according to their relative volatilities. It seems possible that as a result of selective condensation at different temperatures alkali metals may become separated from magnesium and silicon.

It is believed that as a net result of selective volatilization and selective condensation, the planetesimals which formed in the sediment ring would be highly heterogeneous in composition. The majority of planetesimals probably were composed principally of iron-free magnesian silicates. (Note the similar volatilities of magnesium and silicon in Figure 5.) However it is also probable that smaller numbers of planetesimals rich in calcium and aluminum, and containing some metallic iron, were also formed under more restricted conditions (highest temperatures and $H_2/H_2O$ ratios). Finally it is possible that a substantial proportion of planetesimals in the sediment ring were composed of primitive oxidized dust, rich in volatile components. Such planetesimals must have been abundant in the earth's neighborhood during the final stages of accretion, and it appears likely that many of these did not accrete upon the earth, but instead, became incorporated in the sediment ring.

Uranium and thorium appear to be less volatile than calcium and aluminum in the primitive terrestrial atmosphere. Accordingly it is probable that on the average, the material collecting in the sediment ring was impoverished in these radioactive elements. Also, it is possible that a modest depletion of alkali metals occurred because of separation from magnesian silicates during selective condensation. Thus the mean composition of the planetesimals was characterized by a net deficiency in radioactive elements compared to primordial abundances.

The model proposed above is rather flexible at present because of an inadequate understanding of physical and chemical conditions in the primitive terrestrial atmosphere. Its fundamental property lies in providing a possible explanation for fractionation of iron and silicates thereby leading toward an explanation of the moon's low density. The model also implies that the planetesimals in the parental lunar sediment ring were highly heterogeneous in composition and relatively depleted in radioactive elements. An attempt is made in a forthcoming paper to account for many of the observed properties of the moon in terms of its formation from such material [Ringwood, 1965].

## 9. Concluding Remarks

A principal objective of this paper has been to argue the merits of a single-stage model for the origin of the earth as opposed to the multistage models which are currently in vogue. An attractive property of the single-stage model, if it can be sustained, is its simplicity compared to multistage models. The single-stage model has the earth forming by accretion directly from the low-temperature dust-ice solids of the primitive solar nebula. The author believes that a satisfactory and self-consistent explanation of the composition and properties of the earth emerges from this model.

According to the multistage model, a series of complex high-temperature reduction and chemical fractionation processes occurred in the solar nebula prior to formation of the earth. Accordingly the earth formed from a mixture of silicate and metal phase, which had a most involved prior history. A principal argument against these hypotheses is their complexity and the special pleading which must be introduced to explain particular features. It was once assumed that the material from which the earth formed had an origin somewhat similar to ordinary chondritic meteorites. It is now clear that the chemical composition and constitution of the earth is very different in some major aspects from these objects. Although the chondrites have provided us with an immense amount of vital information about the early history of the solar system, it is clear that they do not provide answers to several major problems posed by the earth's composition—e.g., the loss of alkalis, sulfur, and other volatiles, inferred presence of silicon in the earth's core, and lack of equilibrium between core and mantle. This being so, it is necessary on the multistage model to postulate the occurrence of extremely complex processes occurring in the nebula before formation of the earth, for which we have no direct evidence whatsoever and which are introduced in ad hoc fashion, solely to explain specific characteristics or avoid particular difficulties.

The viability of the single-stage model rests squarely upon the assumption that escape of a primitive atmosphere from the earth is possible. This is denied by its critics. An attempt has been made in this paper to show that the outright rejection of this assumption is hardly justified in the light of present knowledge. On the other hand, if the assumption is sustained by future quantitative investigation, then the way is open for a comparatively simple explanation of the formation, not only of the earth, but also of the other terrestrial planets [Ringwood, 1959, 1965]. We have also seen that a possible dividend might be a workable hypothesis for the origin of the moon.

## 10. Summary

The earth probably formed by accretion in the solar nebula from an initially cold cloud of dust and gas. The chemical composition of the primordial dust cloud may be estimated from elemental abundances in the sun, in Type I carbonaceous chondrites, and from nucleosynthetic arguments. Primordial elemental abundances thus derived place limits on the present over-all composition of the earth.

Thermodynamic considerations show that iron in the primordial dust occurred as oxide and not as metal. Formation of the earth from primordial material therefore involved a high-temperature reduction process resulting in the formation of a metal phase. Formation of the earth was also accompanied by selective loss of a large number of relatively volatile elements from the primordial material. It is suggested, however, that the earth retained essentially the primordial abundances of the elements which are not readily volatile under high-temperature, reducing conditions. On the basis of this hypothesis an earth model is constructed based upon the abundances of nonvolatile elements in Type I carbonaceous chondrites. According to this model, the earth's core contains about 20 per cent of elemental silicon. Independent evidence supporting this implication is considered.

Evidence bearing on the composition of the upper mantle is reviewed, and it is concluded that its mean composition is that of a mixture of approximately 3 parts of peridotite to 1 part of basalt. Comparison of the inferred chemical composition of the upper mantle with the composition of the entire mantle derived from the chondrite abundances reveals a distinctive differentiation pattern. The major rock-forming elements and trace elements possessing similar ionic radii and charges to the major elements have undergone very little differentiation. However, trace elements possessing large ionic radii and/or charges which prevent them from entering the major mantle minerals have been strongly concentrated into the upper-mantle crust system by a differentiation process which has affected the entire mantle. The fractionated elements include U, Th, Ba, Sr, Hf, Zr, Ta, and the rare earths. This contrasting differentiation pattern based upon ionic properties is believed to be caused by crystal-liquid fractionation and implies that the entire mantle has been subjected to partial or complete melting.

Abundances of some siderophile elements (Ni, Co, Cu, Pt, Au) in the mantle are considered in relation to the formation of the earth's core. The observed abundances are far too high to be established by an equilibrium metal-silicate partition. It is concluded that the core was formed under conditions such that chemical equilibrium was not maintained with the mantle. This is also implied by the presence of silicon in the core whereas the mantle contains oxidized iron. Core-mantle disequilibrium places important limitations upon the mechanism by which the core was formed.

The fractionation of volatile elements in the earth is discussed. In addition to H, He, N, Ne, and C in which the earth is well known to be strongly depleted compared to primordial abundances, it appears that the earth has lost most of the primordial Na, K, Rb, Cs, S, Zn, Cd, Hg, Ge, Tl, Pb, F, and Cl by volatilization during the process of its formation.

Chemical and physical processes occurring during accretion of the earth from the primordial dust cloud are considered. Theories of origin of the earth may be classed as multistage or single-stage. Multistage theories hold that reduction leading to formation of metal phase and loss of volatile elements occurred in a complex preplanetary stage of evolution, before the earth accreted. According to these views, the material from which the earth accreted was a mixture of dispersed metal and silicates possessing the same average composition as the present earth. Reasons for rejecting such multistage hypotheses are given.

It is maintained here that the earth formed by a single-stage accretion process directly from the cold parental dust cloud, and that a preplanetary stage of reduction and volatile loss did not occur. During the early stages of accretion when the gravitational energy released was relatively small, the central nucleus remained cool and completely oxidized. These conditions permitted the incorporation of volatile components in the growing earth. As the size of the nucleus increased, the accretion energy also increased, leading to transient heating of infalling dust and planetesimals, degassing, and partial reduction to metal phase. In the later stages of accretion, the temperature, maintained by the liberation of gravitational energy and shielding by the surrounding opaque dust cloud, became very high, leading to melting and production of a highly reduced metal phase. Eventually melting became widespread, and the metal phase segregated to form a core. Segregation occurred comparatively rapidly so that over-all chemical equilibrium between metal and silicates was not maintained. The core and mantle were subsequently homogenized by convection, but remained out of chemical equilibrium with one another. The mantle then crystallized from a completely or partially molten state and underwent crystal-liquid fractionation.

In the later stages of accretion, the temperature became so high that a large proportion of the more volatile elements present in the primordial material remained entirely in the hot, dense atmosphere (mainly $H_2$ and $CO$) which surrounded the earth. This atmosphere subsequently escaped from the earth carrying the volatile elements, so that the earth became depleted in the volatile elements previously mentioned. Possible mechanisms of atmospheric escape are discussed.

In the terminal phase of accretion, the temperature was sufficiently high so that the common oxyphile elements Si, Mg, Al, and Ca were volatilized and entered the atmosphere. At this stage, the condensed matter accreting on the earth consisted principally of metallic iron. When the primitive atmosphere was disrupted and escaped, the accompanying expansion and cooling caused precipitation of the nonvolatile silicate components of the atmosphere as iron-poor planetesimals and smoke. Precipitation of this material occurred in a disk surrounding the earth. Subsequently the disk became unstable and broke up to form a number of accretions which collected together to form the moon.

## Acknowledgments

This paper was written while the author was at the Geophysical Laboratory, Washington. The author wishes to thank Dr. P. H. Abelson and the Trustees of the Carnegie Institution for support and hospitality.

## References

Alder, B., State of matter at high pressure, in *Progress in Very High Pressure Research*, edited by F. P. Bundy, W. R. Hibbard, and H. M. Strong, p. 152, John Wiley & Sons, New York, 1961.

Alder, B., Physical experiments with very strong pressure pulses, in *Solids under Pressure*, edited by W. Paul and D. M. Warschauer, Chap. 13, pp. 385–421, McGraw-Hill Book Company, Inc., New York, 1963.

Anders, E., Origin, age and composition of meteorites, *Space Sci. Revs.*, *3*, 583–714, 1964.

Bate, G. L., and J. R. Huizenga, Abundance of ruthenium, osmium and uranium, in some cosmic and terrestrial sources, *Geochim. Cosmochim. Acta*, *27*, 345–360, 1963.

Benson, W. H., The tectonic conditions accompanying the intrusion of basic and ultra-basic plutonic rocks, *Natl. Acad. Sci. Mem.*, *19* (1), 6, 1926.

Birch, F., Elasticity and constitution of the earth's interior, *J. Geophys. Res.*, *57*, 227–286, 1952.

Birch, F., Differentiation of the mantle, *Bull. Geol. Soc. Am.*, *69*, 483–486, 1958.

Birch, F., Composition of the earth's mantle, *Geophys. J.*, *4*, 295–311, 1961.

Birch, F., Some geophysical applications of high-pressure research, in *Solids under Pressure*, edited by W. Paul and D. M. Warschauer, pp. 137–162, p. 478, McGraw-Hill Book Company, Inc., New York, 1963.

Birch, F., Density and composition of mantle and core, *J. Geophys. Res.*, *69*, 4377–4388, 1964.

Birch, F., and P. Le Compte, Temperature-pressure plane for albite composition, *Am. J. Sci.*, *258*, 209–217, 1960.

Bishop, H. L., H. N. Lander, N. J. Grant, and J. Chipman, Equilibrium of sulfur and oxygen between liquid iron and open hearth-type slags, *J. Metals*, *8*, 862–868, 1956.

Bowen, N. L., *The Evolution of the Igneous Rocks*, Princeton University Press, Princeton, N.J., 1928.

Boyd, F. R., Geological aspects of high-pressure research, *Science*, *145*, 13–20, 1964.

Boyd, F. R., and J. L. England, Pyrope, *Carnegie Inst. Wash. Yearbook*, *58*, 83–87, 1959.

Brown, H., Rare gases and the formation of the earth's atmosphere, in *The Atmosphere of the Earth and Planets*, edited by G. P. Kuiper, Chap. 9, pp. 258–266, 2nd edition, University of Chicago Press, 1952.

Bullard, E. C., Discussion of paper by Revelle and Maxwell, *Nature*, *170*, 200, 1952.

Bullard, E. C., The flow of heat through the floor of the Atlantic Ocean, *Proc. Roy. Soc. London, A*, *222*, 408–429, 1954.

Bullard, E. C., and D. T. Griggs, The nature of the Mohorovicic discontinuity, *Geophys. J.*, *6*, 118–123, 1961.

Burbidge, E. M., G. R. Burbidge, W. A. Fowler, and F. Hoyle, Synthesis of the elements in the stars, *Rev. Mod. Phys.*, *29*, 548–650, 1957.

Cameron, A. G. W., A revised table of abundances of the elements, *Astrophys. J.*, *129*, 676–699, 1959.

Cameron, A. G. W., The origin of the atmospheres of Venus and the Earth, *Icarus*, *2*, 249–257, 1964.

Chamberlin, R. T., Geological evidence on the evolution of the earth's atmosphere, in *The Atmosphere of the Earth and Planets*, edited by G. P. Kuiper, Chap. 6, pp. 248–257, 2nd edition, University of Chicago Press, 1952.

Claas, W. J., The composition of the solar atmosphere, *Rech. Astron. Observ. Utrecht*, *XII, Part I*, 1–52, 1951.

Clark, S. P., and A. E. Ringwood, Density distribution and constitution of the mantle, *Rev. Geophys.*, *2*, 35–88, 1964.

Clayton, D. D., and W. A. Fowler, Abundances of heavy nuclides, *Ann. Phys. N.Y.*, *16*, 51–68, 1961.

Dawson, J. B., Basutoland kimberlites, *Bull. Geol. Soc. Am.*, *73*, 545–560, 1962.

Donn, B., The origin and structure of icy cometary nuclei, *Icarus*, *2*, 396–402, 1963.

Drickhamer, H. G., The electronic structure of solids under pressure, in *Solids under Pressure*, edited by W. Paul and D. M. Warschauer, Chap. 12, pp. 357–384, McGraw-Hill Book Company, Inc., New York, 1963.

DuFresne, E. F., and E. Anders, The record in the meteorites, V, A thermometer mineral in the Mighei carbonaceous chondrite, *Geochim. Cosmochim. Acta*, *23*, 200–208, 1961.

Elsasser, W. M., Early history of the earth, in *Earth Science and Meteoritics*, dedicated to F. G. Houtermans, edited by J. Geiss and E. D. Goldberg, Chap. 1, pp. 1–30, North-Holland Publishing Company, Amsterdam, 1963.

Engel, A. E. J., Geologic evolution of North America, *Science*, *140*, 143–152, 1963.

Fermor, L. L., Preliminary note on garnet as a geological barometer and on an infra plutonic zone in the earth's crust, *Records Geol. Surv. India*, *43*, 31, 1913.

Fermor, L. L., The relationship of isostasy, earthquakes and vulcanicity to the earth's infra-plutonic shell, *Geol. Mag.*, *1*, 65–67, 1914.

Frey, F. A., and L. Hasken, Rare earths in oceanic basalts, *J. Geophys. Res.*, *69*, 775–780, 1964.

Gast, P. W., Limitations on the composition of the upper mantle, *J. Geophys. Res.*, *65*, 1287–1297, 1960.

Goldberg, L., E. A. Muller, and L. H. Aller, The abundances of the elements in the solar atmosphere, *Astrophys. J.*, *Suppl. Ser.*, *45*, vol. 5, 1–138, 1960.

Goldschmidt, V. M., Über die Massenverteilung im Erdinnern, verglichen mit der Struktur gewisser Meteoriten, *Naturwissenschaften*, *10*, 918–920, 1922.

Goles, G. G., and E. Anders, Abundance of iodine, tellurium and uranium in meteorites, *Geochim. Cosmochim. Acta*, *26*, 723–737, 1962.

Grant, N. J., and J. Chipman, Sulphur equilibria between liquid iron and blast furnace slags, *Trans. AIME*, *167*, 134–149, 1946.

Green, D. H., Alumina content of enstatite in a Venezuelan high-temperature peridotite, *Bull. Geol. Soc. Am.*, *74*, 1397–1402, 1963.

Green, D. H., The petrogenesis of the high-temperature peridotite intrusion in the Lizard area, Cornwall, *J. Petrol.*, *5*, 134–188, 1964.

Green, D. H., and A. E. Ringwood, Mineral assemblages in a model mantle composition, *J. Geophys. Res.*, *68*, 937–946, 1963.

Greenland, L., Fractionation of chlorine, germanium and zinc in chondritic meteorites, *J. Geophys. Res.*, *68*, 6507–6514, 1963.

Harris, P. G., Zone refining and the origin of potassic basalts, *Geochim. Cosmochim. Acta*, *12*, 195–208, 1957.

Harris, P. G., and J. A. Rowell, Some geochemical aspects of the Mohorovicic Discontinuity, *J. Geophys. Res.*, *65*, 2443–2459, 1960.

Haskin, L. A., and M. A. Gehl, The rare-earth distribution in sediments, *J. Geophys. Res.*, *67*, 2537–2541, 1962.

Hatch, G. G., and J. Chipman, Sulphur equilibria between iron blast furnace slags and metal, *J. Metals*, *1*, 274–284, 1949.

Heier, K. S., and J. W. Rogers, Radiometric determination of thorium, uranium and potassium in basalts and in two magmatic differentiation series, *Geochim. Cosmochim. Acta*, *27*, 137–154, 1963.

Hess, H. H., Island arcs, gravity anomalies and serpentinite intrusions, *Proc. Internat. Geol. Congress. Moscow, 1937*, Rept. 17, vol. 2, 263–283, 1939.

Hess, H. H., The oceanic crust, *J. Marine Research, Sears Foundation*, *14*, 423–439, 1955a.

Hess, H. H., Serpentines, orogeny and epeirogeny, in *Crust of the Earth*, edited by A. Poldervaart, Geol. Soc. Am. Spec. Paper 62, 391–408, 1955b.

Hess, H. H., Stillwater igneous complex, *Geol. Soc. Am. Mem.*, *80*, 177–185, 1960.

Herbig, G. H., in *Stellar Populations*, edited by D. O'Connell, Part III, p. 127, Interscience Publishers, Inc., New York, 1958.

Herbig, G. H., The properties and problems of T-Tauri stars and related objects, in *Advances in Astronomy and Astrophysics*, edited by Z. Kopal, pp. 47–103, Academic Press, New York, 1962.

Hormann, P. K., Zur Geochemie des Germaniums, *Geochim. Cosmochim. Acta*, *27*, 861–876, 1963.

Hoyle, F., On the condensation of the planets, *Monthly Notices Roy. Astron. Soc.*, *106*, 406–414, 1946.

Hoyle, F., and W. A. Fowler, On the abundances of uranium and thorium in solar system material, in *Isotopic and Cosmic Chemistry*, volume dedicated to H. C. Urey, edited by H. Craig, S. L. Miller, and G. J. Wasserburg, Chap. 30, pp. 516–529, North-Holland Publishing Company, Amsterdam, 1964.

Hurley, P. M., H. Hughes, G. Faure, H. W. Fairbairn, and W. H. Pinson, Radiogenic strontium-87 model of continent formation, *J. Geophys. Res.*, *67*, 5315–5336, 1962.

Kennedy, G. C., The origin of continents, mountain ranges and ocean basins, *Am. Scientist*, *47*, 491–504, 1959.

Knopoff, L., and G. J. F. MacDonald, An equation of state for the core of the earth, *Geophys. J.*, *3*, 68–77, 1960.

Kuiper, G. P., Planetary atmospheres and their origin, in *The Atmospheres of the Earth and Planets*, edited by G. P. Kuiper, Chap. 12, pp. 306–405, 2nd edition, University of Chicago Press, 1952.

Kuno, H., K. Yamasaki, C. Iida, and K. Nagashima, Differentiation of Hawaiian Magmas, *Japan J. Geol. and Geography, Trans.*, *28*, 179–218, 1957.

Kushiro, I., and H. Kuno, Origin of primary basalt magmas and classification of basaltic rocks, *J. Petrol.*, *4*, 75–89, 1963.

Latimer, W. M., Astrochemical problems in the formation of the earth, *Science*, *112*, 101–104, 1950.

Lovering, J. F., Pressures and temperatures within a typical parent meteorite body, *Geochim. Cosmochim. Acta*, *12*, 253–261, 1957.

Lovering, J. F., The nature of the Mohorovicic discontinuity, *Trans. Am. Geophys. Union*, *39*, 947–955, 1958.

Lovering, J. F., The evolution of the meteorites—evidence for the coexistence of chondritic, achondritic and iron meteorites in a typical meteoritic parent body, *Research on Meteorites*, edited by C. B. Moore, pp. 179–198, John Wiley & Sons, New York, 1962.

Lovering, J. F., and J. W. Morgan, Uranium and thorium abundances in stoney meteorites, I, The chondritic meteorites, *J. Geophys. Res.*, *69*, 1979–1988, 1964.

Lovering, J. F., W. Nichiporuk, A. Chodos, and H. Brown, The distribution of gallium, germanium, cobalt, chromium, and copper in iron and stoney-iron meteorites in relation to nickel content and structure, *Geochim. Cosmochim. Acta*, *11*, 263–278, 1957.

MacDonald, G. A., and T. Katsura, Variations in lava of the 1959 eruption in Kilauea Iki, *Pacific Sci.*, *15* (3), 358–369, 1961.

MacDonald, G. J. F., Chondrites and the chemical composition of the earth, *Researches in Geochemistry*, edited by P. H. Abelson, pp. 476–494, John Wiley & Sons, New York, 1959a.

MacDonald, G. J. F., Calculations on the thermal history of the earth, *J. Geophys. Res.*, *64*, 1967–2000, 1959b.

MacDonald, G. J. F., The deep structure of continents, *Rev. Geophys.*, *1*, 587–665, 1963.

MacDonald, G. J. F., and L. Knopoff, The chemical composition of the outer core, *Geophys. J.*, *1*, 284–297, 1958.

MacDonald, G. J. F., and N. F. Ness, Stability of phase transitions within the earth, *J. Geophys. Res.*, *65*, 2173–2190, 1960.

MacKenzie, D. B., High-temperature alpine type peridotite from Venezuela, *Bull. Geol. Soc. Am.*, *71*, 303–318, 1960.

Marshall, R. R., Isotopic composition of common leads and continuous differentiation of the crust of the earth from the mantle, *Geochim. Cosmochim. Acta*, *12*, 225–237, 1957.

Mason, B., The origin of meteorites, *J. Geophys. Res.*, *65*, 2965–2970, 1960.

Mason, B., *Meteorites*, John Wiley & Sons, New York, 1962.

Mason, B., The carbonaceous chondrites, *Space Sci. Revs.*, *1*, 621–646, 1963.

Masuda, A., and Y. Matsui, Geometrically progressional residual model as the explanation of lanthanide pattern variation, *Institute for Nuclear Study, University of Tokyo Rept. INSJ-53*, 1963.

Mueller, R. F., A comparison of oxidative equilibria in meteorites and terrestrial rocks, *Geochim. Cosmochim. Acta*, *27*, 273–278, 1963.

Munk, W. H., and D. Davies, The relationship between core accretion and the rotation rate of the earth, in *Isotopic and Cosmic Chemistry*, volume dedicated to H. C. Urey, edited by H. Craig, S. L. Miller, and G. J. Wasserburg, Chap. 22, North-Holland Publishing Company, Amsterdam, 1964.

Nishimura, M., and E. B. Sandell, Zinc in meteorites, *Geochim. Cosmochim. Acta*, *28*, 1055–1080, 1964.

Nixon, P. H., O. von Knorring, and J. M. Rooke, Kimberlites and associated inclusions of Basutoland: a mineralogical and geochemical study, *Am. Mineralogist*, *48*, 1090–1131, 1963.

O'Keefe, J. A., Two avenues from astronomy to geology, in *The Earth Sciences*, edited by T. W. Donnelly, pp. 43–58, University of Chicago Press, Chicago, 1963.

Ol'shanski, Y. I., System iron sulphide-iron oxide-silica, *Dokl. Akad. Nauk SSSR*, *70*, 246–248, 1950.

Onishi, H., Geochemistry of germanium, *Bull. Chem. Soc. Japan*, *29*, 686–694, 1956.

Öpik, E. J., The origin of the moon, *Irish Astron. J.*, *3*, 245–248, 1955.

Öpik, E. J., *The Oscillating Universe*, p. 16, A Mentor book, New American Library, New York, 1960.
Öpik, E. J., Jupiter: Chemical composition, structure and origin of a giant planet, *Icarus*, *1*, 200–257, 1962a, pp. 221–223.
Öpik, E. J., Surface properties of the moon, in *Progress in the Astronomical Sciences*, edited by S. F. Singer, Chap. 5, pp. 219–260, North-Holland Publishing Company, Amsterdam, 1962b.
Öpik, E. J., Dissipation of the solar nebula, in *Origin of the Solar System*, edited by R. Jastrow and A. G. L. Cameron, pp. 73–75, Academic Press, New York, 1963a.
Öpik, E. J., Selective escape of gases, *Geophys. J.*, *7*, 490–509, 1963b.
Öpik, E. J., Jupiter, *Irish Astron. J.*, *6*, 135–149, 1963c.
Patterson, C., Age of meteorites and the earth, *Geochim. Cosmochim. Acta*, *10*, 230–235, 1956.
Patterson, C., Characteristics of lead isotope evolution on a continental scale in the earth, in *Isotopic and Cosmic Chemistry*, edited by H. Craig, S. L. Miller, and G. J. Wasserburg, Chap. 19, pp. 244–268, North-Holland Publishing Company, Amsterdam, 1964.
Patterson, C., and M. Tatsumoto, The significance of lead isotopes in detrital felspar with respect to chemical differentiation within the earth's mantle, *Geochim. Cosmochim. Acta*, *28*, 1–22, 1964.
Powers, H. A., Composition and origin of basaltic magma of the Hawaiian Islands, *Geochim. Cosmochim. Acta*, *7*, 77–107, 1955.
Pottasch, S. R., The lower solar corona—the abundance of iron, *Monthly Notices Roy. Astron. Soc.*, *125*, 543–556, 1963.
Ramsay, W. H., On the constitution of the terrestrial planets, *Monthly Notices Roy. Astron. Soc.*, *108*, 406–413, 1948.
Ramsay, W. H., On the nature of the earth's core, *Monthly Notices Roy. Astron. Soc., Geophys. Suppl.*, *5*, 409–426, 1949.
Reed, G. W., The distribution of fluorine in stoney meteorites, Abstract, *Trans. Am. Geophys. Union*, *45*, 87, 1964.
Reed, G. W., K. Kigoshi, and A. Turkevich, Determination of concentration of heavy elements in meteorites by activation analysis, *Geochim. Cosmochim. Acta*, *20*, 122–140, 1960.
Revelle, R., and A. E. Maxwell, Heat flow through the floor of the Eastern North Pacific Ocean, *Nature*, *170*, 199–200, 1952.
Ringwood, A. E., Constitution of the mantle, 3, Consequences of the olivine-spinel transition, *Geochim. Cosmochim. Acta*, *15*, 195–212, 1958.
Ringwood, A. E., On the chemical evolution and densities of the planets, *Geochim. Cosmochim. Acta*, *15*, 257–283, 1959.
Ringwood, A. E., Some aspect of the thermal evolution of the earth, *Geochim. Cosmochim. Acta*, *20*, 241–259, 1960.
Ringwood, A. E., Chemical and genetic relationships among meteorites, *Geochim. Cosmochim. Acta*, *24*, 159–197, 1961a.
Ringwood, A. E., Silicon in the metal phase of enstatite chondrites and some geochemical implications, *Geochim. Cosmochim. Acta*, *25*, 1–13, 1961b.
Ringwood, A. E., Present status of the chondritic earth model, in *Researches on Meteorites*, edited by C. B. Moore, pp. 198–216, John Wiley & Sons, New York, 1962a.
Ringwood, A. E., A model for the upper mantle, *J. Geophys. Res.*, *67*, 857–867, 1962b.

Ringwood, A. E., A model for the upper mantle 2, *J. Geophys. Res.*, *67*, 4473–4477, 1962c.
Ringwood, A. E., The origin of high temperature minerals in carbonaceous chondrites, *J. Geophys. Res.*, *68*, 1141–1143, 1963.
Ringwood, A. E., Chemical evolution of the terrestrial planets, *Geochim. Cosmochim. Acta*, in press, 1965.
Ringwood, A. E., and D. H. Green, Experimental investigations bearing on the nature of the Mohorovicic Discontinuity, *Nature*, *201*, 566–570, 1964.
Robertson, E. C., F. Birch, and G. J. F. MacDonald, Experimental determination of jadeite stability relations to 25,000 bars, *Am. J. Sci.*, *255*, 115–137, 1957.
Ross, C. S., M. D. Foster, and A. T. Myers, Origin of dunites and of olivine-rich inclusions in basaltic rocks, *Am. Mineralogist*, *39*, 693–737, 1954.
Rubey, W. W., Geologic history of sea water, *Bull. Geol. Soc. Am.*, *62*, 1111–1147, 1951.
Rubey, W. W., Development of the hydrosphere and atmosphere with special reference to the probable composition of the early atmosphere, in *Crust of the Earth*, edited by A. Poldervaart, pp. 631–650, Geol. Soc. Am. Spec. Paper 62, 1955.
Runcorn, S. K., Palaeomagnetic evidence for continental drift and its geophysical cause, in *Continental Drift*, edited by S. K. Runcorn, Chap. 1, p. 1–39, Academic Press, 1962.
Runcorn, S. K., A growing core and a convecting mantle, in *Isotopic and Cosmic Chemistry*, edited by H. Craig, S. L. Miller, and G. J. Wasserburg, Chap. 21, North-Holland Publishing Company, Amsterdam, 1964.
Schmitt, R. A., R. H. Smith, J. E. Lasch, A. W. Mosen, D. A. Olehy, and J. Vasileoskis, Abundances of the fourteen rare earth elements, scandium and yttrium in meteoritic and terrestrial matter, *Geochim. Cosmochim. Acta*, *27*, 577–622, 1963a.
Schmitt, R. A., R. H. Smith, and D. A. Olehy, Cadmium abundances in meteoritic and terrestrial matter, *Geochim. Cosmochim. Acta*, *27*, 1077–1088, 1963b.
Schmitt, R. A., R. H. Smith, and D. A. Olehy, Rare earth, yttrium and scandium abundances in meteoritic and terrestrial matter II, *Geochim. Cosmochim. Acta*, *28*, 67–86, 1964.
Suess, H. E., Die Häufigkeit der Edelgase auf der Erde und im Kosmos, *J. Geol.*, *57*, 600–607, 1949.
Sumner, J. S., Consequences of a polymorphic transition at the Mohorovicic Discontinuity, *Trans. Am. Geophys. Union*, *35*, 385, 1954.
Talwani, M., G. H. Sutton, and L. H. Worzel, A crustal section across the Puerto Rico trench, *J. Geophys. Res.*, *64*, 1545–1555, 1959.
Taylor, S. R., Trace element abundances and the chondritic earth model, *Geochim. Cosmochim. Acta*, *28*, 1989–1999, 1964a.
Taylor, S. R., Chondritic earth model, *Nature*, *202*, 281–282, 1964b.
Tilton, G. R., and G. W. Reed, Radioactive heat production in eclogite and some ultramafic rocks, in *Earth Science and Meteoritics*, dedicated to F. G. Houtermans, Chap. 2, pp. 31–43, North-Holland Publishing Company, Amsterdam, 1963.
Ter Haar, D., and Wergeland, H., On the temperature of the earth's crust, *Kgl. Norske Videnskab. Selskabs, Forh.*, *20*, 52, 1948.

Thayer, T. P., Some critical differences between alpine-type and stratiform peridotite-gabbro complexes, *Intern. Geol. Congr., 21st Copenhagen, 1960, Rept. Session, Norden*, Part 13, 247–259.
Turekian, K. K., and K. H. Wedepohl, Distribution of the elements in some major units of the earth's crust, *Bull. Geol. Soc. Am.*, 72, 175–186, 1961.
Turner, F. J., and J. Verhoogen, *Igneous and Metamorphic Petrology*, p. 694, 2nd edition, McGraw-Hill Book Company, Inc., New York, 1960.
Urey, H. C., *The Planets*, Yale University Press, New Haven, 1952.
Urey, H. C., Discussion on Nuclear Processes in Geologic Settings, *National Research Council Publ. 400*, p. 49, Washington, D.C., 1953.
Urey, H. C., Diamonds, meteorites and the origin of the solar system, *Astrophys. J.*, 124, 623–637, 1956.
Urey, H. C., Boundary constitutions for the origin of the solar system, in *Physics and Chemistry of the Earth*, edited by L. H. Ahrens, F. Press, K. Rankama, and S. K. Runcorn, vol. 2, pp. 46–76, Pergamon Press, Ltd., London, 1957a.
Urey, H. C., Meteorites and the origin of the solar system, 41st Guthrie Lecture, *Yearbook of the Physical Society*, London, 14–29, 1957b.
Urey, H. C., The early history of the solar system as indicated by the meteorites, *Proc. Chem. Soc.*, 68–78, March, 1958.
Urey, H. C., On the chemical evolution and densities of the planets, *Geochim. Cosmochim. Acta*, 18, 151–153, 1960.
Urey, H. C., Evidence regarding the origin of the earth, *Geochim. Cosmochim. Acta*, 26, 1–13, 1962.
Urey, H. C., The origin and evolution of the solar system, in *Space Science*, edited by D. P. LeGalley, Chap. 4, pp. 123–168, John Wiley & Sons, New York, 1963.
Urey, H. C., A review of atomic abundances in chondrites and the origin of meteorites, *Rev. Geophys.*, 2, 1–34, 1964.
Vening Meinesz, F. H., Major tectonic phenomena and the hypothesis of convection currents in the earth, *Quart. J. Geol. Soc. Lond.*, 103, 191–207, 1947.
Verhoogen, J., Petrological evidence on temperature distributions in the mantle of the earth, *Trans. Am. Geophys. Union*, 35, 85–92, 1954.
Verhoogen, J., Temperatures within the earth, in *Physics and Chemistry of the Earth*, vol. 1, edited by L. H. Ahrens, K. Rankama and S. K. Runcorn, McGraw-Hill Book Company, Inc., New York, 1956.
Vincent, E. A., and J. H. Crocket, Studies in the geochemistry of gold I, *Geochim. Cosmochim. Acta*, 18, 130–142, 1960.
Vincent, E. A., and J. H. Crocket, Studies in the geochemistry of gold II, *Geochim. Cosmochim. Acta*, 18, 143–148, 1960.
Vincent, E. A., and A. A. Smales, The determination of palladium and gold in igneous rocks by radioactivation analysis, *Geochim. Cosmochim. Acta*, 9, 154–160, 1956.
Vinogradov, A. P., The origin of the material of the earth's crust, Communication I, *Geochemistry USSR English Transl.*, 1–32, 1961.
Vinogradov, A. P., K. P. Florensku, and V. F. Volynets, Ammonia in meteorites and igneous rocks, *Geochemistry USSR English Transl.*, No. 10, 905–916, 1963.
Wager, L. R., Beneath the earth's crust, Presidential address to Section C, British association for the Advancement of Science, *Advan. Sci.*, 58, 1–14, 1958.

Wager, L. R., and R. L. Mitchell, The distribution of trace elements during strong fractionation of basic magma—a further study of the Skaergaard intrusion, East Greenland, *Geochim. Cosmochim. Acta*, *1*, 129–208, 1951.

Wagner, P. A., The evidence of the kimberlite pipes on the constitution of the outer parts of the earth, *S. African J. Sci.*, *25*, 127–148, 1928.

Wagner, P. A., The diamond fields of Southern Africa, *The Transvaal Leader*, Johannesburg, 1914.

Wardani, El S. A., On the geochemistry of germanium, *Geochim. Cosmochim. Acta*, *13*, 5–19, 1957.

Wasserburg, G., G. J. F. MacDonald, F. Hoyle, and W. A. Fowler, Relative contributions of uranium, thorium, and potassium to heat production in the earth, *Science*, *143*, 465–467, 1964.

Wiik, H. B., The chemical composition of some stoney meteorites, *Geochim. Cosmochim. Acta*, 279–289, 1956.

Williams, A. F., *The Genesis of the Diamond*, two vols., Ernest Benn Ltd., London, 1932.

Wilshire, H. G., and R. A. Binns, Basic and ultrabasic xenoliths from volcanic rocks of New South Wales, *J. Petrol.*, *2*, pp. 185–208, 1961.

Wilson, J. T., The development and structure of the crust, in *The Earth as a Planet*, edited by G. P. Kuiper, pp. 138–214, The University of Chicago Press, 1954.

Wise, D. U., An origin of the moon by rotational fission during formation of the earth's core, *J. Geophys. Res.*, *68*, 1547–1551, 1963.

Wood, J. A., Chondrites and the origin of the terrestrial planets, *Nature*, *194*, 127–130, 1962.

Wood, J. A., On the origin of chondrules and chondrites, *Icarus*, *2*, 152–180, 1963.

Wright, T. L., and M. Fleischer, The geochemistry of the platinum metals, Unpublished report, U.S. Geol. Survey, 1964.

Yoder, H. S., and G. A. Chinner, Almandite-pyrope-water system at 10,000 bars, *Carnegie Inst. Wash. Yearbook*, *59*, 81, 1960.

Yoder, H. S., and C. E. Tilley, Origin of basalt magma: an experimental study of natural and synthetic rock systems, *J. Petrol.*, *3*, 362–532, 1962.

# MINERALOGY OF THE MANTLE

*A. E. Ringwood*

Department of Geophysics and Geochemistry,
Australian National University, Canberra, Australia

## Introduction

The past decade has witnessed some major advances in our knowledge of the phase constitution of the mantle. These have arisen in part from an improved understanding of its chemical composition, but more especially from the development of new experimental high-pressure techniques which have enabled the properties and stability fields of mantle minerals to be studied directly. The pressures and temperatures developed in the mantle range up to 1.3 million atmospheres and 3000 to 5000°C. Current static high-pressure apparatus can maintain pressures of 130,000 atmospheres simultaneously with temperatures up to 2500°C. These $P$-$T$ conditions cover the range found in the upper mantle (Figure 1). Within a few years there is little doubt that the experimental pressure limit will be extended to 200,000 atm and beyond. Another technique involving the generation of enormous transient pressures by shock waves is capable of covering the entire pressure range of the mantle. From the few applications of shock-wave techniques which have so far been made to geophysical problems, the potential of the method, particularly for studying equations of state of high-pressure phases, is clearly very great.

The most important source of physical information on the mantle is provided by seismology and in particular, by the variations of $P$ and $S$

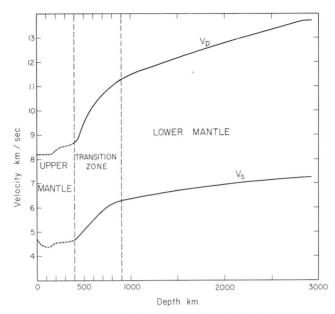

FIG. 1. Seismic velocity distributions and principal subdivisions of the mantle.

wave velocities with depth. These are shown in Figure 1. While there is general agreement concerning the nature of the depth-velocity curves below 600 km, their location at shallower depths is subject to appreciable uncertainty owing partly to regional variation in depth-velocity profiles. Accordingly, they are shown as broken lines. Despite the uncertainty concerning details, the general trend is clear. The depth-velocity curves divide the mantle into three well-defined regions. We shall see subsequently that these divisions are of a fundamental nature, being based directly upon general properties of the mineral phases present. It is therefore appropriate to classify the mantle into regions on the basis of these curves (Figure 1). This was originally done by Bullen [1940] who called them Regions $B$, $C$, and $D$ of the earth. The more descriptive usage—Upper Mantle, Transition Zone, and Lower Mantle—is preferred in this paper. We shall return to the question of mantle subdivisions and nomenclature in Section 2e.

The Upper Mantle is characterized by generally low velocity-depth gradients. The $S$ wave distribution passes through a shallow minimum between 100 and 150 km, whereas for $P$ waves the minimum is less pronounced or absent. The lower limit of the Upper Mantle (as defined above) is not finally established. However, the latest evidence

[Anderson, 1964a, b; Kovach and Anderson, 1964] suggests that relatively low velocity gradients persist to a depth of approximately 400 km. Below this depth and extending to approximately 900 km, the velocity gradients are very high. This is the Transition Zone. Finally, from 900 km to the core at 2900 km, seismic velocities increase at a uniform and moderate rate, thus defining the Lower Mantle.

## 2. The Upper Mantle

*a. Composition*

The chemical composition of the upper mantle was extensively discussed in the accompanying chapter, "The Chemical Composition and Origin of the Earth" (Section 4), and a brief summary only is

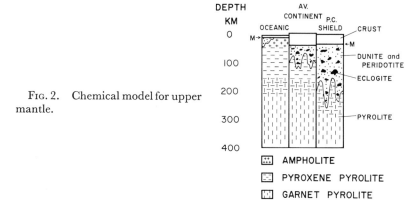

FIG. 2. Chemical model for upper mantle.

required. The upper mantle is believed to be of ultramafic composition and chemically zoned. Beneath continents there is a layer of dunite-peridotite of variable thickness but perhaps averaging between 100 to 200 km. This passes downward into a more primitive rock-type—"pyrolite" or pyroxene-olivine rock. The composition of pyrolite is defined by the property that when it is fractionally melted, it yields a basalt magma together with an unmelted refractory residue equivalent to a normal dunite or alpine peridotite. The composition of pyrolite thus falls between basalt and peridotite. A basalt:peridotite ratio in the vicinity of 1:3 is indicated by various lines of evidence. Beneath oceans, the dunite-peridotite layer is thin or it may be absent so that the primitive pyrolite extends to the suboceanic Moho. To a first approximation, the mean chemical composition of the upper mantle and crust over any extensive region of the earth and down to depths of a few hundred kilometers is identical, whether the surface is continental or oceanic. Continents are considered to have segregated essentially vertically owing to differentiation by fractional melting of the primitive pyrolite. The suggested chemical model is shown in Figure 2.

The occurrence of eclogite in the upper mantle was discussed in Section 4a of the accompanying paper. As is seen in Figure 2, eclogite is believed to occur in local segregations within the peridotite layer. The distribution is probably sporadic, and on the average, its abundance compared to peridotite is comparatively minor. However, it is possible that on a local scale there may be regions where eclogite is a more abundant constituent. The eclogite segregations have probably formed during fractional melting processes leading to the production of basaltic magma. A substantial portion of the magma probably does not reach the surface but instead crystallizes as widespread pockets and segregations of eclogite. Experimental investigations on individual components of eclogite [Robertson et al., 1957; Boyd and England, 1960; Yoder and Chinner, 1960] and also on natural basalt and eclogite [Boyd and England, 1960; Yoder and Tilley, 1962; Ringwood and Green, 1964] have clearly demonstrated that eclogite is the stable assemblage for material of basaltic composition over most of the $P$-$T$ range encountered in the upper mantle (under conditions of low water vapor pressure).

*b. Pyrolite Stability Fields*

Having set up a model for the chemical composition of the upper mantle, we must next investigate the effects of the range of $P$-$T$ conditions occurring in the upper mantle on the mineralogy of dunite, peridotite, and pyrolite. From evidence to be discussed later, it appears that the dominant components of the dunite-peridotite zone, olivine and orthopyroxene (low in Al), are stable throughout the $P$-$T$ range of the upper mantle. On the other hand, the mineralogical assemblages displayed by pyrolite are sensitively dependent on temperature and pressure.

Rocks approaching the pyrolite composition occur rarely at the surface of the earth. They are found most commonly as inclusions in diamond pipes, as inclusions in basalts, and as high-temperature peridotites. Green and Ringwood [1963] observed that naturally occurring rocks approaching the pyrolite composition displayed four distinct mineral assemblages (Table 1), clearly indicative of different $P$-$T$ conditions of crystallization and equilibration. These four assemblages were as follows:

1. Olivine + amphibole: Ampholite
2. Olivine + Al-poor pyroxenes + Plagioclase: Plagioclase pyrolite
3. Olivine + Aluminous pyroxenes ± Spinel: Pyroxene pyrolite
4. Olivine + Al-poor pyroxenes + Pyrope-rich garnet: Garnet pyrolite

If the upper mantle is indeed largely composed of rocks approaching the pyrolite composition, it appears that large-scale mineralogical

### Table 1

COMPARISON OF CHEMICAL COMPOSITIONS AND MINERAL ASSEMBLAGES OF ROCKS APPROACHING THE PYROLITE COMPOSITION

|  | Model Pyrolite Composition 1 | Olivine +* Amphibole 2* | Olivine + Pyroxenes + Plagioclase 3* | Olivine + Aluminous Pyroxenes ± Spinel 4 | Olivine + Pyroxenes + Garnet. 5 |
|---|---|---|---|---|---|
| $SiO_2$ | 45.16 | 44.89 | 44.72 | 44.69 | 45.58 |
| $MgO$ | 37.47 | 38.62 | 40.48 | 39.80 | 42.60 |
| $FeO$ | 8.04 | | | 7.54 | 6.41 |
| Total Fe as FeO | 8.45 | 8.49 | 8.23 | 7.63 | 6.65 |
| $Fe_2O_3$ | 0.46 | | | 0.09 | 0.27 |
| $Al_2O_3$ | 3.54 | 3.99 | 3.52 | 3.19 | 2.41 |
| $CaO$ | 3.08 | 2.82 | 2.03 | 2.97 | 2.10 |
| $Na_2O$ | 0.57 | 0.35 | 0.18 | 0.18 | 0.24 |
| $K_2O$ | 0.13 | 0.05 | 0.07 | 0.02 | nil |
| $Cr_2O_3$ | 0.43 | 0.40 | 0.45 | 0.45 | 0.09 |
| $NiO$ | 0.20 | 0.20 | 0.20 | 0.26 | n.d. |
| $CoO$ | 0.01 | 0.01 | 0.01 | — | n.d. |
| $TiO_2$ | 0.71 | 0.28 | 0.18 | 0.08 | 0.15 |
| $MnO$ | 0.14 | 0.11 | 0.14 | 0.14 | 0.12 |
| $P_2O_5$ | 0.06 | | | 0.04 | 0.03 |
| $H_2O$ | | | | 0.43 | |
| $CO_2$ | | | | 0.17 | |
|  | 100.00 | 100.00 | 100.00 | 100.05 | 100.00 |

* Recalculated to 100 per cent anhydrous.
1. Table 5, accompanying chapter.
2, 3. Lizard peridotite, Green [1964].
4. Tinaquillo peridotite, Green [1963].
5. Kimberlite xenolith, Dawson [1962].

zoning controlled by the *P-T* stability fields of these assemblages must be present.

The next step is to locate the *P-T* stability fields of these assemblages by direct experiment. This task is currently in progress. A preliminary outline of pyrolite stability fields is given in Figure 3. The boundary

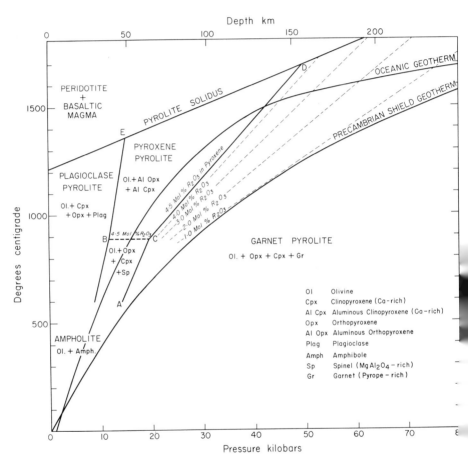

Fig. 3. Preliminary stability fields for pyrolite mineral assemblages.

between pyroxene pyrolite and garnet pyrolite has been reasonably well established by Boyd and England [1964] and by MacGregor and Ringwood [1964]. See also Ringwood, MacGregor, and Boyd [1964]. The pyrolite solidus and limits of the plagioclase pyrolite field are poorly known and considerable further work will be necessary before they can be finally established. Nevertheless, the preliminary estimates in Figure 3 based upon Kushiro and Yoder [1964] and Green and Ringwood [unpublished results] are unlikely to be seriously in error.

The stability field of the olivine + amphibole assemblage is mainly controlled by the partial pressure of water vapor and temperature. Under conditions where $P_{H_2O}$ = load pressure, the upper limit to the thermal stability of this assemblage may exceed 1000°C. However, in an environment where water pressure is much smaller than load pressure, the stability field will be displaced to lower temperatures. Since quantitative data on the stability of amphiboles under conditions where $P_{H_2O} < P_{Load}$ are not available, the stability field of the ampholite assemblage can be indicated only in very general terms in Figure 3.

The plagioclase pyrolite field in Figure 3 consists of the minerals olivine, diopsidic pyroxene (low in aluminum), orthopyroxene (low in aluminum), and plagioclase. With increasing pressure, plagioclase becomes unstable and breaks down according to the following basic equilibria.

$$\underset{\text{Albite}}{NaAlSi_3O_8} \rightarrow \underset{\text{jadeite}}{NaAlSi_2O_6} + \underset{\text{quartz}}{SiO_2}$$

$$\underset{\text{anorthite}}{CaAl_2Si_2O_8} \rightarrow \underset{\substack{\text{Ca-Tschermak's}\\\text{molecule}}}{CaAl_2SiO_6} + \underset{\text{quartz}}{SiO_2}$$

$$\underset{\text{quartz}}{SiO_2} + \underset{\text{forsterite}}{Mg_2SiO_4} \rightarrow \underset{\text{enstatite}}{2MgSiO_3}$$

The jadeite and Tschermak's molecule enter into solid solution in pyroxenes. Thus the net result is the formation of omphacite and aluminous pyroxene at the expense of plagioclase. Experimental investigations dealing with the foregoing equilibria have been carried out by Birch and Le Compte [1960] and Kushiro and Yoder [1964]. Green [1964] has discussed the stability relationships of plagioclase in the Lizard peridotite.

The pyroxene pyrolite field consists of the assemblage olivine + aluminous enstatite + aluminous diopside ± spinel. Its boundaries depend critically upon the chemical composition assumed for pyrolite, particularly the ratio of trivalent oxides:

$$(Al_2O_3 + Cr_2O_3 + Fe_2O_3) = R_2O_3$$

to total pyroxene. For the pyrolite composition derived in Section 4c of the accompanying paper, the maximum content of $R_2O_3$ in the pyroxene, exclusive of that associated with sodium as $NaFeSi_2O_6$, $NaCrAl_2O_6$ or $NaAlSi_2O_6$, is 4.5 molecular per cent or approximately 8 weight per cent. The stability boundaries in Figure 3 apply to this composition.

The pyroxene pyrolite field is divided into two subfields by the line $BC$ according to the presence or absence of spinel. Below $BC$, pyroxenes contain less than 4.5 mole per cent $R_2O_3$ and the remainder of the $R_2O_3$ appears in the form of spinel. As temperature increases, the solubility of $R_2O_3$ in pyroxene in equilibrium with spinel increases, and the amount of spinel therefore decreases. Along $BC$ spinel is finally consumed according to the reaction

$$x\text{MgAl}_2\text{O}_4 + (1 + x)\text{MgSiO}_3 = \text{MgSiO}_3 x\text{Al}_2\text{O}_3 + x\text{Mg}_2\text{SiO}_4$$
$$\text{spinel} \quad\quad \text{enstatite} \quad\quad \text{aluminous enstatite} \quad\quad \text{forsterite}$$
$$(x = 0.045)$$

This equilibrium has been investigated by MacGregor [1964], and the boundary $BC$ is based upon his results. In the natural case, it is probable that the presence of $(\text{MgFe})\text{Cr}_2\text{O}_4$ in solid solution in the spinel will permit the stability of chrome-rich spinel at somewhat higher temperatures than indicated. Above $BC$, the phases present are olivine + aluminous pyroxenes.

At higher pressures pyroxene pyrolite becomes unstable and transforms to a denser assemblage consisting of olivine, orthopyroxene, diopsidic pyroxene (lower in aluminum), and pyrope-rich garnet. The transformation involves two distinct equilibria. Along the boundary $AC$, spinel and pyroxene react to produce garnet and olivine.

$$4\text{MgSiO}_3 + \text{MgAl}_2\text{O}_4 = \text{Mg}_3\text{Al}_2\text{Si}_3\text{O}_{12} + \text{Mg}_2\text{SiO}_4$$

This equilibrium has been investigated by MacGregor [1964], and the boundary $AC$ is based upon his results for the simple system. Above $AC$, and along the boundary $CD$, pyrope garnet is formed by the direct breakdown of aluminous pyroxene according to a different equilibrium.

$$3\text{MgSiO}_3 \cdot x\text{Al}_2\text{O}_3 = x\text{Mg}_3\text{Al}_2\text{Si}_3\text{O}_{12} + 3(1 - x)\text{MgSiO}_3 \,(x \leq 0.045)$$
$$\text{Aluminous enstatite} \quad\quad \text{Pyrope} \quad\quad \text{Enstatite}$$

An analogous reaction can be written for aluminous diopsidic pyroxene. Since garnet and pyroxene have similar basic formulas of the $R_2O_3$ type, we are dealing essentially with a $P$-$T$ controlled solid solubility of garnet in pyroxene. As the pressure increases, and as temperature decreases, the solid solubility of garnet in pyroxene falls. The light broken lines subparallel and to the right of $CD$ define the alumina (or dissolved garnet) content of orthopyroxene in equilibrium with pyrope garnet.

The boundary in Figure 2 between pyroxene pyrolite and garnet pyrolite is based upon the work of Boyd and England [1964] on the system $\text{MgSiO}_3$—$\text{Mg}_3\text{Al}_2\text{Si}_3\text{O}_{12}$ and MacGregor and Ringwood

[1964] on solid solubility relationships of naturally occurring garnet, orthopyroxene, and clinopyroxene. The subject was further discussed by Ringwood, MacGregor, and Boyd [1964].

### c. *Mineralogical Zoning in the Upper Mantle*

The occurrence of the various pyrolite mineral assemblages in the mantle is largely determined by the intersection of geotherms with the stability fields of the assemblages. This subject has previously been discussed by Ringwood [1962a, b], Ringwood et al. [1964], and Clark and Ringwood [1964]. In Figure 3 two characteristic geotherms have been drawn representing possible temperature distributions beneath oceanic regions and Precambrian shields. The reasons for the differences between the two geotherms were extensively discussed by Clark and Ringwood [1964]. The geotherms in Figure 3 extend to somewhat higher temperatures than the curves calculated by these authors corresponding to a higher assumed opacity for the mantle.

The occurrence of ampholite is probably limited to the uppermost layer of the suboceanic mantle. This would require a water content of 0.5 to 0.8 per cent in this region and an environment in which the water vapor pressure was much smaller than the load pressure [Ringwood, 1962b]. These requirements do not seem unreasonable in view of the abundance of water above the crust. Recently, Oxburgh [1964] has suggested that the olivine + amphibole assemblage persists to depths greater than 100 km and is responsible for the low-velocity zone. This does not appear plausible to the author since it implies the presence of an extremely high partial pressure of water vapor in this region where the temperature probably exceeds 1000°C. It then becomes difficult to understand the absence of large quantities of hydrated phases like talc and serpentine in the cooler, uppermost regions of the suboceanic mantle as shown by the seismic velocity. Furthermore basalt produced by partial melting of an olivine-amphibole assemblage below 100 km would contain an excessive amount of water.

From Figure 3 it appears improbable that plagioclase pyrolite occurs in normal oceanic regions. However, in local areas of high heat flow, where the crust is thin and the partial pressure of water low, plagioclase pyrolite may be the stable assemblage. This might be the case in part of Japan and the western United States, where the seismic velocity at the Moho is smaller than 8 km/sec. Plagioclase pyrolite may also occur beneath some island arc areas characterized by low mantle seismic velocities and beneath mid-oceanic ridges.

From Figure 3 we see that pyroxene pyrolite is probably a major component of the upper mantle beneath oceans, where its stability field extends 130 to 140 km. Pyroxene pyrolite probably also occurs beneath normal continental areas characterized by geotherms intermediate between those on Figure 3. However, the presence of a refractory zone of residual dunite-peridotite as the upper layer of the subcontinental mantle reduces the region occupied by pyroxene pyrolite.

Garnet pyrolite possesses an extensive stability field. Along the oceanic geotherm, garnet first appears around 135 km. With increasing depth, more and more garnet is exsolved from the pyroxenes, which become less aluminous, being reduced to < 1 per cent $Al_2O_3$ at around 220 km. The Precambrian shield geotherm is in the garnet pyrolite field throughout its entire course in the upper mantle. This is consistent with the frequent occurrence of garnet peridotite nodules in diamond pipes penetrating stable crustal regions, including shields [MacGregor and Boyd, 1964]. However, as discussed previously, it is believed that the upper mantle beneath shields has been deeply fractionated and deprived of its low-melting components. Accordingly, rocks of garnet pyrolite composition are much less abundant than dunite-peridotite in this region.

Olivines and low alumina pyroxenes appear to be stable phases throughout the range of $P$-$T$ conditions present in the upper 400 km of the mantle (Section 3), and there appears little likelihood that garnet will transform to a denser phase in this depth interval. Accordingly, unless some presently unsuspected phase is formed by a combination of these minerals, it appears that the garnet pyrolite stability field also extends to approximately 400 km.

Possible occurrences of the various pyrolite mineral assemblages in the mantle as discussed previously are shown in Figure 2.

### d. Influence of Mineralogy on Seismic Profiles of Upper Mantle

The pyrolite mineral assemblages previously discussed possess distinctly different physical properties. Calculated densities and seismic velocities ($V_p$) for the pyrolite composition derived in Section 4c of the accompanying paper are given in Table 2. (The densities differ slightly from those given by Green and Ringwood [1963] for a pyrolite composition differing in minor respects from the present one.) Densities were calculated from the mineralogical composition using the data for mineral components and solid solutions given by Tröger [1959]. The seismic velocities were also calculated from the mineralogical composition using the data of Birch [1960, 1961a], Verma [1960], and Birch et al. [1960]. The characteristic velocities of pyroxenes, amphi-

**Table 2**

CALCULATED DENSITIES AND $P$-WAVE VELOCITIES
FOR PYROLITE MINERAL ASSEMBLAGES AND
PERIDOTITE

|  | Density gm/cc | $Vp$ km/sec |
|---|---|---|
| Dunite | 3.32 | 8.48 |
| Peridotite* | 3.31 | 8.32 |
| Ampholite† | 3.27 | 7.98 |
| Plagioclase pyrolite | 3.26 | 8.01 |
| Pyroxene pyrolite | 3.33 | 8.18 |
| Garnet pyrolite | 3.38 | 8.38 |

\* 20 per cent orthopyroxene.
† 35 per cent amphibole.

boles, and plagioclase were found by extrapolating Birch's data obtained in the 4 to 10 kb range back to zero pressure to avoid the effect of porosity. The velocities of olivine and garnet were obtained from Verma [1960]. In most cases, the velocities obtained were for minerals which possess different compositions from those involved in the pyrolite mineral assemblages. The compressibility ($\beta$) data of Birch et al. [1942] and the density ($\rho$) data of Tröger [1959] on members of mineral solid solution series were therefore used to apply compositional corrections to the observed velocities. It was assumed that $Vp \propto (1/\sqrt{\beta\rho})$ for members of a solid solution series. This approximation is adequate for what amounts in most cases to a minor correction. The velocities obtained refer to normal temperatures and pressures. Accordingly, their absolute values are of limited significance in interpreting seismic velocity–depth profiles where the velocities measured correspond to a range of high temperatures and pressures. Nevertheless, the values in Table 2 provide a reasonable estimate of the relative velocities of the upper mantle mineral assemblages, which cover a range of about 0.5 km/sec. It is clear that the velocity profile in the mantle will be strongly influenced by the stability fields of the pyrolite mineral assemblages and the depths at which they are intersected by geotherms.

It is generally agreed by seismologists that a low-velocity zone is a widespread feature of the upper mantle. This is caused in part by high-temperature gradients, particularly beneath oceans, which cause the velocity to decrease at a greater rate than the normal increase caused by pressure. Birch [1952] and MacDonald and Ness [1961] have shown that temperature gradients in the uppermost mantle may be sufficiently high to cause such a velocity inversion.

However, this offers at best only a partial explanation of the low-velocity zone. It implies that the inversion should be greatest where the temperature gradient is greatest—immediately beneath the Moho. This is not observed. Gutenberg [1959a] and Shurbet [1964] believe that seismic velocity increases initially for the first few tens of kilometers beneath the oceanic Moho. Virtually all the current preferred surface wave solutions for the oceanic mantle possess a zone of either constant or increasing velocity between the Moho and the low-velocity zone. If confirmed this implies the presence of inhomogeneity in this region [MacDonald and Ness 1961]. In the model under discussion (Figures 2 and 3), the inhomogeneity is caused by the presence of a layer of ampholite overlying the pyroxene pyrolite. Dehydration of ampholite with increasing depth and transition to pyroxene pyrolite would cause a small velocity increase (Table 2) sufficient to offset the decrease which would otherwise be caused by the temperature gradient. Within the homogeneous pyroxene pyrolite region, the temperature gradients for the first 100 km are probably sufficient to cause velocity to decrease [MacDonald and Ness, 1961], and the position of the velocity minimum in the suboceanic mantle is therefore primarily determined by the temperature gradient in this region. According to the subcontinental velocity distributions of Lehmann [1959, 1961] and Takeuchi et al. [1962], the velocity does not decrease until a depth of approximately 120 km, and the low-velocity zone lies between 120 and 220 km. This cannot be explained by high-temperature gradients which would lead to serious thermal difficulties in this region (Ringwood, 1962a). Again, mineralogical and chemical inhomogeneity seem indicated. According to the model (Figure 2), the uppermost layer of the mantle beneath continents consists of dunite-peridotite, which has a higher intrinsic velocity than pyroxene pyrolite (Table 2). Below this zone, garnet pyrolite occurs and velocity increases.

According to the CANSD model [Brune and Dorman, 1963] for the $S$-wave velocity distribution beneath the Canadian shield, the low-velocity zone for $S$ waves extends between 120 and 310 km. Again, temperature gradients cannot be the sole cause. Since this region is believed to consist dominantly of residual, refractory dunite-peridotite, and pyroxene pyrolite is absent (Figure 2), another cause must be sought. It is possible that an increase in orthopyroxene/olivine ratio with depth could be responsible. Orthopyroxene has a lower velocity than olivine [Birch, 1960]. The zoning might be the result of a change in the fractional melting equilibria with depth leading to a greater tendency of orthopyroxene to remain in the refractory residue. There is some experimental evidence supporting this trend [Green and Ringwood, 1964].

From Figure 3 it is seen that the depth of the transition from pyroxene pyrolite to garnet pyrolite is sensitively dependent upon temperature and upon the ratio of trivalent oxides ($R_2O_3$) to pyroxene as set by the assumed composition of pyrolite. In Figure 3 the oceanic geotherm enters the garnet pyrolite field at 135 km. The transition is gradual, and the amount of garnet exsolved from aluminous pyroxene increases with depth until about 220 km, when nearly all alumina in the pyroxene has been exsolved as pyrope-rich garnet. This transition causes an increase in velocity of approximately 0.20 km/sec (Table 2). An additional velocity increase occurs because of self-compression (within this depth interval the effect of pressure on velocity is greater than that of temperature). As mentioned earlier, it appears that garnet pyrolite will be stable between 220 and 400 km, hence the increase of velocity in this region is at the comparatively small rate caused by self-compression.

It is interesting to compare qualitative aspects of the seismic velocity distribution appropriate to the pyrolite model with some recent depth-velocity distributions. In the Gutenberg [1959b] model, S-wave velocity begins to increase rapidly at 140 km, and uniformly high gradients are maintained to 900 km. This cannot be reconciled with the pyrolite model, where there is a sharp rise between 135 to 220 km followed by a much slower rate of velocity increase. Recently, Anderson [1964a, b] has carried out an elaborate surface wave investigation of the suboceanic mantle and finds that a velocity increase of about 3 per cent is required between 150 and 200 km, below which the gradient remains low or zero until 400 km. This velocity distribution is in remarkable harmony with the consequences of the pyrolite model. The transition from pyroxene to garnet pyrolite involves a 2.5 per cent velocity increase (Table 2), and self-compression between 135 and 220 km would contribute another 1 per cent to the velocity increase. The seismic velocity distribution qualitatively appropriate to the pyrolite model is shown in Figure 1. Lehmann [1959, 1961] has also advocated a *P*- and *S*-wave velocity increase of 3 per cent occurring discontinuously at 220 km. Lehmann mentions that her solution is not unique and that appreciable variation in the sharpness, magnitude, and position of the discontinuity may be allowable.

The sharp velocity increase between 150 and 200 km was not found beneath the Canadian shield by Brune and Dorman [1963]. According to the model Figures 2 and 3, this is explicable since the pyroxene pyrolite is not encountered along the shield geotherm. The residual dunite-peridotite zone passes directly downward into garnet pyrolite with a negligible change of velocity (Table 2).

*e. Definition and Extent of Upper Mantle*

The usage of the term "Upper Mantle," particularly within the context of the International Upper Mantle Project, has become somewhat confused. Some investigators regard the upper mantle as extending down to 1000 km and thereby including the Transition Zone (Figure 1). Others limit the upper mantle to the greatest depths to which deep focus earthquakes have been recorded—approximately 700 km. In the present chapter [see also Clark and Ringwood, 1964], the Upper Mantle is regarded as a region characterized by "normal" mineralogy, mainly olivine, pyroxenes, and garnet in which the silicon atom is in fourfold coordination. As we shall see, the mineralogies of the transition zone and lower mantle are fundamentally different. Thus, both from the mineralogic and seismic point of view, the mantle is best divided into three major subdivisions as in Figure 1. Whether the term "upper mantle" should be used for the uppermost region can be debated. In the author's opinion, it is a satisfactory term if some agreement can be reached on its definition. The principal objection appears to be the more extended usage of the term in connection with the International Upper Mantle Project. Nevertheless, when the programs associated with this project are examined, it is found that the vast majority are aimed at elucidating phenomena occurring within the upper 400 km of the mantle, so that this objection does not appear unduly serious.

The lower boundary of the upper mantle is not yet established. The seismic evidence [Clark and Ringwood, 1964, p. 64; Anderson, 1964a, b] suggests a depth in the vicinity of 400 km, and this is supported by the mineralogical evidence (Section 3). However, a somewhat greater depth is not excluded.

## 3. The Transition Zone

The Transition Zone as outlined in Figure 1 extends between 400 km and 900 km and is characterized by a rapid increase of seismic velocity with depth. According to the Gutenberg and Jeffreys velocity distributions, the velocity increase between 400 and 900 km occurs at a uniform rate. More recently, surface wave investigations by Anderson

[1964a, b] suggest that most of the increase may be concentrated within two separate regions within the 400 to 900 depth interval.

In a series of classical investigations, Bullen [1936, 1937, 1940] established the existence of an inhomogeneous region in the earth's outer mantle in which density increased more rapidly than would occur in a uniform, self-compressed layer. Birch [1939, 1952] examined the nature of the inhomogeneity in considerable detail. He compared the observed variation of seismic velocity with depth with the variation which would be expected on thermodynamic grounds in a homogeneous medium. From this comparison, Birch concluded that the rate of increase of velocity with depth between 400 and 900 km was much higher than would occur in a homogeneous self-compressed layer. Accordingly, he inferred that the inhomogeneity which had been established by Bullen occurred in this region. However, the rate of increase of velocity with depth between 900 and 2900 km agreed closely with expectations for a homogeneous, self-compressed layer, and accordingly, Birch concluded that this region was probably homogeneous. When the elastic ratio of the material of this region was extrapolated to surface conditions using an equation of state, Birch found that it was far higher than those of common silicates like olivine and pyroxene but agreed well with the elastic ratios of close-packed oxide phases such as spinel, rutile, corundum, and periclase. Accordingly, he suggested that the region below 900 km consisted of dense, closely packed polymorphs of normal ferromagnesium

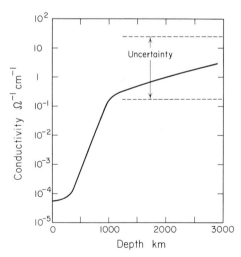

FIG. 4. Variation of electrical conductivity with depth in the mantle. After McDonald [1957].

silicates possessing "oxide" structures. The inhomogeneous region between 400 and 900 km was then interpreted as a region over which the gradual transition from normal to closely packed silicates took place.

A further line of evidence supporting the view that the region between 400 and 900 km is inhomogeneous comes from a study of the electrical conductivity distribution in the mantle [Runcorn, 1955; Tozer, 1959]. As seen from Figure 4, the shape of the conductivity distribution curve qualitatively resembles that of the seismic velocities. Runcorn concludes that the rapid rise of conductivity in the vicinity of 700 km cannot be explained by the effect of plausible temperature distributions upon the conductivity of olivine. He concludes that a phase change to a more closely packed structure in this region could explain the data. The high-pressure phase is assumed to have a higher electronic conductivity than the olivine modification. This assumption has since been experimentally verified for the olivine-spinel transition in fayalite [Bradley et al., 1962]. Tozer [1959] has further developed the interpretation of the electrical conductivity distribution of the mantle in terms of phase changes between 400 and 900 km.

*a. Chemical Composition*

The chemical composition of the mantle was discussed in some detail in the accompanying paper, and only a broad outline is necessary here. From the solar and meteorite elemental abundances and on various geochemical grounds, it seems clear that the dominant constituents of the mantle are $MgO$, $SiO_2$, and perhaps $FeO$. The first two of these components are probably more abundant by an order of magnitude than $Al_2O_3$, $CaO$, and $Na_2O$, while these in turn considerably exceed the abundances of the next group $Cr_2O_3$, $NiO$, $K_2O$, $TiO_2$, $MnO$, and $P_2O_5$. There is strong evidence that the $FeO/(FeO + MgO)$ ratio in the mantle is between 0.1 and 0.2. This is the range inferred for the upper mantle on petrological grounds (Section 4 of accompanying paper). This range is also indicated by the investigations of Birch [1961$b$] on the mean atomic number of the mantle, using an empirical relationship between velocity and density, and from the elastic ratio of the lower mantle [Birch 1952; Clark and Ringwood, 1964].

The foregoing broad limitations on chemical composition of the mantle are sufficient to permit a discussion on phase constitution. Because of the inferred overwhelming preponderance of $MgO$ and $SiO_2$, the phase assemblages will be dominated by equilibria involving these components. After the behavior of the fundamental $MgO$-$SiO_2$ system at high temperatures and pressures has been elucidated, the influence of less abundant components can be considered, more or less as second order effects.

In the previous section, three independent lines of physical evidence on the mantle suggested that important phase changes occurred in the mantle between 400 and 900 km and the detailed hypothesis of Birch [1952] regarding the nature of these phase changes was discussed. The dominant minerals of the upper mantle are magnesian olivine and pyroxene. Birch's hypothesis therefore implies that these minerals should transform to close-packed phases possessing the required elastic properties within a specific *P-T* range corresponding to conditions between 400 and 900 km.

Experimental evidence relating to this hypothesis is discussed in the following sections.

*b. Crystal Chemical and Thermodynamic Relationships between Silicates and Germanates*

The maximum pressure attained in high-pressure–temperature apparatus currently being used for geophysical studies is about 130 kb, which is approximately the pressure at a depth of 400 km in the mantle. This is just the beginning of the transition zone, which extends down to 900 km where the pressure exceeds 300,000 atmospheres. Although some important transitions have been discovered directly at pressures below 130 kb, it is clear that the investigation of mineral stability relations within much of the transition zone necessitates the application of indirect techniques. One such indirect technique which has yielded valuable results is the study of germanate model systems at high pressure.

Crystal chemical relationships between germanates and silicates were first elucidated in a classical paper by Goldschmidt [1931]. Both silicon and germanium readily form tetravalent ions which possess similar outer electronic structures and radii ($Si^{4+}$ 0.42 A, $Ge^{4+}$ 0.49 A [Ahrens, 1952; and personal communication]). Accordingly, the crystal chemistry of silicates is very closely related to that of germanates, particularly for oxy-compounds. Corresponding silicates and germanates are usually isostructural with one another and capable of forming continuous solid solutions. Over 40 examples are now known of complex silicates which possess isostructural germanate analogues, and many more remain to be discovered. In general terms it appears that if a germanate with a new structure should be synthesized, there would be a reasonable probability that under some appropriate *P-T* conditions a corresponding isostructural silicate would be stable.

A second important relationship between germanates and silicates has also emerged: it appears that germanates often behave as high-pressure models for the corresponding silicates. If a germanate is found to display a given phase transformation at a particular pressure, the corresponding silicate usually displays the same transformation but at a much higher pressure (not strictly correct for jadeite, see Table 2). The reverse of this relationship has not been observed. Furthermore, in cases where high-pressure polymorphism is displayed by a silicate, the corresponding germanate, if it does not display the same polymorphism, is found to display only the structure of the high-pressure silicate polymorph. Examples of this relationship are given in Table 3.

The reason for this behavior has been indicated by Bernal [1936] and Wentorf [1962a, b]. The structures of ionic compounds are largely governed by ionic radius ratios of constituent ions. This applies

### Table 3

COMPARATIVE STABILITIES OF ISOSTRUCTURAL GERMANATE AND SILICATE PHASES

| Structure Type | Germanate Compound | Stability Conditions $P$ (kb) | Silicate Compound | Stability Conditions $P$ (kb) | Ref.‡ |
|---|---|---|---|---|---|
| Rutile | $GeO_2$ | $P = 0$ < 1007°C | $SiO_2$ | > 120 kb, 1000°C  > 105 kb, 530°C | 1, 2  3 |
| Garnet | $Ca_3Al_2Ge_3O_{12}$ | $P = 0$ ~ 1000°C | $Ca_3Al_2Si_3O_{12}$ | 800–1000°C  > 15 kb | 4, 5 |
| Spinel | $Ni_2GeO_4$ | $P = 0$, 650°C | $Ni_2SiO_4$ | > 18 kb, 650°C | 6 |
| Spinel | $Co_2GeO_4$ | $P = 0$, 700°C | $Co_2SiO_4$ | > 70 kb, 700°C | 7 |
| Spinel | $Fe_2GeO_4$ | $P = 0$ 800°C | $Fe_2SiO_4$ | > 38 kb, 600°C | 8 |
| Spinel | $Mg_2GeO_4$ | $P = 0$ 600°C | $Mg_2SiO_4$ | > 130 kb,† 600°C | 9, 10 |
| Jadeite* | $NaAlGe_2O_6$ | > 12 kb, 600°C | $NaAlSi_2O_6$ | > 11.2 kb, 600°C | 11, 12 |

\* The differences between pressures required for stability of germanium and silicon jadeites are much smaller than the experimental uncertainty range.
† Extrapolated.
‡ 1. Robbins and Levin [1959].  2. Stishov and Popova [1961].  3. Ringwood and Seabrook [1962a].  4. Tauber et al., [1958].  5. Pistorius and Kennedy [1960]. 6. Ringwood [1962a].  7. Ringwood [1963].  8. Ringwood [1958b].  9. Dachille and Roy [1960].  10. Ringwood and Seabrook [1962b].  11. Dachille and Roy [1962].  12. Robertson et al. [1957].

particularly to oxide compounds in which the critical parameters are usually the radius ratios between the small cations and the large oxygen anions. When a silicate or germanate is subjected to high pressure, the large oxygen ions tend to contract relatively more than the small $Si^{4+}$ and $Ge^{4+}$ ions; hence the radius ratios $R_{Si}/R_O$ and $R_{Ge}/R_O$ increase. Transformation into a new high-pressure phase occurs when these radius ratios attain some critical value. Since the zero-pressure radius of $Ge^{4+}$ (0.48 A) is already slightly larger than that of $Si^{4+}$ (0.42 A), germanates require smaller pressures to achieve the critical radius ratios required for given transitions than the corresponding silicates do. Alternatively, because of their initially higher radius ratios, germanates may crystallize at zero pressure in a structure which is only attained by the silicate at high pressure.

For these reasons, the study of germanates as high-pressure models of silicates offers us the possibility of obtaining useful information about phase transformations which may occur in silicates at pressures beyond the range of currently available experimental techniques. It is also possible to take advantage of solid solubility relationships between silicates and germanates in systems in which the germanate forms a high-pressure phase but not the silicate. The study of such equilibria may provide thermodynamic data from which the stability field of the high pressure form of the silicate can be calculated. Quantitative information may also be obtained on a more empirical basis, simply by extrapolating to higher pressures phase boundaries determined in such systems over a limited range of pressures.

Examples of all of these uses of germanates will be given in the following sections.

### c. *The Olivine-Spinel Transition*

Olivine-spinel polymorphism was discovered by Goldschmidt (1931) in the compound $Mg_2GeO_4$. The olivine-spinel transition in $Mg_2GeO_4$ was confirmed by Dachille and Roy [1960], who also determined the *P-T* dependence of the transition. Bernal [1936] suggested by analogy with $Mg_2GeO_4$ that common olivine might transform in the mantle under a sufficiently high pressure to a spinel phase with a substantial increase in density. This suggestion was adopted by Jeffreys [1937] as the basis for an explanation of the 20 degree discontinuity.

Bernal's hypothesis was strongly supported by the results of Ringwood (1958b, 1959, 1962c, d, 1963) who discovered that the olivines $Fe_2SiO_4$, $Ni_2SiO_4$, and $Co_2SiO_4$ transformed to spinel structures at pressures between 15 and 70 kilobars in the temperature interval 600 to 700°C. The transformations were accompanied by a density

### Table 4

PARAMETERS OF OLIVINE-SPINEL TRANSITIONS

| Compound | Temperature (°C) | Transition Pressure (kb) | Increase in Density (%) |
|---|---|---|---|
| $Mg_2GeO_4$ | 820 | 0 | 8.5 |
| $Fe_2SiO_4$ | 600 | 38 ± 3 | 10.5 |
| $Ni_2SiO_4$ | 650 | 18 ± 5 | 9 |
| $Co_2SiO_4$ | 700 | 70 ± 20 | 9.5 |
| $Mg_2SiO_4$ | 600 | 130 ± 20* | ~9 |

* Extrapolated, see text.

increase of about 10 per cent (Table 4). Both $Fe_2SiO_4$ and $Ni_2SiO_4$ are minor components of natural olivines. The pressure for the olivine-spinel transition in fayalite was determined by Boyd and England [1960b] as 60 kb at 1600°C. This point together with that of Ringwood (Table 4) indicates a transition curve gradient of 45°C per kilobar. Dachille and Roy [1960] measured a gradient of 25° C/kb for the olivine-spinel transition in $Mg_2GeO_4$.

An olivine-spinel transition has not yet been found directly in $Mg_2SiO_4$ by static experiments, apparently because the high pressure required is beyond the range of existing apparatus. Accordingly, the olivine-spinel transition in $Mg_2SiO_4$ has been studied using germanate model systems as discussed earlier.

A quantitative estimate of the pressure required for the $Mg_2SiO_4$ olivine-spinel transition was obtained by Ringwood [1956, 1958a, c] from a thermodynamic investigation of solid solution equilibria at atmospheric pressure in the system $Ni_2GeO_4$—$Mg_2SiO_4$. He found that the transition should occur within the pressure interval 175 ± 55 kilobars at 1500°C. This pressure corresponded to a depth of approximately 500 ± 140 km which is within the transition zone. Confidence in the efficacy of the thermodynamic method for predicting transition pressures was encouraged when an analogous application to solid solubility equilibria in the system $Ni_2GeO_4$—$Ni_2SiO_4$ yielded a pressure for the olivine-spinel transition in $Ni_2SiO_4$ which agreed closely with the directly measured value [Ringwood, 1962c].

Another method for estimating the pressure of the olivine-spinel transition in $Mg_2SiO_4$ was used by Dachille and Roy [1960] and Ringwood and Seabrook [1962b]. This was based upon the direct extrapolation of phase boundaries determined over a wide range of pressures in the systems $Mg_2GeO_4$–$Mg_2SiO_4$, and $Ni_2GeO_4$–$Mg_2SiO_4$ (Figures 5 and 6). Dachille and Roy obtained a pressure of 100,000 ± 15,000 bars for the transition at 530°C, whereas Ringwood and Seabrook obtained 130 ± 20 kb at 600°C. As is seen in the figures, the latter work covered a more extensive solid solution range than the former, so that less extrapolation was required. Accordingly, the higher estimate of transition pressure is probably more accurate. This view is supported by the results of Sclar *et al*. [1964] who found that olivine was the stable form of $Mg_2SiO_4$ at pressures up to 115 kb at 500 to 800°C. The higher value is not inconsistent with the experimental data of Dachille and Roy if the curvature which they arbitrarily introduced into their phase boundaries is eliminated or reduced.

Densities of $Mg_2SiO_4$ spinel were obtained by calculation from extrapolated lattice parameters in the systems $Mg_2GeO_4$–$MgSiO_4$ and $Ni_2GeO_4$–$Mg_2SiO_4$. Dachille and Roy [1960] found that the spinel was 4.7 per cent denser than the olivine, whereas Ringwood and Seabrook estimated that the spinel was at least 9 per cent denser than the olivine. The differing results arise from the fact that the lattice parameters of the germanate-silicate solid solutions do not vary linearly with molar composition, hence an uncertainty is introduced into the extrapolation. Deviations from Vegard's rule are apparently caused by complex order-disorder equilibria involving the partition

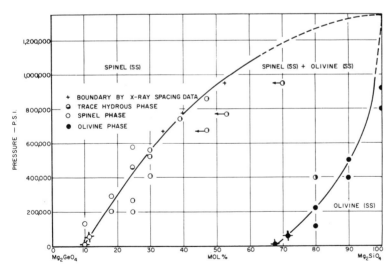

Fig. 5. The system $Mg_2GeO_4$–$Mg_2SiO_4$ at 542°C and 0 to 70 kb. After Dachille and Roy [1960].

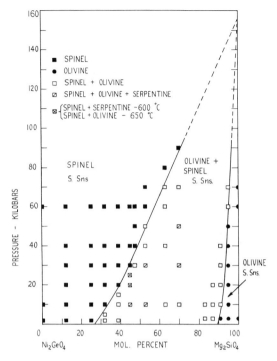

Fig. 6. The system $Ni_2GeO_4$—$Mg_2SiO_4$ at 600°C and 0 to 90 kb. After Ringwood and Seabrook [1962b].

of $Mg^{2+}$ and $Ge^{4+}$ ions at inverse and normal sites in the spinel lattice. The estimate by Ringwood and Seabrook is probably more accurate since it is based upon more extensive data and requires a smaller extrapolation. From Table 4, it is seen that the density change so obtained is in good agreement with those associated with other olivine-spinel transitions. McQueen et al. [1964] have recently found a phase transition in olivine by shock wave techniques. The transition occurs around 400 kb and involves a density increase of about 10 per cent. McQueen et al., suggest that this may be the olivine-spinel transition. They suggest that the high pressure at which it is encountered may be caused by the lower temperature attained during the shock compared with the temperatures in the earth. This explanation fails because they assume that the transition pressure has a negative temperature gradient, whereas it is almost certainly positive. All of the olivine-spinel transitions are sluggish, and it appears more likely that a considerable overpressure is needed before the transition is reached in the few microseconds available during the shock. The pressure measured was therefore probably much higher than the equilibrium transition pressure.

Dachille and Roy [1960] obtained evidence that $Mg_2SiO_4$ would form a normal spinel with $Si^{4+}$ ions in tetrahedral coordination and $Mg^{2+}$ in octahedral coordination. Tarte and Ringwood [1962] found that $Ni_2SiO_4$ spinel was also normal. Thus the transformations of silicate olivines into spinels do not involve changes in coordination of constituent ions.

Dachille and Roy [1960] concluded that $Mg_2GeO_4$ formed an inverse spinel while Tarte [personal communication] found it to be normal. It seems that $Mg_2GeO_4$ is close to the stability borderline for inverse and normal spinels and the structure may be dependent upon the $P$-$T$ conditions under which it was prepared.

*d. Quartz-Coesite-Stishovite Transitions*

The first high-pressure polymorph of quartz—"coesite"—was discovered by Coes [1953] in the course of his pioneering investigations in high-pressure mineralogy. The equilibrium between quartz and coesite was studied by MacDonald [1956], Boyd and England [1960a], and Dachille and Roy [1959]. The density of coesite is about 2.9, and its structure shows that the silicon ions remain in fourfold coordination [Zoltai and Buerger, 1959].

A major discovery was made by Stishov and Popova [1961], who showed that at a pressure of approximately 130 kb at 1600°C coesite transformed to a new phase possessing the rutile structure. The density of the new phase, "stishovite," was close to 4.3 gm/cc. The enormous increase in density was caused by the change in coordination of silicon from 4 to 6.

A natural occurrence of stishovite was described from Meteor Crater, Arizona by Chao et al. [1962], where it had been formed by shock during meteoritic impact. Laboratory syntheses of stishovite were subsequently reported by Wentorf [1962], Sclar et al. [1962], and Ringwood and Seabrook [1962a]. The latter authors found that the coesite-stishovite transition occurred close to 105 kb at 530°C. Sclar et al. [1962] synthesized minute amounts of stishovite at pressures as low as 80 kb. It is doubtful whether the stishovite field extends to pressures as low as this, since coesite was by far the most abundant phase produced in these experiments. It appears possible that small-scale inhomogeneity in pressure distribution within the charge may have been responsible for the production of stishovite at this low pressure.

Stishov [1963] estimated the standard entropy of stishovite by extrapolation of an observed regular density-entropy relationship displayed by the rutile type compounds $PbO_2$—$SnO_2$—$GeO_2$. Using this estimate together with experimental data previously obtained by himself on the coesite-stishovite transition and by others on the quartz-coesite transition, Stishov [1963] calculated an equilibrium $P$-$T$ curve for the coesite-stishovite transition (Figure 7). It is seen that the point at 530°C and 105 kb previously determined by Ringwood and Seabrook [1962a] falls close to the theoretical curve.

The importance of the discovery by Stishov and Popova is twofold. First, it is possible that stishovite may occur as a distinct phase in the mantle. The pressure required for its formation is reached in the upper regions of the transition zone. Second, the demonstration that the coordination of silicon can change from 4 to 6 under high pressure greatly increases the range of transformations which are possible in the deeper regions of the mantle. This is particularly important in the case of the pyroxenes and will be discussed in detail in the next section.

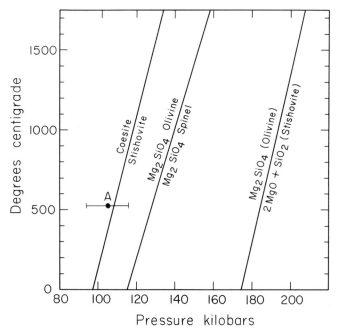

Fig. 7. Estimated equilibrium lines for the coesite-stishovite, $Mg_2SiO_4$ olivine-spinel and $Mg_2SiO_4$ olivine-oxides transformations [Stishov, 1963; Ringwood and Seabrook, 1962b]. The latter transformation involves a metastable equilibrium. The point A denotes an experimental determination of the coesite-stishovite transition by Ringwood and Seabrook [1962a].

As seen in Table 3, germanium dioxide displays both quartz and rutile polymorphs at zero pressure and thus behaves as a model for $SiO_2$.

### e. High-Pressure Transitions in Pyroxenes

Apart from some preliminary results by Sclar et al. [1964] on $MgSiO_3$, attempts to detect major high-pressure phase transitions in silicate pyroxenes have so far been unsuccessful. However, an investigation of the stability of several germanate pyroxenes at high pressure has provided considerable insight into the probable behavior of silicate analogs [Ringwood and Seabrook, 1963]. In every case, the germanate pyroxenes were found to be unstable at high pressures and transformed to one or more denser phases (Table 5). Accordingly, from the

### Table 5

HIGH-PRESSURE PHASE CHANGES IN GERMANATE PYROXENES AND RELATED COMPOUNDS AT $700 \pm 10°C$

| Compound | Initial Structure | Transformation Pressure (kb) | High-Pressure Phases | Density Increase due to Phase Trans. (approx.) (%) |
|---|---|---|---|---|
| $MgGeO_3$ | Orthopyroxene | $28 \pm 3$ | $MgGeO_3$ (ilmenite structure) | 15 |
| $MnGeO_3$ | Orthopyroxene | $25 \pm 5$ | $MnGeO_3$ (ilmenite structure) | 18 |
| $FeGeO_3$ | Clinopyroxene | $10 \pm 3$ | $Fe_2GeO_4$ (spinel) + $GeO_2$ (rutile) | 11 |
| $CoGeO_3$ | Clinopyroxene | $10 \pm 3$ | $Co_2GeO_4$ (spinel) + $GeO_2$ (rutile) | 11 |
| $(Mg_{0.75}Ni_{0.25})GeO_3$ | Clinopyroxene | $22 \pm 3$ | $(Mg_{0.75}Ni_{0.25})_2GeO_4$ (spinel) + $GeO_2$ (rutile) | 10 |
| $Mg(GeSi)O_3$* | Orthopyroxene | $\sim 50$ | $Mg_2(GeSi)O_4$ (spinel) + $(GeSi)O_2$ (rutile) | 15 |
| $CaGeO_3$ | Wollastonite? | $\sim 40$ | $CaGeO_3$ (garnet) | |
| $CaMgGe_2O_6$ | Diopside | $\sim 80$ | $CaGeO_3$ (garnet) + $MgGeO_3$ (ilmenite) | |

* Compositions of solid solutions ranged between $Mg(Ge_{0.9}Si_{0.1})O_3$ and $Mg(Ge_{0.5}Si_{0.5})O_3$

model relationships between silicates and germanates discussed in Section 3c, we might reasonably expect silicate pyroxenes to display analogous transformations at higher pressures.

All of the transformations in germanate pyroxenes involve either a complete or partial change in coordination of the germanium ions from fourfold to sixfold. In pure $GeO_2$, this change in coordination occurs at atmospheric pressure and at a temperature of 1007°C [Robbins and Levin, 1959]. However, in complex germanium oxycompounds, germanium almost invariably occurs in tetrahedral coordination at atmospheric pressure. It appears that the combination of a basic oxide with $GeO_2$ to form a germanate tends to stabilize germanium in tetrahedral coordination. There are straightforward reasons for this behavior which have been well treated in general terms by Weyl [1951, 1958].

Because of the relative stabilization of $GeO_4^{-4}$ groups in germanates as compared to $GeO_2$, higher pressures are required to cause changes in Ge coordination from 4 to 6 in germanates than in $GeO_2$. This generalization may reasonably be applied to silicates. It implies that high-pressure transformations in silicates which lead to octahedral coordination for silicon are likely to require pressures which are higher than those required for the 4 to 6 transformation in pure $SiO_2$. Thus, the equilibrium curve for the coesite-stishovite transition (Figure 7) will constitute a lower pressure limit for 4 to 6 transitions in silicates.

Possible transformations in magnesia-rich silicate pyroxenes, e.g., enstatite, may now be considered in the light of these generalizations. From the foregoing discussion it is improbable that enstatite will undergo a major* transformation below the stishovite stability region. This inference, drawn by Ringwood [1962c] and Ringwood and Seabrook [1963], has been experimentally verified by Sclar et al. [1964].

It is seen from Table 5 that germanate pyroxenes display two distinct modes of high-pressure transformation. They may either transform directly to an ilmenite structure, or they may break down into a denser mixture of orthogermanate and $GeO_2$ rutile. The probable mode of transformation of $MgSiO_3$ is seen from the behavior of $MgGeO_3$ and $MgGeO_3$—$MgSiO_3$ solid solutions (Table 5). Whereas $MgGeO_3$ transforms directly into an ilmenite structure, the substitution of only 10 per cent of the germanium by silicon completely changes the nature of the transformation. Solid solutions of $MgGeO_3$ and $MgSiO_3$ containing between 10 and 50 per cent of $MgSiO_3$ breakdown to

* This does not include transformations involving minor symmetry and volume changes such as the enstatite-clinoenstatite transition.

orthogermanate-silicate spinel solid solutions plus rutile solid solutions.

$$2\text{Mg}(\text{GeSi})\text{O}_3 \rightarrow \text{Mg}_2(\text{GeSi})\text{O}_4 + (\text{GeSi})\text{O}_2$$
orthopyroxene  spinel  rutile

The rutile solid solution contains very little silicon at pressures below 80 kb. At higher pressures, however, the solid solubility of $SiO_2$ in $GeO_2$ (rutile) is quite pronounced [Ringwood and Seabrook, unpublished results].

From a consideration of the results on germanates together with thermodynamic data and data on other transformations, Ringwood [1962c] and Ringwood and Seabrook [1963] concluded that enstatite would break down into forsterite plus stishovite around 120 kb and that the transformation pressure would be relatively insensitive to temperature:

$$2\text{MgSiO}_3 \rightarrow \text{Mg}_2\text{SiO}_4 + \text{SiO}_2 \qquad \Delta v = -5.2 \text{ cc}$$
enstatite  forsterite  stishovite

A measure of confirmation of this prediction has recently been provided by Sclar et al. [1964] in a study of the reaction of MgO and $SiO_2$ in a 1 to 1 ratio at pressures between 100 and 125 kb and temperatures between 450 and 800°C. At pressures below 115 kb, the reaction products were clinoenstatite plus traces of forsterite and stishovite. Above 115 kb, however, forsterite and stishovite became much more abundant amounting to 30 per cent of the charge. Sclar et al. suggest that their synthesis fields are indicative of chemical equilibrium and that the stability field of clinoenstatite is replaced by that of forsterite plus stishovite at 115 kb. The question of whether equilibrium was attained is not yet finally settled; nevertheless, their results are most suggestive.

From Table 5 it is seen that $CaGeO_3$, which possesses a structure similar to wollastonite ($CaSiO_3$) at low pressures, transforms to a garnet structure with the formula $Ca_3^{VIII}CaGe^{VI}Ge_3^{IV}O_{12}$ at high pressure. It is possible that wollastonite may behave likewise. Ringwood and Seabrook [1962a] found a transformation in wollastonite, but the volume change was only a few per cent and did not involve a change in silicon coordination. The garnet form did not appear when wollastonite glass devitrified at pressures up to 70 kb. An alternative mode of transformation of wollastonite might be to a perovskite structure. From previous discussion, this would require pressures in excess of 120 kb. Germanium diopside was observed to disproportionate at high pressure into $MgGeO_3$ (ilmenite) + $CaGeO_3$ (garnet) (Table 5). Because of the differing high-pressure behavior of $MgGeO_3$ and

MgSiO$_3$, it is possible that ordinary diopside may break down as follows:

$$2CaMgSi_2O_6 \rightarrow Mg_2SiO_4 + SiO_2 + 2CaSiO_3$$
$$\text{diopside} \qquad \text{forsterite} \quad \text{stishovite} \quad \text{garnet or perovskite}$$

*f. High-Pressure Breakdown of Mg$_2$SiO$_4$ Spinel*

Considering only the components of the mantle MgSiO$_3$ and Mg$_2$SiO$_4$, we have inferred in previous sections that MgSiO$_3$ would break down to forsterite plus stishovite around a depth of 400 km, and forsterite would transform to a spinel structure in the vicinity of 500 km. The stable phases at this depth would therefore be Mg$_2$SiO$_4$ spinel + stishovite. The density of the spinel is approximately 3.53 gm/cc. Since the silicon atoms in spinel are tetrahedrally coordinated, it is far from close-packed. It is therefore to be expected that the spinel will in turn transform at higher pressure to denser phases characterized by octahedral coordination of silicon [Ringwood, 1962c; Stishov, 1962].

Two such reactions have been proposed. Firstly, spinel might decompose into its constituent oxides

$$Mg_2SiO_4 \rightarrow 2MgO + SiO_2 \qquad \Delta v = -3.1 \text{ cc}$$
$$\text{spinel} \qquad \text{periclase stishovite}$$

We will discuss silicate-oxide decomposition reactions in the next section. Although they are thermodynamically possible and yield an import class of upper stability limits in the mantle, it is concluded that this type of breakdown is less likely to occur in the transition zone than the type which we discuss next.

An alternative mode of breakdown of spinel would be into one or more new binary phases composed of MgO and SiO$_2$, which possess approximately the same density as the equivalent mixture of periclase and stishovite. This type of breakdown is favored by the strong compound-forming tendency of MgO and SiO$_2$.

Once again, the high-pressure behavior of germanates and germanate-silicate solid solutions provides useful insight into the probable behavior of the silicates. Ringwood and Seabrook [1963] obtained evidence that Mg$_2$(GeSi)O$_4$ spinel solid solutions in equilibrium with (GeSi)O$_2$ rutile solid solutions became unstable above approximately 90 kb and reacted to form an ilmenite phase:

$$Mg_2(GeSi)O_4 + (GeSi)O_3 \rightarrow 2Mg(GeSi)O_3$$
$$\text{spinel} \qquad \text{rutile} \qquad \text{ilmenite}$$

This suggests that below 500 km, the following reaction might occur:

$$Mg_2SiO_4 + SiO_2 \rightarrow 2MgSiO_3 \qquad \Delta v = -2.3 \text{ cc}$$
$$\text{spinel} \quad \text{stishovite} \quad \text{ilmenite}$$

Furthermore, the same authors observed a breakdown of $Mg_2GeO_4$ spinel above 90 kb at 570°C into $MgGeO_3$ (ilmenite) and a phase which appeared to have the composition $5MgO \cdot GeO_2$. This suggests the possibility that $Mg_2SiO_4$ spinel might break down in an analogous manner, either to a mixture of periclase and $MgSiO_3$ ilmenite [Ringwood, 1962c], or to some new binary phase $xMgO \cdot SiO_2$ plus $MgSiO_3$ ilmenite:

$$Mg_2SiO_4 \rightarrow MgO + MgSiO_3 \qquad \Delta v = -2.7 \text{ cc}$$
$$\text{spinel} \quad \text{periclase} \quad \text{ilmenite}$$

The possibility that $MgSiO_3$ might ultimately transform into a corundum structure (essentially a disordered ilmenite structure) was first suggested by J. B. Thompson [Birch, 1952, p. 234]. The corundum and ilmenite structures are based on hexagonal close-packing of oxygen ions with the octahedral interstices occupied by the cations. Such structures are commonly formed between end members possessing rutile and rock salt structures. For example, the end members MgO, FeO, NiO, CaO, MnO, and CdO (rock-salt structures) all form binary ilmenite type compounds with $TiO_2$ (rutile structure). The density of the ilmenite compounds is generally between 1 and 4 per cent smaller than the mean of the end members.

The above relationships between end-members possessing rutile and rocksalt structures, such as MgO and $SiO_2$, strongly suggest that the formation of a binary compound $MgSiO_3$ with the ilmenite structure will be possible at sufficiently high pressure. In view of the model relationships previously discussed between silicates and germanates, the synthesis of an ilmenite form of $MgGeO_3$ also strongly supports the view that the ilmenite form of $MgSiO_3$ may play an important part in the lower parts of the transition zone according to the equilibria discussed.

*g. Decomposition of Silicates to Oxides under Pressure*

Another class of transitions which has been frequently discussed is the breakdown of silicates into their denser simple oxide components under high pressure, e.g.:

$$Mg_2SiO_4 \rightarrow 2MgO + SiO_2 \qquad \Delta v = -7.1 \text{ cm}^3$$
$$\text{forsterite} \quad \text{periclase} \quad \text{stishovite}$$

Pressures required for such transitions can be readily estimated from thermodynamic considerations. The free energy change $\Delta G$ associated with the previous reaction can be written

$$\Delta G = \Delta G_0 + P\,\Delta v \tag{1}$$

where $P$ is the pressure, $\Delta G_0$ is the free energy of the reaction at zero pressure, and $\Delta v$ is the molar volume change for the reaction. Because of the low compressibilities of the phases, $\Delta v$ can be assumed independent of pressure—at least in a first approximation. For equilibrium, $\Delta G = 0$; therefore, the equilibrium pressure is simply:

$$P = -\frac{\Delta G_0}{\Delta v} \tag{2}$$

$\Delta G_0$ can be calculated from the one-atmosphere thermodynamic properties of forsterite, quartz, and periclase, together with free energies for the quartz-coesite and coesite-stishovite transitions obtained from the experimental equilibrium curves and measured densities. Pressures required to cause breakdown according to the above reaction have been calculated by MacDonald [1962], Ringwood [1962c], and Stishov [1963]. The most precise calculations are those of Stishov, and his equilibrium curve for the breakdown of forsterite into periclase and stishovite is given in Figure 7.

The principal value of such calculations is that they set upper limits to the stability of upper-mantle minerals like forsterite and enstatite. Thus, if these minerals do *not* transform into other denser polymorphs at pressures smaller than those required for complete dissociation into oxides, then they will decompose into oxides at the calculated pressure. However, if transformation into a high-pressure polymorph of intermediate density occurs, the pressure for ultimate dissociation into oxides will be moved upward. The reason for this is seen in Figure 8. The curve *ABD* represents the equilibrium for the transition of Phase I into a second phase assemblage II. If, however, a phase or phase assemblage III with an intermediate density becomes stable at a lower pressure (curve *BC*) than that required for the I–II equilibrium, it is seen from Figure 8 that the pressure required for the ultimate stability of assemblage II is increased, and the calculated I–II equilibrium *BD* represents a metastable condition. This situation is realized in the case of $Mg_2SiO_4$ where, as seen in Figure 7, the pressure for the olivine-spinel transition is much smaller than the oxide-decomposition pressure. Accordingly, the pressure required for the latter will be much higher than calculated. It can be shown that in

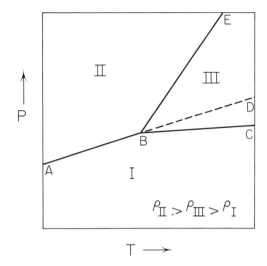

FIG. 8. Stable and metastable fields near a triple point.

general, because of the Relationship 2, the smaller the density difference between II (dense mixture of oxides) and III (binary compound with intermediate density), the higher will be the pressure required to reach the stability field of II.

This situation renders it doubtful whether the pressure required for oxide decomposition would be reached in the mantle. According to previous discussion, it is probable that with increasing pressure, $Mg_2SiO_4$ spinel breaks down into $MgO + MgSiO_3$ (ilmenite). When a binary compound with the ilmenite structure is formed between end members possessing rocksalt and rutile structures, its density is usually slightly less than the mean densities of its components—the difference amounting to between 1 and 4 per cent. Thus it appears possible that the combination $MgO + MgSiO_3$ ilmenite will be very slightly less dense than $MgO + SiO_2$ (stishovite). In principle, therefore, the latter combination could be formed by the decomposition of the former with increasing pressure. However, because of the small $\Delta v$ involved in such a breakdown, it appears from the previous discussion that the pressure required would be very much higher than required for stability of $MgSiO_3$ (ilmenite). A further effect favoring the stability of binary compounds like $MgSiO_3$ rather than discrete simple oxides arises from entropy considerations. At the high temperature attained in the mantle, Mg and Si ions in $MgSiO_3$ ilmenite will probably become disordered, leading to further stabilization because of the free energy of mixing. Also, as discussed in the next section, an additional contribution to the stability of $MgSiO_3$ (ilmenite) in the mantle would be made because of its probable ability to dissolve $Al_2O_3$ and $Cr_2O_3$.

Finally, there is the possibility of further polymorphic changes leading to structures which are slightly denser than $MgSiO_3$ ilmenite. The ultimate transition of $MgSiO_3$ into a perovskite structure is a possibility. In the ideal perovskite structure, Mg ions would be in eightfold coordination and silicon ions in octahedral coordination. The perovskite structure is frequently formed between oxide components possessing rocksalt and rutile structures and is usually slightly denser than the mean density of the oxide components. Transformation of $MgSiO_3$ (ilmenite) into a perovskite in the lower mantle would bypass an otherwise possible decomposition into oxides.

For the preceding reasons, the author does not believe it probable that oxide decomposition transitions which have been considered by many authors, e.g., Birch [1952, 1963], MacDonald [1956, 1962], Shimazu [1958], Stishov [1962], and Ringwood [1962c], actually occur in the mantle. Investigations of such transitions are, however, of considerable value since they set upper stability limits for the common mineral phases of the mantle and also provide limits to the density of the phases in the lower mantle.

*h. The Transition Zone—Synthesis*

We will now consider the sequence of transitions which, in the light of the preceding discussions, appear likely to occur in the simple system $Mg_2SiO_4$—$MgSiO_3$ over the *P-T* range in the earth's outer mantle. It seems reasonably certain that $Mg_2SiO_4$ olivine and $MgSiO_3$ enstatite would be stable phases down to a depth of approximately 400 km. Around this depth, enstatite would break down to forsterite plus stishovite. Next, around 500 km, forsterite would transform into a spinel structure, so that the stable phases would be $Mg_2SiO_4$ (spinel) and stishovite. At greater depth, $Mg_2SiO_4$ spinel becomes unstable, breaking down to phases characterized by octahedral coordination of silicon. The most probable reactions are believed to involve transformations leading to the formation of an $MgSiO_3$ phase possessing the ilmenite structure. The suggested sequence given in Table 6 and estimated densities of phase assemblages are given in Table 7.

The depths at which transitions (3) and (4) (Table 6) occur is not known, but it is plausible to assume that they occur in the interval 500 to 900 km. An upper limit to the pressure at which such transitions must occur is provided by an investigation of the equilibrium

$$\underset{\text{spinel}}{Mg_2SiO_4} \rightarrow 2MgO + \underset{\text{stishovite}}{SiO_2} \qquad \Delta v = -3.1 \text{ cc}$$

### Table 6

PROBABLE SEQUENCE OF HIGH-PRESSURE TRANSITIONS IN SYSTEM $Mg_2SiO_4$—$MgSiO_3$

---

1. $2MgSiO_3$ (enstatite) → $Mg_2SiO_4$ (forsterite) + $SiO_2$ (stishovite)
   $\Delta v = -5.2$ cc
2. $Mg_2SiO_4$ (forsterite) → $Mg_2SiO_4$ (spinel)   $\Delta v = -4.0$ cc
3. $Mg_2SiO_4$ (spinel) + $SiO_2$ (stishovite) → $2MgSiO_3$ (ilmenite)   $\Delta v = -2.3$ cc
4. $Mg_2SiO_4$ (spinel) → $MgSiO_3$ (ilmenite) + $MgO$ (periclase)   $\Delta v = -2.7$ cc

---

### Table 7

PROBABLE ZERO PRESSURE DENSITIES OF HIGH-PRESSURE EQUIVALENTS OF $Mg_2SiO_4$ AND $MgSiO_3$

| | gm/cm³ |
|---|---|
| A. $Mg_2SiO_4$ Composition | |
| $Mg_2SiO_4$ (forsterite) | 3.21 |
| $Mg_2SiO_4$ (spinel)* | 3.53 |
| $MgO + MgSiO_3$ (ilmenite) | 3.79 |
| $2MgO + SiO_2$ (stishovite) | 3.83 |
| B. $MgSiO_3$ Composition | |
| $MgSiO_3$ enstatite | 3.19 |
| $Mg_2SiO_4$ (olivine + $SiO_2$ (stishovite) | 3.47 |
| $Mg_2SiO_4$ (spinel) + $SiO_2$ (stishovite) | 3.73 |
| $MgSiO_3$ (ilmenite)† | 3.90 |
| $MgO + SiO_2$ (stishovite) | 3.96 |

\* Ringwood and Seabrook [1962].
† Density estimated from average density changes for equilibria of the type
AO (rocksalt structure) + $BO_2$ (rutile structure) → $ABO_3$ (ilmenite structure)

The pressure for the breakdown can be roughly estimated according to the procedure given in Section 3g. Ringwood [1962c] estimated that it would occur at a depth of approximately 1000 km. As discussed earlier, it is more probable that the calculation refers to a metastable equilibrium, since transformation to other structures of intermediate density occurs at a lower pressure.

As seen from Table 7, the series of transformations involve an increase in zero-pressure density from 3.2 to between 3.8 and 3.9 gm/cm³. A most interesting result is that in such a simple system as $MgO$—$SiO_2$, the ultimate attainment of a close-packed state should involve successive transformations through several states of intermediate density. This is responsible for the considerable width of the phase transition zone in the mantle, according to the discussion in Section 3$h$. The explanation of this wide transition interval in terms of phase transitions had previously been cited as a difficulty in the phase change hypothesis [Verhoogen, 1954].

The influence of the less abundant components of the mantle (FeO, NiO, $Al_2O_3$, $Cr_2O_3$, $Fe_2O_3$, CaO, MnO, $Na_2O$, and $TiO_2$) upon the basic equilibria involving MgO and $SiO_2$ remains to be discussed. All of these components, except $Na_2O$ and CaO, are readily able to enter the spinel and ilmenite structures. It is therefore highly probable that at the high temperatures of the mantle these components will enter into solid solutions with the predominating spinel and ilmenite phases. In addition, FeO and NiO can enter the periclase structure. The radius of $Ca^{++}$ (0.99 Å) is probably too large to permit its ready substitution for $Mg^{++}$ (0.66 Å) in the close-packed and highly compressed phases under consideration. It seems more likely that Ca occurs in a separate phase possibly $CaSiO_3$ (perovskite) as previously discussed. If such a phase were formed, manganese ($Mn^{++}$ 0.80 Å) and perhaps much of the available $Fe^{++}$ (0.75 Å) would probably enter, replacing $Ca^{++}$. In addition, a perovskite structure would form a satisfactory locus for $Na^+$ (0.97 Å), which could enter via a coupled substitution of $Na^+ + Al^{+++}$ (or $Cr^{+++}$ or $Fe^{+++}$) for 2 $Ca^{++}$, as is common in minerals of the upper mantle and crust.

The principal effect of solid solution formation is to spread the individual transitions over a substantial depth range. Accordingly, the increase in density due to phase transitions between 400 and 1000 km will not occur in a series of discrete jumps accompanied by first-order seismic discontinuities, but will be somewhat smeared out. Although the density increase in the transition zone is probably smooth, it is not necessarily uniform. It is not possible to choose between the uniform increase of velocity in the transition zone [Gutenberg, 1959$b$; and Jeffry, 1939] (Figure 1) and a model of the type proposed by Anderson (1964$a$, $b$), in which most of the velocity increase in the transition zone is concentrated in two narrower regions. If the latter distribution should prove to be preferable, it might be explained by occurrence of transitions (1) and (2) of Table 6 within a narrow zone around 400 to 500 km, and the transitions (3) and (4) in a limited region in the vicinity of 800 to 900 km.

## 4. The Lower Mantle: 900 to 2900 Kilometers

Below 900 km the seismic velocities increase at a uniform rate with depth until 2700 km, when they flatten off or even decrease slightly until the boundary of the core is reached at 2900 km.

Birch [1952] discussed the constitution of the lower mantle in detail. He concluded that the rate of increase of seismic velocity with depth is consistent with the region being homogeneous on a large scale. There is no evidence suggesting *major* phase changes or significant changes in chemical composition. Continuing the argument of the previous section, it appears that silicates achieve a state of closest atomic packing at 1000 km and that once this state is reached, further major transitions involving changes in crystal structure are not possible, so that the region remains homogeneous on a large scale.

Birch [1952] estimated that the density of the mineral phases of the lower mantle, when reduced to zero pressure, was 4.0 gm/cm$^3$. Clark and Ringwood [1964], using a different method, obtained 4.1 gm/cm$^3$. An allowance for the probable effect of temperature leads to zero $P$-$T$ densities in the range 4.1 to 4.25 gm/cm$^3$. The density of close-packed MgO—SiO$_2$ phases at the bottom of the transition zone was estimated as 3.85 gm/cm$^3$ (Table 7). This value would be increased to about 3.90 if allowance is made for the effect of other components such as Ca, Al, and Cr. (It was previously suggested that Ca may occur in a dense perovskite structure; the density of corundum Al$_2$O$_3$ (4.0) exceeds that of MgSiO$_3$ ilmenite (3.9) while the higher atomic weight of Cr will also produce a small effect.) However, the principal additional component affecting density is iron. If the FeO/(FeO + MgO) ratio of high-pressure phases previously inferred to be present at the lower mantle is between 0.1 and 0.2, then their densities (reduced to low temperature and pressure) are in the range 4.05 to 4.2 gm/cm$^3$ [Clark and Ringwood, 1964]. These values are in essential agreement with the reduced densities for the lower mantle mentioned earlier.

The elastic ratio ($K/\rho$ (where $K$ = incompressibility and $\rho$ = density) of the phases of the lower mantle when reduced to low temperature and pressure, is about 60 (km/sec)$^2$. Birch [1952] and Clark and Ringwood [1964] have shown that the elastic ratio of the high-pressure phase assemblage inferred to be present on the basis of previous discussions is in close agreement with this value. Accordingly, it appears that the sequence of phase changes investigated is capable of providing a rather complete explanation of the presently known physical properties of the lower mantle.

The sequence of phase transitions believed to occur in the transition zone has been based upon transitions which are known or which are inferred indirectly from the behavior of model germanate systems. The limitations of this approach should not be ignored. It is always possible that presently unknown and unsuspected phase transitions may play a significant role and alter the sequence given in Table 6. Accordingly, it is most desirable to develop and apply apparatus capable of maintaining static pressures in the vicinity of 300,000 atmospheres so that this question can be studied directly. Nevertheless, the fact that the presently inferred series of transitions is capable of providing such a complete and satisfying explanation of the properties of the transition zone and lower mantle suggests that we are not far from the correct answer.

On the other hand, additional *minor* transitions involving comparatively small volume changes ( < 5 per cent) cannot be ruled out. Transitions of $MgSiO_3$ from an ilmenite to a perovskite structure and MgO to a cesium chloride structure involve comparatively small overall density differences and must be considered. It is also possible that minor transitions involving rearrangement of outer electronic orbitals may occur without accompanying changes in crystal structure. Indeed Gutenberg [1958] has obtained some evidence for minor irregularities in the lower mantle, which, however, are not large enough to shake Birch's basic conclusion.

A further suggestion regarding the phases of the lower mantle has been made by Stishov [1962]. He suggests that, because of the high temperature in the deep mantle and the accompanying tendency towards extensive solid solution of phases, there might be only one phase present in the lower mantle. It would be characterized by octahedral coordination of all cations and would consist of a highly defective NaCl structure. The suggestion is not inconsistent with the model advocated herein if a defective ilmenite structure (for which there is experimental support) is substituted for the hypothetical defective NaCl type structure.

The cation coordinations in both structures are similar, but the ilmenite is likely to be closer packed than the proposed NaCl structure. It was mentioned previously that such an ilmenite structure would consist of a complex solid solution. However the author is inclined to doubt whether the trend toward extensive solid solubility would be sufficient to result in a single homogeneous lower-mantle phase. The enormous pressure in this region would not favor the existence of a high concentration of lattice defects (vacancies) in the phases present.

At the base of the lower mantle, between 2700 km and 2900 km, the seismic velocities flatten off and may even decrease slightly. The effect could be explained by the presence of some metallic iron in this region. In the accompanying paper it was inferred that the core was out of equilibrium with the mantle. Ringwood [1961] suggested that the metallic iron was formed by reactions arising out of core/mantle disequilibrium.

## 5. Prospects

One of the chief objectives of the solid earth sciences is to determine the variation of a wide variety of physical properties with depth and position in the earth and then to explain and interpret these over-all properties in terms of the nature, composition, and properties of the individual mineral phases present. Major progress has been made during the past ten years toward achievement of this objective. In the upper mantle it is now possible to present a qualitative interpretation of the nature of the seismic velocity distribution in terms of the stability fields and properties of minerals and of mineral assemblages as determined in the laboratory by high-pressure–temperature investigations. Solutions which have so far been proposed are nonunique, and the objective of investigators has been toward setting up self-consistent models. The next steps must be toward a quantitative and unique understanding of the physical and chemical constitution of the upper mantle. This objective is within reach using presently existing techniques or modest advances thereon.

The entire field of $P$-$T$ conditions occurring in the upper mantle is within the range of laboratory static high-temperature–pressure apparatus. Thus, it is possible to investigate directly all significant phase relationships occurring in the upper mantle, to prepare the phases, and to measure their properties. During the coming years, we may hope that seismologists will produce more refined velocity-depth distributions characteristic of the different major geologic provinces of the earth. It should then be possible to interpret more rigorously the seismic structure in terms of mineralogical and chemical constitution. A field which urgently requires expansion is the laboratory determination of the elastic properties of individual mantle minerals over a wide range of temperatures and pressures. When sufficient information on these properties has accumulated, we shall be able to interpret the seismic velocity profiles quantitatively in terms of mineralogical constitution and temperature. Such are the prospects for the upper mantle.

Within the next few years, apparatus capable of developing the temperature and pressure occurring throughout most of the transition zone will be developed. Thus, it will be possible to prepare directly the high-pressure phases which are responsible for the transition zone and to measure their properties. A more refined picture of the constitution of the transition zone and lower mantle will then emerge, although even now some of the major properties of these regions are reasonably well understood. The use of shock-wave techniques to determine the equations of state of high-pressure phases is also certain to enlarge our understanding of the physical state of the lower mantle.

Before this recital of prospects is condemned for overoptimism, it should be mentioned that the objective of achieving a detailed understanding of the present constitution of the mantle is but one of many major objectives of the solid earth sciences. Other outstanding problems include the evolution of the earth as a function of time, thermal history, the mechanics of the upper mantle, the ultimate causes of mountain building, crustal uplift and horizontal displacement and volcanism. Solutions to these problems are likely to prove much more difficult, partly because the geologic time dimension is difficult to simulate in the laboratory. Nevertheless, it is the author's belief that solutions of these fundamental and difficult problems will be closely linked to and dependent upon advances in our knowledge of the earth's present constitution. This statement may appear trite. It should be remembered, however, that in the past, many of the conjectures regarding the ultimate causes of orogeny have been based upon proposed mechanisms, e.g., large convection cells extending throughout the entire depth of the mantle, possessing consequences which are in direct conflict with a great deal of the evidence concerning the present constitution of the mantle. In the author's opinion, a great deal of wasted effort might be spared if those who are concerned with theories of orogeny and crustal displacement worked within the framework set by knowledge and inferences of its present constitution.

## 6. Summary

The mantle may be subdivided into 3 distinct regions on the basis of the seismic velocity distributions. These regions are the Upper Mantle ($M$ to 400 km), the Transition Zone (400 to 900 km) and the Lower Mantle (900 to 2900 km). Recent evidence on the phase constitution of each of these regions is reviewed. The upper mantle is characterized by "normal" mineralogy, being composed dominantly of olivine, pyroxenes, and garnet. It is believed to be chemically zoned, with dunite-peridotite immediately below the continental $M$ discontinuity passing downward into a more primitive rock "pyrolite," which is chemically equivalent to a 3:1 mixture of peridotite and basalt. Beneath oceans the dunite-peridotite may be absent so that the primitive pyrolite extends downward from the $M$ discontinuity. The effect of the range of $P$-$T$ conditions in the upper mantle on the mineralogy of pyrolite is considered. Pyrolite may crystallize in four distinct mineral assemblages each characterized by a specific stability field. This gives rise to large-scale mineralogical zoning in the upper mantle which in turn exercises an important influence on the seismic velocity distribution and, particularly, on the origin of the low-velocity zone. The subject is discussed in some detail, and a close correspondence between recent seismic velocity distributions and the consequences of the pyrolite model is demonstrated.

The transition zone between 400 and 900 km is characterized by a series of major phase transformations. The stabilities of olivines and pyroxenes at high pressure are discussed in the light of recent experimental evidence. This indicates that, at a depth of approximately 400 km, magnesian pyroxenes break down into forsterite plus stishovite. Around 500 km, forsterite transforms to a spinel structure with a density increase of 10 per cent. Between 600 and 900 km, the spinel in turn transforms to denser phases characterized by octahedral coordination of silicon. It is probable that the principal phase so produced is $MgSiO_3$ (ilmenite), and that complete breakdown into simple oxides does not occur. The series of phase transformations results in a total density increase of about 20 per cent. Because of solid solution effects, individual transitions occur over distinct depth intervals so that first-order seismic discontinuities are not to be expected.

At 900 km, the phases present are in a state of close packing, and further major phase transformations are unlikely on structural grounds. Accordingly the lower mantle is essentially homogeneous between 900 and 2900 km. The density and elastic ratio of the material of the lower mantle resulting from the phase transformations discussed agree closely with estimates for these quantities obtained from independent geophysical evidence.

## Acknowledgment

This paper was written at the Geophysical Laboratory, Washington, while the author was the recipient of a Carnegie Fellowship. He wishes to express his gratitude to the Director of the Geophysical Laboratory and the Carnegie Institution for hospitality and support.

## References

Ahrens, L. H., The use of ionization potentials 1, *Geochim. Cosmochim. Acta*, *2*, 155–169, 1952.

Anderson, D. L., Universal dispersion tables 1, Love waves across oceans and continents in a spherical earth, *Bull. Seismol. Soc. Am.*, *54*, 681–726, 1964a.

Anderson, D. L., Recent evidence concerning the structure of the upper mantle from the dispersion of long period surface waves, preprint, 1964b.
Bernal, J. D., Discussion, *Observatory*, *59*, 268, 1936.
Birch, F., The variation of seismic velocities within a simplified earth model in accordance with the theory of finite strain, *Bull. Seismol. Soc. Am.*, *29*, 463–479, 1939.
Birch, F., Elasticity and constitution of the earth's interior, *J. Geophys. Res.*, *57*, 227–286, 1952.
Birch, F., The velocity of compressional waves in rocks to 10 kilobars 1, *J. Geophys. Res.*, *65*, 1083–1102, 1960.
Birch, F., The velocity of compressional waves in rocks to 10 kilobars 2, *J. Geophys. Res.*, *66*, 2199–2224, 1961a.
Birch, F., Composition of the earth's mantle, *Geophys. J.*, *4*, 295–311, 1961b.
Birch, F., Some geophysical applications of high pressure research, in *Solids Under Pressure*, edited by W. Paul and D. M. Warschauer, Chap. 6, pp. 137–162, McGraw-Hill Book Company, Inc., New York, 1963.
Birch, F., J. F. Schairer, and H. C. Spicer, Handbook of Physical Constants, *Geol. Soc. Am. Spec. Papers*, *36*, 1942.
Birch, F., and P. Le Compte, Temperature-pressure plane for albite composition, *Am. J. Sci.*, *258*, 209–217, 1960.
Boyd, F. R., and J. L. England, Pyrope, *Carnegie Inst. Wash. Yearbook*, *58*, 83–87, 1959.
Boyd, F. R., and J. L. England, The quartz-coesite transition, *J. Geophys. Res.*, *65*, 749–756, 1960a.
Boyd, F. R., and J. L. England, Minerals of the mantle, *Carnegie Inst. Wash. Yearbook*, *59*, 48–52, 1960b.
Boyd, F. R., and J. L. England, The system enstatite-pyrope, *Carnegie Inst. Wash. Yearbook*, *63*, 157–161, 1964.
Bradley, R. S., A. K. Jamil, and D. C. Munro, Electrical conductivity of fayalite and spinel, *Nature*, *193*, 965–966, 1962.
Brune, J., and J. Dormann, Seismic waves and earth structure in the Canadian shield, *Bull. Seismol. Soc. Am.*, *53*, 167–210, 1963.
Bullen, K. E., The variation of density and the ellipticities of strata of equal density within the earth, *Monthly Notices Roy. Astron. Soc., Geophys. Suppl.*, *3*, 395–401, 1936.
Bullen, K. E., Note on the density and pressure inside the earth, *Trans. Roy. Soc. New Zealand*, *67*, 122–124, 1937.
Bullen, K. E., The problem of the earth's density variation, *Bull. Soc. Seismol. Am.*, *30*, 235–250, 1940.
Chao, E. C. T., J. J. Fahey, J. Littler, and D. J. Milton, Stishovite, $SiO_2$, a very high pressure new mineral from meteor crater, Arizona, *J. Geophys. Res.*, *67*, 419, 1962.
Clark, S. P., and A. E. Ringwood, Density distribution and constitution of the mantle, *Rev. Geophys.*, *2*, 35–88, 1964.
Coes, L., A new dense crystalline silica, *Science*, *118*, 131–133, 1953.
Dachille, F., and R. Roy, High-pressure region of the silica isotypes, *Z. Krist.*, *111*, 451–462, 1959.
Dachille, F., and R. Roy, High pressure studies of the system $Mg_2GeO_4$—$Mg_2SiO_4$ with special reference to the olivine-spinel transition, *Am. J. Sci.*, *258*, 225–246, 1960.

Dachille, F., and R. Roy, Chapter 9, in *Modern Very High Pressure Techniques*, edited by R. H. Wentorf, Butterworth and Company, London, 1962.

Dawson, J. B., Basutoland Kimberlites, *Bull. Geol. Soc. Am.*, *73*, 545–560, 1962.

Goldschmidt, V. M., Zur Kristallchemie des Germaniums, *Nachr. Ges. Wiss. Goettingen, Math.-Physik. Kl. I*, *2*, 184–190, 1931.

Green, D. H., Alumina content of enstatite in a Venezuelan high-temperature peridotite, *Bull. Geol. Soc. Am.*, *74*, 1397–1402, 1963.

Green, D. H., The petrogenesis of the high-temperature peridotite intrusion in the Lizard area, Cornwall, *J. Petrol.*, *5*, 134–188, 1964.

Green, D. H., and A. E. Ringwood, Mineral assemblages in a model mantle composition, *J. Geophys. Res.*, *68*, 937–945, 1963.

Green, D. H., and A. E. Ringwood, Fractionation of basalt magmas at high pressures, *Nature*, *201*, 1276–1279, 1964.

Gutenberg, B., Velocity of seismic waves in the earth's mantle, *Trans. Am. Geophys. Union*, *39*, 486–489, 1958.

Gutenberg, B., *Physics of the Earth's Interior*, vol. 1, International Geophysics Series, edited by J. V. Miegham, Academic Press, New York, 1959a.

Gutenberg, B., The asthenosphere low-velocity layer, *Ann. Geofis. Rome*, *12*, 439–460, 1959b.

Jeffreys, H., On the materials and density of the earth's crust, *Monthly Notices Roy. Astron. Soc., Geophys. Suppl.*, *4*, 50–61, 1937.

Jeffreys, H., The Times of P, S, and SKS and the velocities of P and S, *Monthly Notices Roy. Astron. Soc., Geophys. Suppl.*, *4*, 498–533, 1939.

Kovach, R. L., and D. L. Anderson, Higher mode surface waves and their bearing on the structure of the earth's mantle, *Bull. Seismol. Soc. Am.*, *54*, 161–182, 1964.

Kushiro, I., and H. S. Yoder, Experimental studies on the basalt-eclogite transformation, *Carnegie Inst. Wash. Yearbook*, *63*, 108–114, 1964.

Lehmann, I., Velocities of longitudinal waves in the upper parts of the earth's mantle, *Ann. Geophys.*, *15*, 93–118, 1959.

Lehmann, I., S and the structure of the upper mantle, *Geophys. J.*, *4*, 124–138, 1961.

McDonald, K. L., Penetration of the geomagnetic secular variation through a mantle with variable conductivity, *J. Geophys. Res.*, *62*, 117–130, 1957.

MacDonald, G. J. F., Quartz-coesite stability relations at high temperatures and pressures, *Am. J. Sci.*, *254*, 713–721, 1956.

MacDonald, G. J. F., On the internal constitution of the inner planets, *J. Geophys. Res.*, *67*, 2945–2974, 1962.

MacDonald, G. J. F., and N. F. Ness, A study of the free oscillations of the earth, *J. Geophys. Res.*, *66*, 1865–1911, 1961.

MacGregor, I. D., The reaction 4 enstatite + spinel = forsterite + pyrope, *Carnegie Inst. Wash. Yearbook*, *63*, 156–157, 1964.

MacGregor, I. D., and F. R. Boyd, Ultamafic rocks, *Carnegie Inst. Wash. Yearbook*, *63*, 152–156, 1964.

MacGregor, I. D., and A. E. Ringwood, The natural system enstatite-pyrope, *Carnegie Inst. Wash. Yearbook*, *63*, 161–163, 1964.

McQueen, R. G., J. N. Fritz, and S. P. Marsh, On the composition of the earth's interior, *J. Geophys. Res.*, *69*, 2947–2978, 1964.

Oxburgh, E. R., Upper mantle inhomogeneity and the low-velocity zone, *Geophys. J.*, *8*, 456–462, 1964.

Pistorius, W., and G. C. Kennedy, Stability relations of grossularite and hydrogrossularite at high temperatures and pressures, *Am. J. Sci.*, *258*, 247–257, 1960.
Ringwood, A. E., The olivine-spinel transition in the earth's mantle, *Nature*, *178*, 1303–1304, 1956.
Ringwood, A. E., The constitution of the mantle 1. Thermodynamics of the olivine-spinel transition, *Geochim. Cosmochim. Acta*, *13*, 303–321, 1958a.
Ringwood, A. E., The constitution of the mantle 2. Further data on the olivine-spinel transition, *Geochim. Cosmochim. Acta*, *15*, 18–29, 1958b.
Ringwood, A. E., The constitution of the mantle 3, Consequences of the olivine-spinel transition, *Geochim. Cosmochim. Acta*, *15*, 195–212, 1958c.
Ringwood, A. E., The olivine-spinel inversion in fayalite, *Am. Mineralogist*, *44*, 659–661, 1959.
Ringwood, A. E., Silicon in the metal phase of enstatite chondrites and some geochemical implications, *Geochim. Cosmochim. Acta*, *25*, 1–13, 1961.
Ringwood, A. E., A model for the upper mantle, *J. Geophys. Res.*, *67*, 857–866, 1962a.
Ringwood, A. E., A model for the upper mantle, 2, *J. Geophys. Res.*, *67*, 4473–4477, 1962b.
Ringwood, A. E., Mineralogical constitution of the deep mantle, *J. Geophys. Res.*, *67*, 4005–4010, 1962c.
Ringwood, A. E., Prediction and confirmation of olivine-spinel transition in $Ni_2SiO_4$, *Geochim. Cosmochim. Acta*, *26*, 457–469, 1962d.
Ringwood, A. E., Olivine-spinel transformation in cobalt ortho-silicate, *Nature*, *198*, 79–80, 1963.
Ringwood, A. E., and D. H. Green, Experimental investigations bearing on the nature of the Mohorovicic Discontinuity, *Nature*, *201*, 566–567, 1964.
Ringwood, A. E., I. D. MacGregor, and F. R. Boyd, Petrological constitution of the upper mantle, *Carnegie Inst. Wash. Yearbook*, *63*, 147–152, 1964.
Ringwood, A. E., and M. Seabrook, Some high-pressure transformations in pyroxenes, *Nature*, *196*, 883–884, 1962a.
Ringwood, A. E., and M. Seabrook, Olivine-spinel equilibria at high pressure in the system $Ni_2GeO_4$—$Mg_2SiO_4$, *J. Geophys. Res.*, *67*, 1975–1985, 1962b.
Ringwood, A. E., and M. Seabrook, High-pressure transformations in germanate pyroxenes and related compounds, *J. Geophys. Res.*, *68*, 4601–4609, 1963.
Robertson, E., F. Birch, and G. J. F. MacDonald, Experimental determination of jadeite stability relations to 25,000 bars, *Am. J. Sci.*, *255*, 115–137, 1957.
Robbins, L. R., and E. M. Levin, The system magnesium oxide-germanium dioxide, *Am. J. Sci.*, *257*, 63–70, 1959.
Runcorn, S. K., The electrical conductivity of the earth's mantle, *Trans. Am. Geophys. Union*, *36*, 191–198, 1955.
Sclar, C. B., L. C. Carrison, and C. M. Schwartz, High-pressure reaction of clinoenstatite to forsterite plus stishovite, *J. Geophys. Res.*, *69*, 325–330, 1964.
Sclar, C. B., A. P. Young, L. C. Carrison, and C. M. Schwartz, Synthesis and optical crystallography of stishovite, a very high pressure polymorph of $SiO_2$, *J. Geophys. Res.*, *67*, 4049–4054, 1962.
Shimazu, Y., A chemical phase transition hypothesis on the origin of the C-layer within the mantle of the earth, *J. Earth Sci. Nagoya Univ.*, *6*, 12–30, 1958.
Shurbet, D. H., The high-frequency $S$ phase and the structure of the upper mantle, *J. Geophys. Res.*, *69*, 2065–2070, 1964.

Stishov, S. M., On the internal structure of the earth, *Geokhimiya*, No. 8, 649–659, 1962.

Stishov, S. M., Equilibrium line between coesite and the rutile-like modification of silica (In Russian), *Dokl. Akad. Nauk SSSR*, *148*, no. 5, 1186–1188, 1963.

Stishov, S. M., and S. V. Popova, New dense polymorphic modification of silica, *Geokhimiya*, no. 10, 837–839, 1961.

Takeuchi, H., M. Saito, and N. Kobayashi, Study of shear velocity distribution in the upper mantle by mantle Rayleigh and Love waves, *J. Geophys. Res.*, *67*, 2831–2839, 1962.

Tarte, P., and A. E. Ringwood, Infrared spectra of the spinels $Ni_2SiO_4$, $Ni_2GeO_4$, and their solid solutions, *Nature*, *193*, 971–972, 1962.

Tauber, A., E. Banks, and H. Kedesy, Synthesis of germanate garnets, *Acta Cryst.*, *11*, 893–894, 1958.

Tozer, D. C., The electrical properties of the earth's interior, in *Physics and Chemistry of the Earth*, edited by L. H. Ahrens, F. Press, K. Rankama, and S. K. Runcorn, Chap. 8, pp. 414–437, Pergamon Press, New York, 1959.

Tröger, W. E., *Optische Bestimmung det gesteinsbildenden Minerale*, vol. 1, E. Schwiezenbart'sche Verlag, Stuttgart, 1959.

Verhoogen, J., Elasticity of olivine and constitution of the earth's mantle, *J. Geophys. Res.*, *58*, 337–346, 1954.

Verma, R. K., Elasticity of some high-density crystals, *J. Geophys. Res.*, *65*, 757–766, 1960.

Wentorf, R. H., Stishovite synthesis, *J. Geophys. Res.*, *67*, 3648, 1962*a*.

Wentorf, R. H., Chemistry at high pressure, preprint, Conference on Physics and Chemistry of High-Pressure, Society of Chemical Industry, London, 1962*b*.

Weyl, W. A., *Colored Glasses*, Society of Glass Technology, Sheffield, England, 1951.

Weyl, W. A., Acid-base relationship in glass systems, *Glass Ind.*, Nos. 5 and 6, 1956.

Yoder, H. S., and Chinner, Almandite-pyrope-water system at 10,000 bars, *Carnegie Inst. Wash. Yearbook*, *59*, 81–84, 1960.

Yoder, H. S., and C. E. Tilley, Origin of basalt magma: an experimental study of natural and synthetic rock systems, *J. Petrol.*, *3*, 342–532, 1962.

Zoltai, T., and M. J. Buerger, The structure of coesite, the dense high-pressure form of silica, *Z. Krist.*, *111*, 129–141, 1959.

# PART V

# THE "SOLID" EARTH II

# EARTH HEAT FLOW MEASUREMENTS IN THE LAST DECADE

*Francis Birch*

*Harvard University, Cambridge, Massachusetts*

My original intention was to review recent studies of heat flow, radioactivity, and the earth's thermal history. It quickly became apparent that this was overambitious, and after eliminating one topic after another, I finally reduced the subject to heat flow alone, and even this will be very sketchily presented. The amount of material has increased at an unprecedented rate over the last 10 years; in 1954, I could tabulate 20 values at sea and 27 on land; recent compilations and unpublished results total over 1000, about nine-tenths at sea. With these large numbers of observations, we encounter new problems of interpretation and of statistical treatment which needed little attention when every measurement was a rare event and could be given careful, individual study. I should like to discuss a few examples of recent work and to consider particularly the reliability of heat flow determinations, having in mind not the instrumental or observational errors but rather the environmental or accidental disturbances which affect these measurements. This discussion is far from satisfactory but possibly it will inspire someone to examine the question more fully.

In geophysics, we understand by the term "heat flow" the rate of heat conduction per unit of area and of time to the earth's surface. The geothermal flux is measured in the tiny unit of microcalories per

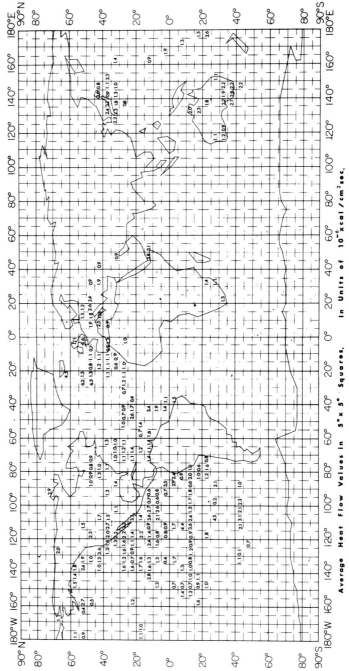

FIG. 1. Distribution of heat flow determination, averaged for 5° "squares" [Wang, unpublished].

cm² second. Nevertheless, integrated over geological periods and over large areas, this feeble flow appears to dominate, by a large margin, the transport of heat through the outer layer to the surface. Even in so-called "geothermal areas," where the heat transport is mainly by moving water or steam, conduction may account for one-tenth to one-quarter of the heat brought to the surface [Banwell, 1963, p. 85; Benseman et al., 1963, p. 57]; in the hotter parts of these areas, the intensity of heat loss is thousands of times greater than the average, but the areas are small and the duration of activity geologically short. For most of the earth, most of the time, the only direct clue to the underlying thermal distribution is the conducted heat.

The outstanding development of the last decade has been the determination of heat flow through the sea floor. The distribution of heat flow determinations over the earth's surface is shown in Figure 1, where the values have been averaged for 5° "squares." The rapid progress of the work at sea and the many unexpected features of the heat flow distribution have stimulated new work on the continents, including many countries where no determinations existed hitherto. The work up to 1963 has been compiled by Lee [1963] [see also Lee and MacDonald, 1963]. The first measurements in South America and in India have just been made; except for Australia and Japan, nearly all of the measurements are in the Western Hemisphere.

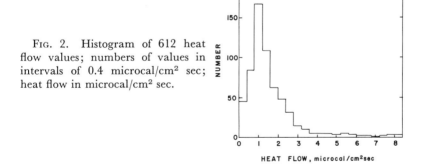

Fig. 2. Histogram of 612 heat flow values; numbers of values in intervals of 0.4 microcal/cm² sec; heat flow in microcal/cm² sec.

The distribution of values is shown in a histogram (Figure 2) which includes about 600 determinations. The unit is the microcal/cm² sec, which will be assumed in the following without further mention. The most frequent value is 1.1, the mean of all, 1.62; the mean for continents, 1.43, for oceans, 1.65 [Lee, 1963]. If we exclude values greater than 4 from the oceanic mean, this is reduced to 1.37. In any case, the oceanic and continental values are remarkably close together; this is the major consequence of this work so far. It has revolutionized our conceptions of the distribution of radioactivity and of the thermal regime. Interpretations at this time are likely to be quickly overtaken by new data; they tend to have a subjective element, believers in mantle convection finding support for this hypothesis, and unbelievers finding many features at variance with it. The fact that the oceanic and continental heat flow means are about the same has already reversed the direction of these currents; alternations of high and low heat flow determinations now require currents of a great variety of sizes and shapes, if this is to be the sole explanation.

The frequency distribution can be thought of as combining two effects—the first, a real dispersion of values, and the second, local disturbances for which we would apply corrections if the circumstances were known. I should like to make an effort to separate these effects, by considering various details of the observations themselves.

The concept of heat flow in geophysics contains a number of tacit assumptions. The law of heat conduction gives heat flux $Q$ as the product of local thermal conductivity $K$ and thermal gradient: $Q = K$ grad $T$ (ignoring variations of $K$ with direction and taking the positive direction of the vector $Q$ in the direction of decreasing temperature). These quantities are in general time-dependent, but this relation gives the "real" instantaneous conducted heat flux. For some purposes, this may be what is needed, but if we are interested in the loss of heat from the earth's interior, many time-dependent processes which affect the temperature are thought of as local, irrelevant, or temporary "disturbances." Now over geological periods, nothing is permanent, and our problem is to separate long-period, quasi-stationary effects from short-period ones; we also wish to eliminate the effects of such local accidents as irregularities of terrain, local flows of water, and recent erosion or deposition. The daily and annual variations of surface temperature may be avoided by finding the gradient at a sufficient depth below the surface, but this is not feasible for the fluctuations of temperature associated with glacial periods or uplift and erosion in mountainous regions. Thus we must often estimate corrections for geological happenings for which precise data are lacking or leave our measurements uncorrected for conceivable significant disturbances. Corrections of this kind suggested for various continental measurements range from 0.1 or less to more than 1.0, but there is no general agreement on what should be taken into account. The problem is not usually serious in the stable regions where most continental measurements have been made, but it can lead to significant uncertainties for areas of Tertiary mountain formation or sedimentation [Birch, 1950; Clark and Niblett, 1954; Clark, 1961].

Heat flow measurements require openings below the ground surface for the introduction of thermometers; measurement of thermal gradient means in practice the measurement of temperature differences over finite intervals of depth, below the level of disturbance by annual fluctuation (roughly 100 feet). The simplest situation is a vertical borehole of small diameter in horizontally uniform rock, with horizontal isothermal surfaces; the equation of heat conduction may then be applied in the integrated form $\Delta T \big|_{z_1}^{z_2} = Q \int_{z_1}^{z_2} dz/K(z)$, where $K(z)$ is found by sampling. The degree to which the required conditions are met is often not known. Where boreholes are deep enough to permit independent determinations over a number of intervals and the values of heat flow are closely distributed about the mean value,

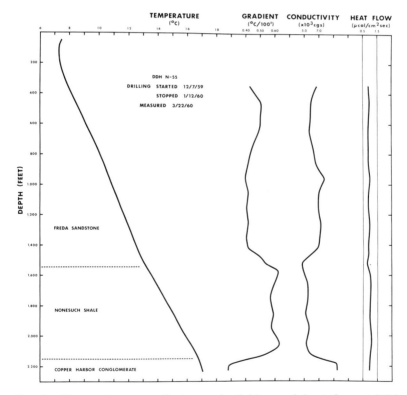

Fig. 3. Temperatures, gradients, conductivities, and heat flow at White Pine, Michigan (drill hole N-55). After Roy [1963].

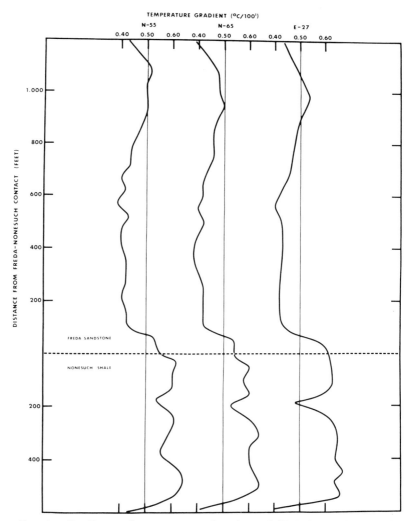

FIG. 4. Gradients of temperature for three drill holes at White Pine, Michigan; the profiles have been aligned on the contact between Freda sandstone and Nonesuch shale. After Roy [1963].

it may be concluded that important lateral disturbances are absent. The uncertainties are greatly reduced if a number of closely spaced holes are available, permitting the construction of isothermal surfaces.

Much of the older work was accomplished in fairly deep borings; while this had the advantage of giving large temperature differences and, at least toward the bottom, considerable diminution of surface disturbances, the gains were offset by a requirement of mean conductivity for large vertical intervals, and the most common source of uncertainty was inadequacy of sampling. Recent continental work has moved toward the oceanic method in using sensitive instruments for measuring temperature differences over smaller intervals with more complete sampling. As contrasted with oceanic measurements, however, in place of a single or perhaps two values of mean gradient, even a relatively shallow hole can provide ten or more independent observations, each at least as good as an oceanic one. I return to this point later.

As an example of nearly ideal conditions, let us look at the measurements for a drill hole (N-55) at White Pine, Michigan, given by Roy [1963] (Figure 3). This passes through three formations of hard, dense rocks having nearly horizonal boundaries; gradients are strongly correlated with conductivities, sampled at roughly 20-foot intervals. The 18 values of heat flow determined for 100-foot intervals between depths of 300 and 2100 feet are all within a few per cent of the mean value. Two other holes nearby show closely similar profiles of gradient (Figure 4), and the values of heat flow for the three holes and three formations are in good agreement (Table 1). The mean value for the

Table 1

HEAT FLOW AT WHITE PINE, MICHIGAN [ROY, 1963]
($microcal/cm^2\ sec$)

| Drill Hole | Formation | | |
|---|---|---|---|
| | Freda | Nonesuch | Copper Harbor |
| N-55 | $1.05 \pm 0.08$ | $1.09 \pm 0.05$ | $1.06 \pm 0.16$ |
| N-65 | $1.04 \pm 0.09$ | $1.08 \pm 0.06$ | |
| E-27 | $1.10 \pm 0.10$ | | |
| Mean | $1.07 \pm 0.05$ | | |

three holes, $1.07 \pm 0.05$, is probably one of very few entitled to 3 significant figures; the standard deviation for a single determination is rarely less than 0.1, and more than 2 significant figures can seldom be justified.

A more typical situation was encountered (Figure 5) by Diment and Robertson [1963] for a 3000-foot hole at Oak Ridge, Tennessee. Here the rocks are much more variable: the dip approaches 30°; the

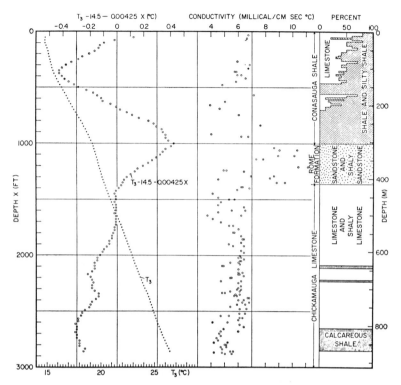

FIG. 5. Geologic section, temperatures, and conductivities for a drill hole at Oak Ridge, Tennessee. After Diment and Robertson [1963].

uppermost 1000 feet consists of shales and limestones which were anisotropic, inhomogeneous, and troublesome in every respect; sampling tends to be biased, in these conditions, toward the harder rocks of higher conductivity. Below 1000 feet, the heat flows for 100-foot intervals become reasonably self-consistent (Figure 6). In this case, if we had only the first 1000 feet, we should evidently be left in doubt about the heat flow. The mean for the lower part has a standard error of about 0.1.

Many sorts of artificial disturbance have been recognized; Figure 7 shows temperatures in a drill hole in Cambridge, which have been affected down to about 300 feet by the presence, during the last 50 years, of a large heated building close by. Determinations in mines are often disturbed by ventilation, temperatures in boreholes by water circulation during drilling, and so on. There is ordinarily no difficulty in principle in correcting for or avoiding these short-period effects,

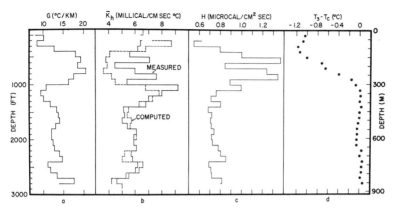

FIG. 6. Mean gradients, conductivities, and heat flow values for 100-foot intervals for the hole at Oak Ridge [Diment and Robertson, 1963].

especially where it is possible to return at intervals to observe progressive changes. A disturbance caused by flow of water in a borehole is shown in Figure 8. Circulation was lost during the drilling of this hole, and subsequent logging showed that water flowed down the hole to a depth of about 1100 feet, carrying with it the low near-surface temperature. When the flow was cut off by grouting, the temperatures began to return toward the original rock temperatures. In this case, the water circulation came into existence with the drilling of the hole. On the other hand, a natural aquifer below the drill hole may redistribute the heat coming from below without giving any evidence of its presence. The movement of ground water frequently dominates the uppermost few hundred feet of boreholes; if we were to use the gradients in the first 300 feet, we should find zero or negative heat flows in many cases. It is to some extent a matter of judgment to decide at what level the observed gradients are to be taken for the

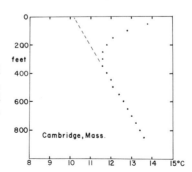

FIG. 7. Temperatures in a drill hole in Cambridge, Massachusetts, showing disturbance by artificial surface heating [Roy, unpublished].

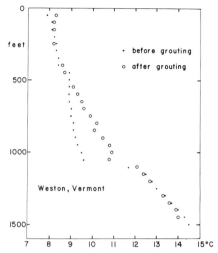

FIG. 8. Temperatures in a drill hole in Weston, Vermont, showing disturbance by flowing water, before and after grouting [Roy, unpublished].

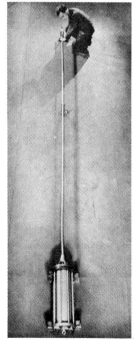

FIG. 9. Probe and recorder for measuring temperature gradients in the sea floor [Bullard, 1954].

determination of heat flow, and many shallow holes are rejected for this purpose simply because experience has shown that consistent results are obtainable only at greater depths.

The large amount of information supplied by measurements of this kind, together with the supplementary geological and topographical information, should be kept in mind as we turn to the very different circumstances of oceanic heat flow determinations.

The first effort to determine a temperature gradient in the ocean bottom appears to have been that of Pettersson [1949], with an alcohol reversing thermometer. Shortly thereafter, Bullard [1954] introduced

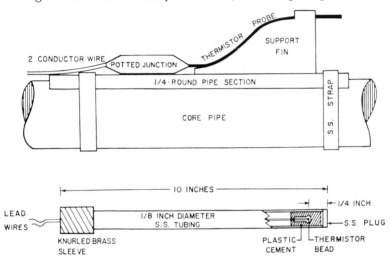

Fig. 10. Outrigger thermal probes, mounted on core barrel [Gerard, Langseth, and Ewing, 1962].

a probe with thermocouples and a photographically recording galvanometer (Figure 9); further developments have included the use of thermistors and self-balancing recorders of various types, and the mounting of small outrigger probes on the barrels of bottom-coring devices [Figure 10, Gerard, Langseth, and Ewing, 1962]. Initial disadvantages of long time constants and of sampling at points which might be several miles from the place of gradient determination have thus been overcome. The temperature difference is found for intervals of depth of from 1.7 to 5 meters and is typically of the order of 0.1°C. This must be combined with the conductivity of the mud in which the measurement is made, which ranges from a value close to that of water alone to about twice this value. The conductivity of the core is sometimes measured at once by a transient method [Von Herzen and

Maxwell, 1959] and sometimes estimated from water content, according to relations based on laboratory studies [Ratcliffe, 1960]. The range is relatively small, and large errors from this source seem unlikely. Ordinarily, the only major correction is for the transient disturbance accompanying the penetration by the probe.

The possibility of working at such shallow depths in the sea floor depends upon the absence of short-term fluctuations of temperature in the bottom water; unlike the land, this receives no solar radiation and is everywhere within a few degrees of 0°C. Changes are believed to be extremely slow. Much of the ocean floor is covered by soft muds which can be penetrated by weighted probes or coring tubes. Combined with seaborne mobility and a great deal of hard work on the part of seagoing geophysicists, these two circumstances have made possible a rapid accumulation of oceanic measurements.

A disadvantage of work at sea is that circumstances which are obvious on land, such as nearby topography, local variations of material, and recent movements of sediment, are thoroughly obscured by a few miles of sea water. For temperature measurements at depths of a few meters, the local topography within a similar radius can generate sizable corrections, but this is still beyond the resolution of sounding equipment. The configuration of the harder, underlying rocks is usually not known, though subsurface profilers are

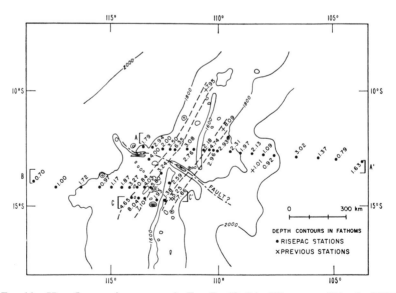

FIG. 11. Heat flow stations across the East Pacific Rise [Herzen and Uyeda, 1963].

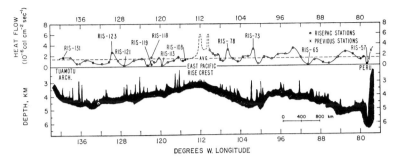

FIG. 12. Heat flow and topography across East Pacific Rise [Herzen and Uyeda, 1963].

beginning to reveal a great deal of complexity [Ewing, Ludwig, and Ewing, 1964]. Various possible sources of disturbance have been considered by Bullard, Revelle, and Maxwell, Von Herzen and Uyeda, and others, but it is not yet clear how much of the variation in oceanic determinations is real and how much arises from local disturbances which we are still unable to evaluate.

The East Pacific Rise has been intensively studied by Von Herzen and Uyeda [1963]. Several traverses cross the Rise along parallels, with observations spaced in some parts at about ½ degree intervals (Figure 11). Heat flow and gross bottom topography along the track are shown in Figure 12. The distribution is perhaps more easily seen on a pin plot, where the length of the pin is proportional to heat flow (Figure 13). The high values are localized, occupying much narrower zones than the broad topographic rise. Aside from the fact that the high values are on the Rise, at a depth of about 3 km, there appears to be little correlation with depth (Figure 14). There is a remarkable pattern of alternating highs and lows east of the Rise, shown in Figure

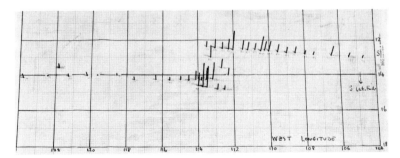

FIG. 13. Pin plot of heat flow across the East Pacific Rise; the length of pin is proportional to heat flow. Values from Von Herzen and Uyeda [1963].

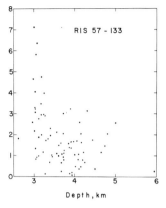

Fig. 14. Heat flow, in microcal/cm² sec, versus depth, in the neighborhood of the East Pacific Rise; station RIS 57 to RIS 133. After Von Herzen and Uyeda [1963].

15. The average is close to the world average, but the horizontal gradients are large, with changes of heat flow from nearly zero to 3 microcal/cm² sec in a distance of a degree, or roughly 100 km. An explanation in terms of recent sedimentation might be offered for the low value in the South American Trench. Von Herzen and Uyeda [1963] have made a special study of the topography in the vicinity of a pair of stations, 2 minutes of longitude apart, which gave the values 1.60 and 0.25, the high one on an abyssal hill and the low value occurring on a flat area, about 25 meters deeper. These writers suggest that the higher value is probably more representative; the preferred explanation for the lower one is irregularity of the underlying rocks of higher conductivity. They suggest that many of the other lows may result from similar situations. For the pair of stations just mentioned, the difference between 1.60 and 0.25 is not considered significant, but is attributed to disturbances at the "low" station. If these are to agree, we must apply a correction for the disturbance

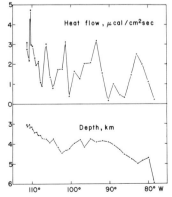

Fig. 15. Heat flow and water depth east of the East Pacific Rise along 12°S latitude. After Von Herzen and Uyeda [1963].

of 0.7 to both values, 1.35 to the low value alone, or some such combination. In any case a large correction is implied.

Still another approach is by comparison with continental values; below the level of ground water disturbance these ought to be comparable with ocean bottom values, if we make an appropriate allowance for the difference in gradients. On the average, the oceanic gradients are about three times as high as continental ones, and thus a 1.7-meter interval in mud may be comparable with a 5-meter interval on land. Most continental gradients are based on intervals of at least 10 meters, and usually there are 10 or more such intervals. Thus the standard error of an oceanic determination, other things being equal, should be at least three times as great as that of a continental one. The best continental measurements may have a standard error of 0.1, but 0.2 is more typical. Thus we find an estimate of from 0.3 to 0.6 for oceanic measurements.

Another way of stating this is to say that we require a group of ten or more closely spaced oceanic determinations as a rough equivalent of one 1000-foot continental hole, and this is still conceding a superior virtue to mud as compared with rock. Ten oceanic holes, with 1.7-meter probes, sample 17 meters of mud; even if we use only the lower half of our 1000-foot hole, we sample 170 meters of rock but with an average gradient only one-third that of the mud. The stations which give low values of heat flow of course show gradients as low or lower than those usually found in rock.

The closest approach to a test of these estimates is provided by the work of Von Herzen and Maxwell [1964] in the neighborhood of the preliminary Mohole site near Guadelupe Island. Measurements in the hole over a vertical interval of 154 meters gave a mean heat flow of $2.8 \pm 0.3$ (error estimated). Five measurements nearby with the usual short probe technique gave values between 2.5 and 2.8, but a sixth gave 4.2. These were all within a horizontal distance of not more than 10 km, and the high value is attributed to a local irregularity of the bottom, about 100 meters high. The mean of all six is 2.9, in good agreement with the deeper hole, but the standard deviation of a single observation is 0.57 or, nearly enough, 0.6. If the high value is discarded, the standard deviation is reduced to 0.1, but the mean is reduced to 2.6. Regions of relatively high heat flow, such as this, may be expected to give more consistent results than those of low heat flow.

Lee [1963] has given averages and standard deviations for various groups of heat flow measurements (Table 2). The average for 561 stations at sea is 1.65 with a standard deviation of 1.27; the average

## Table 2
### [After Lee, 1963, p. 453]

| Sample Region | Number of Observations | Mean Heat Flow (microcal/cm² sec) | Standard Deviation |
|---|---|---|---|
| World | 634 | 1.62 | 1.21 |
| Continents | 73 | 1.43 | 0.57 |
| Oceans | 561 | 1.65 | 1.27 |
| Pacific Ocean | 471 | 1.71 | 1.35 |

### [Lee and MacDonald, 1963]

| | | | |
|---|---|---|---|
| World | 757 | 1.61 | |
| Continents | | 1.48 | |
| Oceans | | 1.63 | |

for 73 stations on land, 1.43, with a standard deviation of 0.57. The high values over the oceanic rises contribute greatly to the large deviation of the oceanic values; if we take only oceanic values between zero and 4.0, comparable with the continental range, then for the 366 stations in Lee's tabulation meeting this requirement, we find a mean of 1.37 with a standard deviation of 0.80. Suppose that this is composed of two parts, of which one is the standard deviation associated with the unevaluated corrections and assessed above as 0.6; this leaves another part equal to 0.53 for the "real" variation, close to the standard deviation for continental determinations.

Even with a standard deviation of 0.5 and a mean of 1.4, a Gaussian distribution would show about 2 per cent of the determinations below 0.4, that is, 7 in 366; in fact, there are 29. It seems likely that the unevaluated disturbing factors increase the dispersion of values in both directions, and that the real variations are somewhat smoother than those of the raw data. The high values will be relatively less affected than the low ones; a value of 8 will probably lie between $8 \pm 0.6$ say, and thus still be high, but we cannot be sure that a value of 0.6 is different either from the mean value or from zero.

Other oceanic areas show much smoother variations; Figure 16, [Reitzel, 1963] shows highly uniform values over a 10° by 10° section of the deep North Atlantic. Evidently there are real differences of texture in the oceanic heat flow patterns, this one resembling a

HEAT FLOW 419

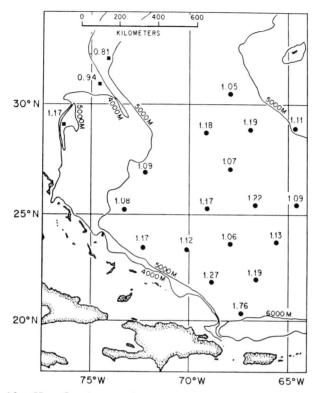

FIG. 16. Heat flow in a portion of the North Atlantic [Reitzel, 1963].

continental area much more closely than it resembles the eastern Pacific.

In addition to the larger amount of information applicable for determination of heat flow, workers on continents also have more

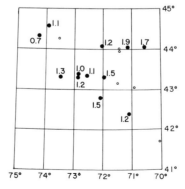

FIG. 17. Heat flow in New England and New York [Roy, preliminary unpublished values].

information useful for interpretation. For example, Figure 17 shows a set of measurements as they might appear if they were obtained at sea; the only additional information might be the depth of water. When we see these stations on a geological map (Figure 18) and log the radioactivity of the cores, the interpretation becomes fairly obvious. High values are found in the Paleozoic granites, especially those of the White Mountain Magma Series, which are known to have several times as much uranium and thorium as the average granite [Billings and Keevil, 1946; Butler, 1961; Adams et al., 1962]. Low values are found in the anorthosite area of the Adirondacks,

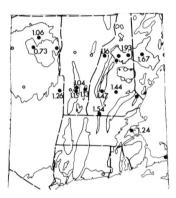

FIG. 18. The values of Fig. 17, with simplified geological map [Roy, unpublished].

and again it is known that the anorthosite makes virtually no contribution to the local heat flow. These contrasts suggest that at least 0.5 of the mean continental heat flow originates in the upper part of the crust, the contribution from the lower crust and mantle amounting, on the average, to about 0.7. Few, if any, continental values are significantly lower than 0.7, though a number of values of about this magnitude are found in the Canadian and Australian shields [Misener and others, 1951; LeMarne and Sass, 1962; Sass, 1964; Howard and Sass, 1964].

The partition of this between lower crust and mantle is still uncertain, but probably at least 0.2 comes from the lower crust. If the distribution of radioactivity assumed by Jeffreys had been correct, we should have found values of 0.3 or so everywhere in the deep oceans, and if the low values which have been found are substantiated, the most straightforward explanation may be extreme depletion of radioactivity in the underlying mantle.

The distribution of values in North America as of 1963 is shown in Figure 19 [after Roy, 1963]; a considerable number of new values will be available in another year. The chief features so far discernible

HEAT FLOW 421

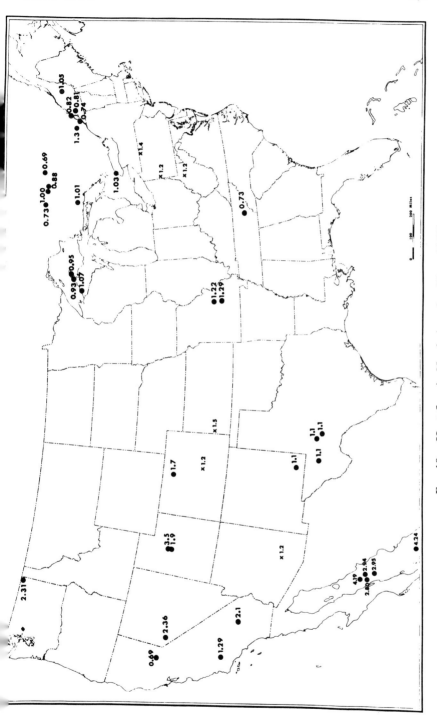

Fig. 19. Heat flow in North America [Roy, 1963].

are the relatively low values in the Canadian Shield and significantly higher values in the Basin and Range Province. These high values seem to be aligned with high values in the Gulf of Lower California, which in turn have been speculatively joined with the East Pacific Rise by Menard [1961]. The Basin and Range Province is thickly dotted with thermal springs [Figure 20, from Stearns and others,

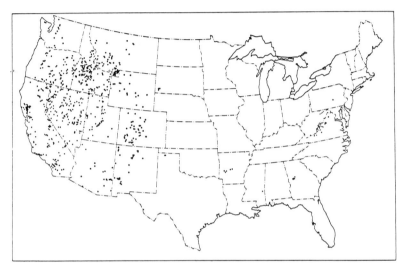

FIG. 20. Hot springs in the United States [Stearns, Stearns, and Waring, 1937].

1937], some of them supplying steam or boiling water as well as much other evidence of recent volcanism. Investigators of heat flow have generally avoided the hot spring areas in this country, but Uyeda and Horai [1964] report values of 15 and 10.8 near Japanese thermal areas. From the data given by Benseman and others [1963, p. 55] for the Waiotapu area of New Zealand, we find a mean value of 130 for an 8-km² area including the warmest part. The transition from these hot spots to normal regions has not been adequately explored, but the areas of moderately high heat flow, say from 4 to 10, must be fairly large; even if the East Pacific Rise is as well supplied with thermal areas as its presumed continental continuation, the large number of high values found by random probing implies a considerable extent of hot ground surrounding each center. Perhaps for a more detailed interpretation of the East Pacific Rise we shall be able to make use of work now in progress in Nevada.

Even where the station spacing is relatively close, contouring of

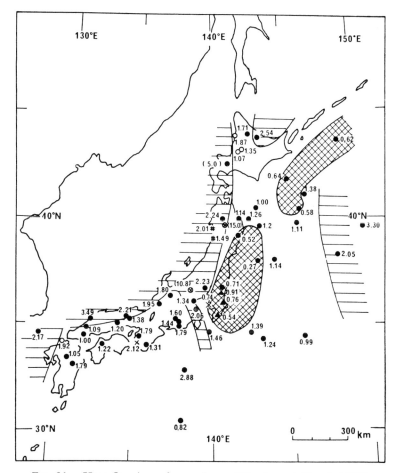

Fig. 21. Heat flow in and near Japan [Uyeda and Horai, 1964].

heat flow is difficult, and this is perhaps another consequence of rather large disturbances. It is often possible to separate low areas from high areas with some arbitrary boundaries, but a rather gerrymandered pattern usually results. A liberal plus or minus allowance in recognition of unassessed corrections might permit more realistic contouring. The Japanese work (Figure 21) shows some of the difficulties, with three areas defined according to the ranges, 0 to 1, 1 to 2, > 2. The low area and the high area have virtually no separation if these boundaries are rigidly adhered to, and high values occur in the intermediate area.

Correlations of heat flow with other kinds of geophysical data, besides the obvious one with local radioactivity, are still relatively undeveloped, mainly because the regions where different kinds of measurements have been made do not often coincide. A few cases where seismic profiles at sea have come close to heat flow stations are shown in Von Herzen's thesis [1960]. It might be expected that under regions of high heat flow and higher mantle temperature, the velocity would be lower, but no great consistency is shown. One can also put together the seismic work of Bunce and Fahlquist [1962] north of the Puerto Rico Trench with some of Reitzel's heat flow stations. The temperature at the base of the oceanic crust may be estimated roughly with the assumption that the conductivity in the oceanic crust has the average value 0.005 cal/cm sec deg. If the oceanic crust is 5 km thick, then the temperature at its base is nearly $100Q$, where $Q$ is the heat flux in microcal/cm² sec (Figure 22). The most likely interpretation is that velocity is not a function of temperaure alone; that is, the

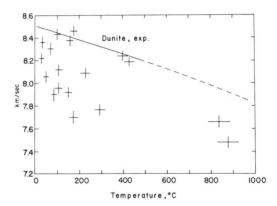

FIG. 22. Upper mantle velocity $V_P$ versus estimated temperature for some Pacific and Atlantic stations.

material below the oceanic crust is not seismically uniform, but a reasonable temperature effect seems to be associated with the East Pacific Rise.

Mantle velocities beneath the United States were contoured by Herrin and Taggart [1962], following the Gnome explosion (Figure 23); these contours have been considerably modified by later work, but a low-velocity region in the Basin and Range Province persists and is at least qualitatively related to the higher heat flow and higher mantle temperatures in this region.

Vacquier and Von Herzen [1964] have investigated heat flow and magnetic field across the Mid-Atlantic Ridge, finding a magnetic pattern, which they take to define the true axis of the Ridge, and an association of heat flow with distance from this axis (Figure 24).

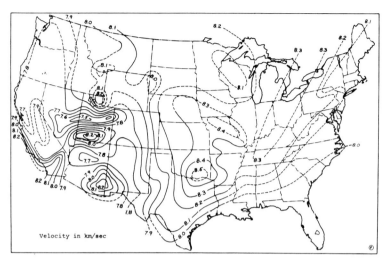

Fig. 23. Upper mantle velocities in the United States [Herrin and Taggart, 1962].

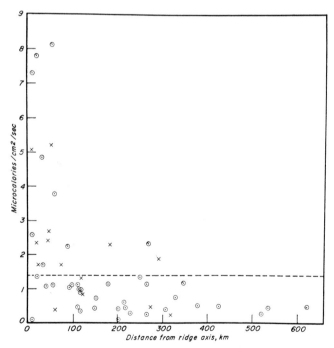

Fig. 24. Heat flow versus distance from the Mid-Atlantic Ridge [Vacquier and Von Herzen, 1964].

While the high values occur on or near the Ridge, no other clear-cut relations exist. Again, some very low values are found close to high ones, especially in places of flat topography resembling others in which low values have been found.

Since high heat flow implies high temperatures and this in turn low densities, other things being equal, correlation with the external gravitational field seems *a priori* likely. Neither heat flow nor gravity seems to respect continental boundaries. Detailed comparison is still not feasible, since most of the determinations of heat flow are in oceanic areas where gravity measurements are scanty, while gravity is well known on the continents where there are still too few values of heat flow. If we simply form averages for $5° \times 5°$ "squares," using only those where at least one value of heat flow exists, we find that in about $\frac{2}{3}$ of the "squares," the departure of heat flow from the mean and of geoid height from the ellipsoid have the expected directions, that is, higher than average heat flow tends to go with depression of the geoid and vice versa. Lee and MacDonald [1963] have used the available heat flow data to find spherical harmonic coefficients up to the

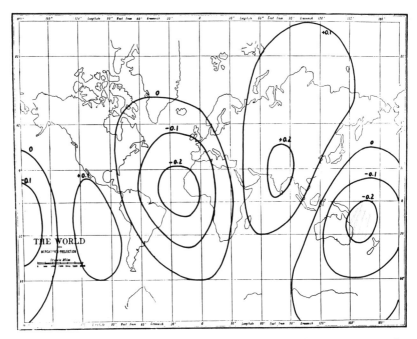

Fig. 25. Variation of heat flow in microcal/cm$^2$ sec from the mean value, for 5° "squares" [Wang, 1964, with harmonic coefficients from Lee and MacDonald, 1963].

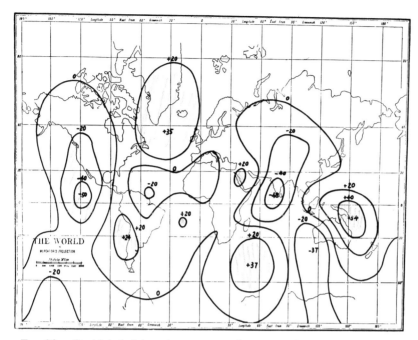

Fig. 26. Geoidal heights, in meters, above a reference ellipsoid with flattening of 1/298.24 [Izsak, 1964].

second order and have plotted the corresponding contours of equal heat flow, remarking the similarities. One of their solutions has been replotted by Wang [1964] to show differences from the mean heat flow (Figure 25). These may be compared with the external gravitational field, which is conveniently expressed in terms of geoidal heights referred to an ellipsoid (Figure 26). Some correspondences may be observed, the predicted low of heat flow near New Guinea corresponding to the high which appears to persist on all of the maps of external potential, and the high of heat flow coinciding with the low of the geoid over India. The disturbances of temperature corresponding to these harmonics of heat flow have been calculated by Wang with various simplifying assumptions; one of a family of such solutions is shown in Figure 27.

Just as the contours of the geoid have been changing, almost from month to month, as new satellite observations are fed into the calculations, the contours of the thermal field are evidently highly tentative, especially for the low harmonics; even with the much greater number of measurements of gravity it was not feasible to obtain the coefficient of the third harmonic in the earth's potential until the advent of the

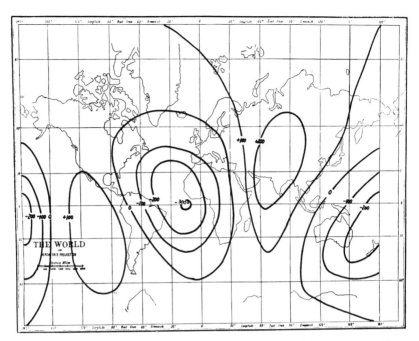

FIG. 27. Disturbance of temperature at a depth of 100 km corresponding to variation of heat flow, of Figure 25, for radioactive sources uniformly distributed through 200 km [Wang, 1964].

artificial satellites. For the present, the expansion of the heat flow in spherical harmonics of low degree may be regarded as a method of smoothing, which eliminates the rapid horizontal variations from which we might hope to obtain the most decisive evidence regarding the sources of heat. In principle, with stations 50 km apart, as over the East Pacific Rise and in New England, we can say something about the variation of heat sources within the upper 50 km or so; this is mainly crust under continents but mantle under the ocean. This application, like most of the others, requires a realistic appraisal of significant differences. Though attention has been centered on the oceanic work in this discussion, the treatment of continental measurements is still far from uniform and much remains to be done before their interpretation can be pushed beyond the most preliminary stage.

## Acknowledgments

I am grateful to Robert F. Roy and Chi-yuen Wang for the unpublished data of Figures 1, 7, 8, 17, and 18.

## References

Adams, J. A. S., M. C. Kline, K. A. Richardson, and J. J. W. Rogers, The Conway granite of New Hampshire as a major low-grade thorium resource, *Proc. Nat. Acad. Sci. U.S.*, *48*, 1898–1905, 1962.

Banwell, C. J., "Programming of temperature surveys," *Waiotapu Geothermal Field, Bulletin 155*, Chapter VIII, New Zealand Department of Scientific and Industrial Research, 1963.

Benseman, R. F., R. G. Fisher, and D. J. Dickinson, "Survey of surface heat output at Waiotapu," *Waiotapu Geothermal Field, Bulletin 155*, Chapter VI, New Zealand Department of Scientific and Industrial Research, 1963.

Billings, M. P., and N. B. Keevil, Radioactivity of the four Paleozoic magma series in New Hampshire, *Bull. Geol. Soc. Am.*, *57*, 797–828, 1946.

Birch, Francis, Flow of heat in the Front Range, Colorado, *Bull. Geol. Soc. Am.*, *61*, 567–630, 1950.

Bullard, Sir Edward, The flow of heat through the floor of the Atlantic Ocean, *Proc. Roy. Soc., London*, *A*, *222*, 408–429, 1954.

Bunce, Elizabeth T., and Davis A. Fahlquist, Geophysical investigation of the Puerto Rico Trench and outer ridge, *J. Geophys. Res.*, *67*, 3955–3972, 1962.

Butler, A. P., Jr., Ratio of thorium to uranium in some plutonic rocks of the White Mountain plutonic-volcanic series, New Hampshire, *U.S. Geological Survey Prof. Paper 424-B*, 67–69, 1961.

Clark, S. P., Jr., Heat flow in the Austrian Alps, *Geophys. J.*, *6*, 55–63, 1961.

Clark, S. P., Jr., and E. R. Niblett, Terrestrial heat flow in the Swiss Alps, *Monthly Notices Roy. Astron. Soc., Geophys. Suppl.*, *7*, 176–195, 1954.

Diment, W. H., and E. C. Robertson, Temperature, thermal conductivity, and heat flow in a drilled hole near Oak Ridge, Tennessee, *J. Geophys. Res.*, *68*, 5035–5047, 1963.

Ewing, Maurice, William J. Ludwig, and John I. Ewing, Sediment distribution in the oceans: the Argentine Basin, *J. Geophys. Res.*, *69*, 2003–2032, 1964.

Gerard, R., M. G. Langseth, Jr., and M. Ewing, Thermal gradient measurements in the water and bottom sediment of the western Atlantic, *J. Geophys. Res.*, *67*, 785–803, 1962.

Herrin, Eugene, and James Taggart, Regional variations in $P_n$ velocity and their effect on the location of epicenters, *Bull. Seismol. Soc. Am.*, *52*, 1037–1046, 1962.

Howard, L. E., and J. H. Sass, Terrestrial heat flow in Australia, *J. Geophys. Res.*, *69*, 1617–1626, 1964.

Izsak, I. G., Tesseral harmonics of the geopotential and corrections to station coordinates, *J. Geophys. Res.*, *69*, 2621–2630, 1964.

Lee, W. H. K., Heat flow data analysis, *Rev. Geophys.*, *1*, 449–479, 1963.

Lee, W. H. K., and Gordon J. F. MacDonald, The global variation of terrestrial heat flow, *J. Geophys. Res.*, *68*, 6481–6492, 1963.

LeMarne, A. E., and J. H. Sass, Heat flow at Cobar, New South Wales, *J. Geophys. Res.*, *67*, 3981–3983, 1962.

Menard, Henry W., The East Pacific Rise, *Sci. Am.*, *205*, 52–61, 1961.

Misener, A. D., L. G. D. Thompson, and R. J. Uffen, Terrestrial heat flow in Ontario and Quebec, *Trans. Am. Geophys. Union*, *32*, 729–738, 1951.

Pettersson, H., Exploring the bed of the ocean, *Nature*, *164*, 468–470, 1949.

Ratcliffe, E. H., The thermal conductivity of ocean sediments, *J. Geophys. Res.*, *65*, 1535–1541, 1960.

Reitzel, John, A region of uniform heat flow in the North Atlantic, *J. Geophys. Res.*, *68*, 5191–5196, 1963.

Roy, R. F., Heat flow measurements in the United States, Ph.D. thesis, Harvard University, 1963.

Sass, J. H., Heat-flow values from the Precambrian shield of Western Australia, *J. Geophys. Res.*, *69*, 299–308, 1964.

Stearns, N. D., H. T. Stearns, and G. A. Waring, Thermal springs in the United States, *U.S. Geological Survey Water-Supply Paper 679-B*, 59–191, 1937.

Uyeda, S., and K. Horai, Terrestrial heat flow in Japan, *J. Geophys. Res.*, *69*, 2121–2141, 1964.

Vacquier, V., and R. P. Von Herzen, Evidence for connection between heat flow and the Mid-Atlantic Ridge magnetic anomaly, *J. Geophys. Res.*, *69*, 1093–1101, 1964.

Von Herzen, R. P., Pacific ocean floor heat flow measurements—their interpretation and geophysical implications, Ph.D. thesis, University of California, 1960.

Von Herzen, R. P., and A. E. Maxwell, The measurement of thermal conductivity of deep-sea sediments by a needle probe method, *J. Geophys. Res.*, *64*, 1557–1563, 1959.

Von Herzen, R. P., and A. E. Maxwell, Measurements of heat flow at the preliminary Mohole site off Mexico, *J. Geophys. Res.*, *69*, 741–748, 1964.

Von Herzen, R. P., and S. Uyeda, Heat flow through the Eastern Pacific Ocean floor, *J. Geophys. Res.*, *68*, 4219–4250, 1963.

Wang, Chi-yuen, Figure of the Earth as obtained from satellite data and its geophysical implications, Ph.D. thesis, Harvard University, 1964.

# GEOCHRONOLOGY, AND ISOTOPIC DATA BEARING ON DEVELOPMENT OF THE CONTINENTAL CRUST*

G. J. Wasserburg

*California Institute of Technology, Pasadena, California*

**Prologue**

I will present some ideas and problems regarding the evolution of the crust of the earth and will also attempt a prognosis of the application of isotopic techniques to the understanding of geological processes. This dissertation might well be entitled "Pride and Prejudice"; it will be my implicit assumption that the use of physicochemical techniques and principles, when used with a sound geologic knowledge, has and will help solve important existing problems and will also create new problems and new fields which are themselves an integral part of the science of geology.

The close dependence of geology on the other physical sciences was emphasized 159 years ago by Sir James Hall [1812], who, in commenting on the lack of success of his contemporaries in explaining the origin of mountains, said—"One principal cause of this failure seems to have lain in the very imperfect state of chemistry, which has only of late years begun to deserve the name of a science. While chemistry was in its infancy, it was impossible that geology should make progress;

---
* California Institute of Technology Contribution No. 1294.

since several of the most important circumstances to be accounted for by this latter science are admitted on all hands to depend on principles of the former."

Sir James Hall also indicated that he abstained from further pursuit of the problem which had excited him in deference to his eminent colleague, Dr. Hutton, and did not resume his investigations until after Hutton's demise. Such gentlemanly deference is no longer characteristic of science, and it is at this point that we depart from Sir James' good example. At the present time the variety and intensity of work is causing continuous revision of our ideas, and today's speculation is often assaulted by tomorrow's fact.

The application of the methods of classical nuclear physics to geologic problems started soon after the discovery of radioactivity by Henri Becquerel in 1896. The use of long-lived natural radioactivities in determining geologic and cosmic time scales was pursued by a number of distinguished workers including Boltwood, Hahn, Rutherford, Soddy, and Strutt. Many of the most advanced researches pursued today were considered and attempted by these workers during the early 1900's. Variations of lead isotope abundances were first reported by Aston [1933].

A series of three classic papers by A. O. Nier appeared in *The Physical Review* in 1939 and 1941 on the variation of isotopic abundances due to radioactive decay and the measurement of geologic time. With the appearance of these works, the modern field of isotopic geochronology was born. Throughout much of this time, Arthur Holmes maintained and stimulated the geologic application of radiometric ages. There was a lull for about ten years followed by an explosion of activity.

Men like Inghram, Urey, Brown, Thode, Houtermans, and Gerling, Tuzo, Wilson, and Gentner were looking for new frontiers and directed their students and colleagues to exciting problems of cosmic and geologic importance. This choice may have been partly based on an aversion to "big machines," but the techniques which were developed during the period 1941 to 1951 permitted the accurate routine measurements of quantities of the order of $10^{-11}$ gram. Concurrent with the development of modern geochronology the study of the abundance variations of stable isotopes has also flourished, particularly due to the activities of H. C. Urey and S. Epstein. The application of these methods to geology by a group of half-breeds who were on the borderland of physics, chemistry, and geology has led to the development of a branch, not a technique, of the earth sciences. There are no old men practicing the art, although there are admittedly some who are prematurely grey and balding.

# GEOCHRONOLOGY

The great activity in the study of nuclear phenomena applied to the study of geologic problems has resulted in some dislocations within the profession and caused some unhappy conservatives to consider this work more or less outside the legitimate field of geology. The practitioners, often referred to as knob twisters or black box operators [J. Hoover Mackin, 1963], are also relegated to a caste of doubtful legitimacy. There have been excessive and erroneous claims made by some isotopic enthusiasts which have caused ill will, but I suppose this is a trait of enthusiasts in general. There appears to be some doubt expressed in some circles over the need for quantification in the science. That is a moot point. The real need is for understanding earth processes; and whatever methods may be applied with advantage will, of necessity, be used. But, in point of fact, the revolution, which is discomfiting to some professionals, is over. The knowledge which is being obtained is in the process of becoming an essential part of the science of geology. These problems range from the evolution of the atmosphere to the evolution of the crust of the earth and the time scale of the solar system and nucleosynthesis.

## Methodology

The methods of geochronology are physical and chemical, and the problems and materials are geological. All of the principal dating schemes (see Table 1) are based on the radioactive decay of a

### Table 1

PRINCIPAL DECAY SCHEMES USED IN GEOCHRONOLOGY FOR TIMES GREATER THAN $10^5$ YEARS

| Parent | Daughter |
|---|---|
| $U^{238} \rightarrow$ intermediate daughters $\rightarrow$ | $Pb^{206}$ |
| $U^{235} \rightarrow$ intermediate daughters $\rightarrow$ | $Pb^{207}$ $\}$ $He^4$ |
| $Th^{232} \rightarrow$ intermediate daughters $\rightarrow$ | $Pb^{208}$ |
| $K^{40}$ | $Ar^{40}$ / $Ca^{40}$ |
| $Rb^{87}$ | $Sr^{87}$ |
| $Re^{187}$ | $Os^{187}$ |

long-lived parent to produce an accumulated stable daughter product. Assuming that the amounts of parent isotope and daughter isotope may be measured with the necessary accuracy, there is a need for other physical constants. These are the half-lives of the radioactive species. In all of the intercomparisons of the various dating methods on minerals, it has been possible to obtain, often well before purely laboratory methods, rather good determinations of these constants. But in order to obtain what are in principle "absolute dates," it is necessary to have refined half-lives determined by direct methods. This necessity still exists, and some of our understanding of geologic phenomena awaits further refinement in the determination of absolute decay constants. Dates by two decay schemes are not directly comparable with each other. This is true for all of the decay schemes listed in Table 1 [Aldrich and Wetherill, 1958]. The uncertainties for $U^{235}$ and $U^{238}$ while small are very significant because of the systematics in this dating method [Wasserburg, Wetherill, Silver, and Flawn, 1962]. The $Rb^{87}$ decay constant is still far too uncertain for the precision needed [see Leutz, Wenninger, and Ziegler, 1962] and the Re decay constant should hardly be discussed.

A variety of ages can be obtained in studies on mineral systems from a single rock fragment. This dispersion can be found even within one mineral species by a single method and is due to differential effects of diffusion, weathering, and metamorphism. The ages obtained will have an uncertainty in meaning which, thus, far exceeds the analytical error. Even in cases of very consistent data, discrepancies exist between the various methods, and it is not always possible to distinguish between regular continuous (?) diffusion loss and errors in decay constants. For example, if the size distribution function in the fine size region ($\lambda D/a^2 \gg 1$) of mosaic units in crystals in igneous rocks is relatively constant from macroscopic crystal to crystal, then diffusion losses would be very regular and would not be easily discernible from a decay constant error. With these difficulties in mind, is still possible clearly to resolve events by the $Sr^{87}$-$Rb^{87}$ and $Ar^{40}$-$K^{40}$ methods which differ by 100 million years at an age of 1000 million years under favorable circumstances [Zartman, 1963]. The resolution ability of a dating method appears to be related to parent-daughter systematics. This exists for the Sr-Rb system but is particularly true for the coupled $U^{235}$-$Pb^{207}$ and $U^{238}$-$Pb^{206}$ systems. The theoretical basis for such systematics (in the absence of intermediate daughter loss) has been given by Wetherill [1956], Tilton [1960], and Wasserburg [1963]. The result is that for this coupled system, regardless of the details of the history of daughter or parent loss, the precise age may be found by

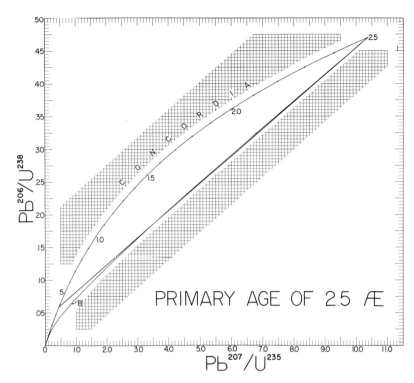

FIG. 1(a). Daughter-parent systematics. Lead evolution diagram for the case of continuous lead diffusion loss controlled by radiation damage ($B'$). The straight line is a tangent to $B'$ at the concordia curve. The linear relationship is always true in the neighborhood of concordia regardless of the law of stable daughter loss.

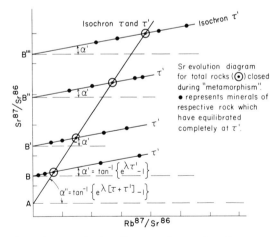

FIG. 1(b). Strontium evolution diagram for total rock systems and mineral phases showing the systematics for a two-stage history.

extrapolation of a very linear array of experimental data. This is illustrated in Figure 1a. Silver has shown [1963a, b] in work on different zircon fractions from a given rock that it is possible to obtain remarkably linear arrays and it appears possible to resolve events differing by 60 million years (m.y.) in 1700 million years. This work appears to offer the greatest resolution which can be obtained by isotopic means. Daughter-parent systematics of a less general type are also found for the Rb-Sr system as shown on Figure 1b. In general, on both a local scale (hand specimen, outcrop) and on a regional scale, the isotopic ages which are obtained will have a dispersion which limits our ability to define simultaneity. In Precambrian terranes the dispersion of ages due to secondary effects and to intrinsic resolution of the methods used causes us to consider dates in a time band as coherent, when in fact they represent distinct events over a duration of time comparable with a good part of all of the Phanerozoic. This myopic vision may force us to recognize some great geologic truth which is confused in the details of the more recent record or it might also lead us astray. The desire to believe in magic numbers—precise episodicity in magmatism and metamorphism—is at present rather strong in the trade and encourages the label of contemporaneity even when the ages are well resolved. This is further confused by the varying quality of the different studies reported in the literature, so that ages which are a "trifle" low for particular magic numbers are assumed to be due to daughter loss and are automatically upgraded in their assignment. This general problem has been previously discussed in the literature from different viewpoints [Gilluly, 1949, 1963; Wasserburg, 1961; Wasserburg, Wetherill, Silver, and Flawn, 1962]. With the preceding *caveat*, I will proceed with the main topic.

**Introduction**

In understanding the evolution of continental masses, the question of the growth of continental bodies through geologic time is of fundamental importance. It appears well demonstrated that the continental masses are intrinsically unstable due to erosional processes, and hence accretionary processes must be active in order to maintain regions of positive relief over long periods of time. With regard to the hypothesis of continental evolution, it is necessary to establish whether there exist major subdivisions which have meaning in terms of evolutionary characteristics and to determine whether the geologically younger parts of the continental mass represent the addition of new material or are in fact the product of metamorphism of older, preexisting provinces or materials derived therefrom. In addition, it is of prime importance to identify new material, derived from depth, which is being added to the crust. The continental and oceanic crusts are the two obvious distinctive features. Seismic and volcanologic observations indicate that the sources of lavas on some oceanic islands are from a depth associated with the (seismic) upper mantle [Eaton and Murata, 1960]. Because of the thinness of the oceanic crust and sediments, it is usually inferred that the oceanic volcanic magmatic rocks represent fractionated upper-mantle material relatively uncontaminated by crustal material. The implicit assumption which usually follows from this is that the oceanic mantle is compositionally a uniform layer. This presumed simplicity may be partly a reflection of our lack of observations but is also deeply tied to the question of the age and stability of the ocean basins themselves and the consequences of crustal evolution and of continental drift. The absence of a thickness of deep-sea sediments as estimated from modern sedimentation rates and the age of the earth may indicate some processes of transformation which could produce an upper mantle of considerable chemical complexity. By contrast with what is surmised about the oceanic crust, the continental crust is known to be of great complexity in both structures and rock types, and it has not been possible to attribute directly the magma sources of continental rocks to the mantle. However, we are able to sample materials in the continental crust of great age, but so far no ancient samples of oceanic crust have yet been recognized. The oldest igneous material from oceanic islands is only 650 million years, while the oldest known sediments are only Cretaceous in age. By contrast, exposures of continental crust as old as 2700 million years are quite common. It follows that we can only investigate the products of relatively recent and intense magmatic activity in oceanic regions [Menard, 1964] which are derived from a presumed "simple" region and from continental rocks which represent a wide range in age but which may often represent a complicated, hybrid history.

I will give examples of isotopic studies made in both types of regions and attempt to indicate what additional knowledge they have contributed. My emphasis will be on the Precambrian because it is in our understanding of this time region that these techniques have had their greatest impact.

### The Grenville Province

The Grenville province is a large subcontinental segment of the Canadian shield (see Figure 2). This province was defined principally on the basis of structural and lithologic criteria which distinguish it from the adjacent Superior province which bounds it on the west. The Grenville province contains extensive areas of marbles intermingled with a variety of gneisses, highly metamorphosed sedimentary and volcanic rocks, and numerous granites. The structural relationships within the Grenville province are often exceedingly complex, and the granitic rocks are commonly gneissic and concordant bodies in a

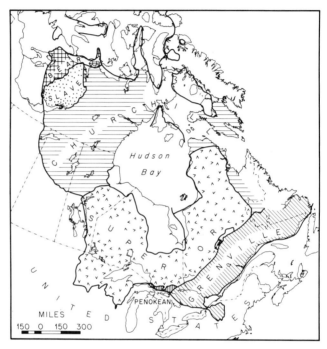

FIG. 2. "Structural" provinces of the Canadian Shield [Stockwell, 1962].

generally migmatitic terrane. The structural trends of the Grenville province transect the structural trends of the so-called Archean and Proterozoic sedimentary rocks of the Superior province which are characteristically of low metamorphic grade. The zone bounding these two provinces is called the Grenville front and is in part a zone of faulting and in part a region of metamorphic transition. The metamorphic and igneous complexes of the Adirondack mountains and the New Jersey highlands are similar in character to that found in the "typical Grenville" province areas and have naturally been associated with this province. The eastern boundary of the Grenville province is uncertain, to say the least, but on lithologic grounds may extend into eastern Canadian Appalachia. The further extent of this great province into the United States has been hazarded by some authors based on the fundamental theorem of tectonics which says that orogenic belts continue along their strike.

Very thorough and sophisticated geologic investigations have been made in many of these areas which have cast great light on the stratigraphy, metamorphism, and local geologic history, but in general no specific ideas have been put forth which would permit a genetic interpretation that could be of use in predicting the character and scale of this province "in the large."

Some measurements of isotopic ages within the Canadian Grenville were obtained which lay within a time band of between 800 million years to 1150 million years. As more measurements were made, it became apparent that the "ages of rocks" within the Grenville province were characteristically in this interval, and that this was distinctively younger than the ages of around 2300 million years which appeared within the Superior province [Leech, Lowden, Stockwell, and Wanless, 1963; Lepp, Goldich, and Kistler, 1963]. Age measurements in the Adirondacks and New Jersey Highlands [Tilton et al., 1960; Long, 1961] also gave ages between 800 to 1150 million years. These data showed that the complex structural and lithologic criteria which were originally used to define the Grenville province also appeared to define a time zone. In Maryland, the Glenarm series and the crystalline rocks which underlie them were of uncertain stratigraphic age. These rocks lie to the east of the major folded Appalachian structures. Age determinations indicated that they were metamorphosed during early Paleozoic time ($\sim 350$ million years), but more complete investigation by Tilton, Wetherill, Davis, and Hopson [1958] showed that the zircons of the Baltimore gneiss indicated an age of 1150 million years, suggesting that part of the deformed basement of the Appalachian mountain belt consisted of gneisses which appeared to be time-correlative with the metamorphism within the "classical" Grenville province. This was a major discovery in its own right since it showed the possibility of recognizing a time of primary crystallization through later periods of metamorphism [Wetherill, Kouvo, Tilton, and Gast, 1962].

Other investigations showed rocks of about 1000 million years in exposures in North Carolina and in some basement cores from Ohio [Tilton, Wetherill, Davis, and Bass, 1960]. In central Texas a large exposure of a Precambrian metamorphic and igneous complex was found to be of about 1050 million years [Aldrich, Wetherill, Davis, and Tilton, 1958]. An age study of the exposed Precambrian and of basement cores in an E-W section across the state of Texas showed what appears to be a continuous band of rocks with ages of about 1000 million years which are bounded to the north by older metamorphic and igneous rocks (1200? to 1450). The regions adjacent to this belt are currently being investigated [Wasserburg, Towell, and Steiger, 1965] and show a distinct age jump from 1090 to 1400 million years between El Paso and the southern end of the San Andres mountains—ages of 1400 million years appear consistently in a N-S traverse from White Sands to Albuquerque.

In some localities in the Precambrian of Texas, the lithologic types and structural characteristics are comparable with what are found in the Grenville province, but in many places the rocks can hardly be correlated on such a basis. It is, obviously, plausible to correlate these rocks which were crystallized or metamorphosed at $\sim 1000$ million years and to make this an extension of the Grenville province. In doing so we obtain a belt of fully continental dimensions (Figure 3). However, this is no longer the Grenville province as defined by lithologic and structural criteria but a continuous (?) region which was subject to contemporaneous (?) crystallization or recrystallization $\sim 10^9$ years ago. It is inferred that this event, or series of events covering about 250 million years, corresponds to an orogenic episode. The term *event* was introduced into the literature to distinguish whatever it is that happened from more specific geologic phenomena. We now frequently refer to regions which show time correlative "dates" as orogenic provinces. This usage, while most probably correct, is an inference. In simple cases the ages which are obtained by the various methods determine a time since the mineral was a closed system with regard to the parent-daughter isotopes. This can, in complicated (but common) circumstances, have very varied meanings as far as "age" is concerned. Assuming that the data are sufficiently coherent so that the age may be taken as representing the last period of metamorphism or intrusion, it is important to make the distinction between these events and orogeny in general. The classical criteria for tectonic activity are structural-stratigraphic, and the time-correlative terms, based on radioactive decay schemes which are now being obtained and used, may have a very different meaning. As emphasized by Gilluly [1963], "radiometric dates yield, not the times of orogenic activity, but of those orogenic episodes which were accompanied by plutonism. Judged by tectonic criteria, these episodes were not more intense than many orogenic episodes not so accompanied." He further states that "there is no obvious correlation of tectonics with metamorphism or with granitic intrusion, though both metamorphism and granite are obviously associated with tectonism." However, I would even prefer to remain uncertain as to the latter correlation. It may be that some of the less tangible "events" which have been reported, such as the

FIG. 3. The belt of igneous and metamorphic rocks which have ages typical of the Grenville time band, which is here taken to be 900 to 1200 million years. Other ages are only shown to indicate the boundary.

so-called 1350 to 1450 million year orogeny, represent tectonic episodes without broad-scale plutonism and (obvious) metamorphism. More recent studies suggest that even the ghostly 1350 to 1450 million year event has extensive associated plutonism.

We have seen from the preceding example that isotopic ages have independently confirmed some of the better-defined structural provinces.* They have also, in some cases, completely upset classifications in Precambrian terranes. The isotopic ages now permit another means of studying orogenic episodes in the Precambrian, but this new technique measures something which is quite different than has been used previously as a basis for describing orogeny. While confirming the coherence of the Grenville province, it has, using a completely different criterion, also added to it a mass of highly varied geologic materials several times its original size. There appear to be even more widespread outliers of this 1000 million year episode: an isolated 1.06 AE rapakivi pluton intrusive into 1.7 AE metamorphic rocks in Nevada, diabase sills in the Sierra Ancha of Arizona and pegmatites in the San Gabriel Mountains of California, 1 AE intrusives at the southern tip of Greenland; and if one really wishes to think big, Grenville ages as far as Norway and Sweden [Polkanov and Gerling, 1961; Kulp and Neumann, 1961]. It is clear that at this rate, the tail may well wag the dog and that what is needed to utilize properly the geologic "age mapping" in the Precambrian, which will undoubtedly be pursued for the next 10 to 20 years, are some theories of orogenesis that can make consequential predictions which are testable by observation and which may prove a format that will permit interpretations to be made; preferably, theories which can correlate the classical structural stratigraphic descriptions with the isotopic ages. This point can be best emphasized by considering the most modern texts on historical geology. One such book recently published, titled *The History of the Earth*, devotes 97 per cent of its contents to a description of 13 per cent of geologic time. This is with good reason; the Precambrian *is* prehistoric in the sense of paleontologic-stratigraphic techniques. And if in fact we were today to expand the coverage to be commensurate with the periods of time involved, while we would have a much thicker book (with many blank pages), the break in methodology would still be seen with the appearance of Olenellus. But the refined morphologic observations which describe parts of earth history (with far finer resolution than absolute age measurements) have not yet provided a genetic-theoretical basis for bridging the gap with "Prehistoric" geology. It is to this point that significant efforts must, in my opinion, be directed.

We will now jump from the broad continental-scale province to

* In the preliminary version of the new tectonic map of Canada (Stockwell, A Tectonic Map of the Canadian Shield, Figure 2, The Tectonics of the Canadian Shield RSC Spec. Pub. No. 4, edited by J. S. Stevenson), it may be seen that the provinces appear to have taken on a strongly time-orogenic character as compared with the previous map. The classification seems to now be based more on a "time" basis than on structural criteria.

consider the boundary between the younger Grenville and the older Superior province within the region delineated on the Canadian shield. This boundary zone is a natural place to determine whether the geologically younger belts represent the addition of new material or are, in fact, the product of metamorphism of the older preexisting geologic provinces. An investigation was made by Grant [1964] across a segment of the Grenville Front south of Lake Timagami, Ontario. By careful field mapping, he was able to demonstrate that the metamorphic transition, which defines the front in this region, is unfaulted

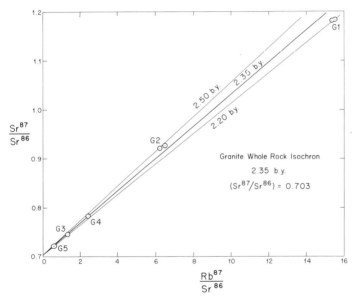

FIG. 4. Total rock samples for granites taken from a traverse across the Grenville Front [Grant, 1964].

for three miles and that it appeared possible to tentatively correlate geological units of the Superior and Grenville provinces across the front. This possibility had also been estimated in other areas by previous workers. The lithologies, structures, and grades of metamorphism found on each side of the front are typical for that province and appear quite distinctive. On the "Grenville side" there is a migmatitic terrane is which granite is a major component of the migmatite and which includes generally concordant foliated homogeneous granitic bodies. This terrane is typical of the northwestern part of the Grenville province. From previous studies using the strontium evolution diagram, it has been shown that the determination of a "total rock

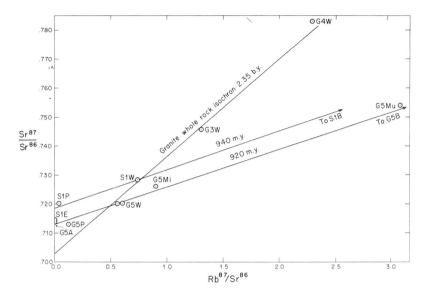

FIG. 5. Mineral and total rock analyses for a granite (G5) and a schist (S1) originally of Superior Province age but "cannibalized" during Grenville metamorphism [Grant, 1964].

isochron" and a "mineral isochron" can be used to determine respectively the primary age of crystallization of a suite of cogenetic igneous bodies and also the time at which they were last recrystallized [Compston, Jeffery, and Riley, 1960; Fairbairn, Hurley, and Pinson, 1961; Lanphere, Wasserburg, Albee, and Tilton, 1964]. Applying this technique to a suite of granitic rocks which were taken from a traverse across the front, Grant found that the total rock samples lie along a fairly well-defined isochron with a slope corresponding to about 2.35 AE (Figure 4). Mineral isochrons for a granite and an adjacent schist well within the Grenville region showed that they had achieved virtually complete Sr isotopic homogenization at about 930 million years ago, indicating this "Grenville" age of metamorphism (Figure 5). A granite north of the front which appeared macroscopically unaffected by the Grenville orogeny shows that a partial strontium isotopic reequilibration took place among the constituent minerals also within Grenville time (Figure 6). This study confirms that granitic rocks and metasediments of the Superior province with primary ages of 2.35 AE or greater were reconstituted during the Grenville orogeny and now form part of the Grenville province. In addition, there is distinct evidence that the Grenville episode produced metamorphic effects within what is macroscopically the Superior province. This study shows that at least part of the younger orogenic belt is older continental material and that the band which defines the Grenville front is, in terms of registering effects of this orogeny, probably several miles wide. A study by Krogh [1964] and Hills and Gast [1963] on Grenville province igneous and metamorphic rocks in the Hastings Basin area and the Adirondacks farther to the southeast has shown that there do not appear in this region any rocks which are older than 1300 million years. In general, it therefore appears that some of the younger material comprising this Grenville orogenic belt is reworked older rocks, and some is "new" material added to the continents. The positive identification which was made of Superior province rocks is more definitive than the results which fail to produce evidence of older ages, and so "new" material must be used with some caution (we will return to this point later). It is clear that more extensive studies across the boundaries of this and similar "orogenic" units will be pursued with great intensity in order to give a more complete answer to these questions. In addition to the obvious important tectonic implications of such studies as the preceding one, it is also clear that these techniques are of fundamental importance

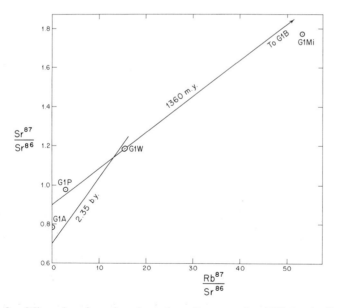

FIG. 6. Mineral and total rock analyses for a granite (G1) in the Superior Province in the neighborhood of the Grenville Front. Note the displacement of the mineral points from the 2.35 AE isochron which shows incomplete Sr isotopic equilibration during Grenville metamorphism [Grant, 1964].

to the field of metamorphic petrology. The use of isotopic redistribution as a means of detecting even very subtle metamorphism and of understanding the nature of the processes which control element redistribution in general will become an inherent part of the subject of metamorphic petrology and *not* an independent activity. Argon loss, lead-uranium diffusion, and strontium homogenization and oxygen isotope ratios will belong in the text of metamorphic petrology along with the other physicochemical methods which have long been a part of the organism of this discipline.

## Common Pb and Sr

In the "dating" methods the usual parameters are the amount of daughter isotope produced by radioactive decay and the amount of parent isotope present today. In general the chemical elements which we observe today represent the materials produced during nucleosynthesis which were incorporated in the earth and the contribution from the decay of natural long-lived activities within the earth. The relative abundances of those isotopes which have long-lived progenitors to those which have none is therefore a function of the chemical abundances of parent and daughter, and a function of time. The study of the isotopic abundances of such elements in systems which are not selected because of a high parent/daughter ratio and which have relatively small radiogenic enrichments constitutes the so-called common lead and common strontium problems.

If continental crust, oceanic crust, and the mantle materials have different chemical abundances of daughter and parent elements, the

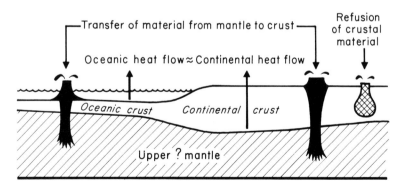

Fig. 7. Schematic diagram showing magmatic rocks derived from upper (?) mantle below the oceanic crust, below the continental crust, and magmatic rocks produced by refusion of continental crust.

study of the relative isotopic composition of common Pb and common Sr gives a means of distinguishing material from these different sources and of following their time evolution and the fractionation of U-Th-Pb, which is so important in the heat balance of the earth (see Figure 7).

After the demonstration by Nier [1938] that the isotopic composition of ore leads was variable and roughly a function of age, and with the

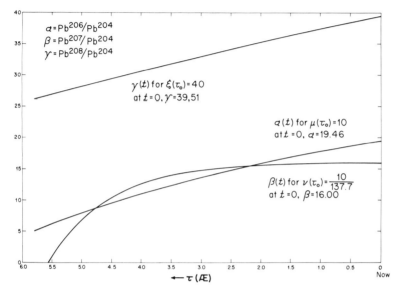

Fig. 8. Lead evolution curves for a closed homogeneous system with $(U^{238}/Pb^{204})_{today} = 10$ and $(Th^{232}/Pd^{204})_{today} = 40$. Modern lead ($t = 0$) as indicated.

relationship of these parameters to the "age of the earth" as developed by Houtermans and Holmes, a large number of measurements were accumulated. If the mantle represents a homogeneous reservoir with given ratios of $U^{238}/Pb^{204}$ and $Th^{232}/Pb^{204}$, then the relationship between age and isotopic composition will be as shown in Figure 8. This model has been investigated with ore leads by Russell [1963] and Russell and Farquhar [1960], and has been presumed by some workers to represent the evolution of mantle lead. In terms of the lead evolution diagram introduced by Houtermans (Figure 9), the evolution of lead is described by the coordinates $\alpha = Pb^{206}/Pb^{204}$, $\beta = Pb^{207}/Pb^{204}$, $\gamma = Pb^{208}/Pb^{209}$ and are parametrized by the time at which lead is removed from the parent reservoir and the abundance of $U^{238}$ relative to $Pb^{204}$ today ($\mu$).

If the source of magmatic rocks is a homogeneous (U-Pb-Sr-Rb) reservoir in which the ratio of $U^{238}/Pb^{204}$ only changes by the effect of radioactive decay, then samples of these rocks will have as their initial Pb isotopic composition the same value as the homogeneous reservoir at that time (and the same value of $\mu$). They will then lie along a constant $\mu$ curve. If there are several magmatic sources which originally had the same Pb isotopic composition but different initial $U^{238}/Pb^{204}$ ratios and which maintained their separate integrity,

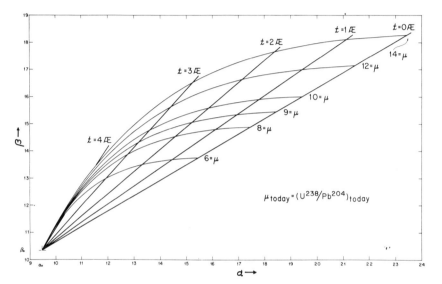

FIG. 9. Lead evolution diagram for closed systems with the same initial lead but with different U/Pb ratios. After Houtermans [1946, 1947].

then at a given time $t$ (years ago) samples of lead from each of the reservoirs would lie along a straight line (isochron) on the $\alpha,\beta$ diagram. The slope (for say reservoirs $i$ and $j$) is given by

$$\frac{\alpha_i - \alpha_j}{\beta_i - \beta_j} = \left(\frac{U^{238}}{U^{235}}\right)_{today} \frac{[e^{\lambda_8 \tau_0} - e^{\lambda_8 t}]}{[e^{\lambda_5 \tau_0} - e^{\lambda_5 t}]}$$

This is a function of the "age" $\tau_0$ of the parent system and the age $t$. If $\tau_0$ is known, then the time "$t$" is determined. For modern samples $t = 0$, and it is possible to calculate $\tau_0$. If $\tau_0$ and the initial isotopic composition are known, it is possible to calculate $\mu$ from the equation for $\alpha$ given in Table 2. A similar set of equations holds for $\gamma = Pb^{208}/Pb^{204}$ and $\alpha$ which, however, depend on the $Th^{232}/U^{238}$ ratio today.

On this diagram (Figure 9) all unfractionating systems which originally had the same Pb isotopic composition would lie on a straight line

## Table 2

EQUATIONS FOR RESERVOIRS CLOSED OR SUBJECT TO LOSS IN WHICH U-TH-PB ARE NOT FRACTIONATED. NOTE THAT $\mu(\tau_0)/\nu(\tau_0)$ IS A CONSTANT (137.8) FOR ALL TERRESTRIAL SAMPLES

$$\alpha(\tau) - \alpha_0 = \mu(\tau_0)\left[e^{\lambda_8 \tau_0} - e^{\lambda_8 t}\right]$$

$$\beta(\tau) - \beta_0 = \nu(\tau_0)\left[e^{\lambda_5 \tau_0} - e^{\lambda_5 t}\right]$$

$$\gamma(\tau) - \gamma_0 = \xi(\tau_0)\left[e^{\lambda_2 \tau_0} - e^{\lambda_2 t}\right]$$

$$\frac{\alpha - \alpha_0}{\beta - \beta_0} = \frac{\mu(\tau_0)}{\nu(\tau_0)} \frac{\left[e^{\lambda_8 \tau_0} - e^{\lambda_8 t}\right]}{\left[e^{\lambda_5 \tau_0} - e^{\lambda_5 t}\right]}$$

whose slope determines the time since they were isotopically identical. It was exactly ten years ago that C. C. Patterson in his classical experiment showed that the lead from iron meteorites (uranium free), stone meteorites, and "modern" lead from basalts and deep-sea sediments lie on a straight line corresponding to a time of 4.5 AE, a time which is often taken as the "age of the earth." These are the first and most direct data which relate other solar bodies and the earth. Ages of between 4.0 and 4.7 AE were obtained on meteorites by the $Ar^{40}$-$K^{40}$ and $Sr^{87}$-$Rb^{87}$ methods. These data established this time as a period of major chemical differentiation in the history of the solar system. The values $\alpha$, $\beta$ for Fe meteorites are taken to be the isotopic composition of lead at the time of formation of the earth.

In studying ore leads (and a suite of volcanic leads), it was found that some were anomalous in that they lie in the region to the right of the "modern isochron" and would yield negative (i.e., future ages) in terms of the parameters used for $\alpha_0$, $\beta_0$, and $\tau_0$. This leads to the question of the correctness of $\alpha_0$, $\beta_0$, and $\tau_0$ and whether or not it is possible to assign characteristic values of $\mu$ to lead from various sources. This problem has been under attack for the past few years by studying primary Pb in U-free minerals in magmatic rocks whose age of crystallization is known (from independent isotopic age measurements on other phases or using other decay schemes) and by studying the lead isotopes in oceanic sediments and beach sands [Patterson, 1964]. Earlier studies in which it was not possible to assign an age independent of the common Pb composition made it difficult, if not impossible, to determine whether Pb was "anomalous" in terms of the simple model of growth in a nonfractionating reservoir.

Houtermans' model is very simple in that it assumes that U and lead in particular are not fractionated during the formation of crustal materials. This is not demonstrable independently and is best determined by testing the model on well-understood systems. An example of such an investigation is the recent work of Zartman on lead in plutonic and metasedimentary rocks of the Llano region in Texas, part of the "Grenville" age province. The igneous rocks have been well dated by the major methods. Primary lead in the feldspars (which are essentially U free) from all the principal igneous rock types showed a rather uniform isotopic composition and gave model ages $t$ (using the constants $\alpha_0$, $\beta_0$, $\tau_0$ given by Patterson) which are in good agreement with the ages determined by other methods and on more radiogenic minerals. The relative uniformity of the primary lead suggests a homogeneous source. The measured values of $\mu$ for the total rocks and the calculated values of $\mu$ from the model are in good agreement, indicating no serious Pb-U fractionation in the formation of the granites and the parent material in this particular case.

Such work, if extended to other igneous bodies of the same age, will determine whether all of these approximately contemporaneous igneous rocks came from the same magma source or from different sources. On the other hand, the Pb from the metasedimentary rocks was isotopically very variable and confirmed the complicated origin which is apparent from geological considerations. A study by Catanzaro and Gast [1960] indicated rather good agreement between model ages and the regular ages for some pegmatites as old as 2.7 AE. From the few published data, it appears that the simple model is grossly correct and that $\alpha_0$, $\beta_0$, and $\tau_0$ cannot be greatly different from the values given by Patterson.* However, such studies are still not complete, and it is particularly important that they be extended to more well-dated plutonic rocks.

The consequences of fractionation between Pb and U in the formation of magmatic bodies can be considered in terms of several exchanging homogeneous reservoirs. This has been approached by Patterson from consideration of the lead in beach sands and deep-sea sediments. A theoretical discussion of the simple two-reservoir case is of interest. The equations are given by Wasserburg [1964]. The solution for a simple two-layer model with constant transport coefficients are shown in Table 3.

* The work of Reed, Kigoshi, and Turkevich [1960] on the U, Th, and Pb concentrations in stony meteorites shows that there remains some difficulty in explaining the material balance in the parent-daughter systems for these bodies. This observation should not be obscured by our enthusiasm for "magic numbers."

### Table 3

SOLUTIONS FOR $\alpha$ AND $\beta$ FOR CRUST (1) AND MANTLE (2) FOR THE CASE OF A CONTINUOUSLY GROWING CRUST.

| Crust | Mantle |
|---|---|
| $\mu_p^1/\mu_p^\Sigma = \dfrac{(1-e^{-H\tau})}{(1-e^{-G\tau})}$ | $\mu_p^2/\mu_p^\Sigma = e^{-(H-G)\tau}$ |
| $\alpha^1 = \alpha_0 + \mu_0 \dfrac{\left[1-e^{-\lambda_8\tau} - \dfrac{\lambda_8}{(G-H-\lambda_8)}(e^{-(\lambda_8+H)\tau} - e^{-G\tau})\right]}{(1-e^{-G\tau})}$ | $\alpha^2 = \alpha_0 + \mu_0 \dfrac{\left[e^{-(\lambda_8+H-G)\tau} - 1\right]}{(G-H-\lambda_8)}$ |
| $\beta^1 = \beta_0 + \nu_0 \dfrac{\left[1-e^{-\lambda_5\tau} - \dfrac{\lambda_5}{(G-H-\lambda_5)}(e^{-(\lambda_5+H)\tau} - e^{-G\tau})\right]}{(1-e^{-G\tau})}$ | $\beta^2 = \beta_0 + \mu_0 \dfrac{\left[e^{-(\lambda_5+H-G)\tau} - 1\right]}{(G-H-\lambda_5)}$ |

$G$ and $H$ are the transport constants for lead and uranium, respectively. The symbol $\Sigma$ indicates the value for the total system $(1 + 2)$, and the subscripts $p$ and $o$ indicate present and original values, respectively.

Here the functions $G^j$ and $H^j$ are the fractional rates of transport of $Pb^{204}$ and $U^{238}$ out of reservoir $j$ and represent the time constants for chemical differentiation of the earth. If the transport is solely from mantle to form crust, then

$$H^{\text{mantle}} = \frac{-\mathscr{E}}{V^{\text{mantle}}} \frac{dV^{\text{mantle}}}{dt}$$

where $\mathscr{E}$ is the enrichment factor for uranium of the magma compared to the mantle, and $(1/V^{\text{mantle}})(dV^{\text{mantle}}/dt)$ is the fractional rate of change of the volume of the mantle. A schematic drawing of this is illustrated in Figure 10. It is possible to estimate this parameter from considerations of crust and mantle volumes and the estimated content of U in these regions. It then follows that $\alpha$ and $\beta$ for the total crust depend not only on the time but on $G^{\text{mantle}}$ and $H^{\text{mantle}}$. For the mantle and hence for the materials added to the crust from the mantle $\alpha$ and $\beta$ only depend on $G^{\text{mantle}}-H^{\text{mantle}}$, and so a mantle reservoir with different time constants $G$, $H$, but the same value $G-H$ will yield leads of identical composition. An evolution curve for the mantle (and crust) is similar in form to the nonfractionating evolution model. In fact a mantle consisting of a family of systems with the same value of $G-H$ but with different $\mu$ values will also generate isochrons, the only difference being that the slopes will depend on $G\text{-}H$, and hence the ages and $\mu$ values calculated assuming the nonfractionating model will be in error (Figure $11a, b$). Independent age assignments are therefore absolutely necessary to understand the Pb isotopic evolution. The extent to which this type of fractionation is important can only be answered by more precise data of the type mentioned earlier. The value of $\tau_0$, the so-called age of the earth, is also tied up in these parameters so that the effects of fractionation are important for a number of reasons. The variety of important work currently

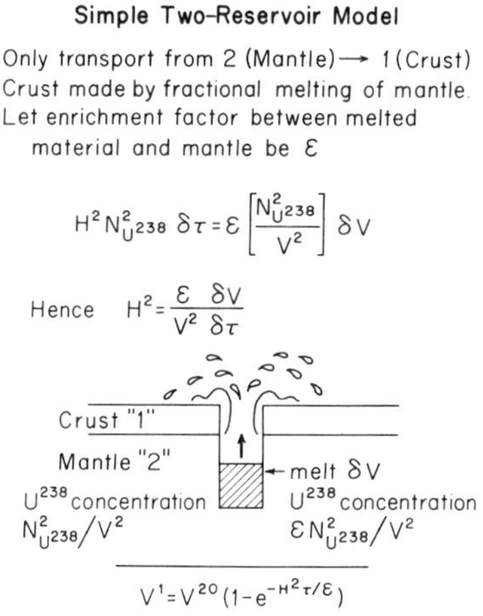

**Simple Two-Reservoir Model**

Only transport from 2 (Mantle) → 1 (Crust)
Crust made by fractional melting of mantle.
Let enrichment factor between melted
material and mantle be $\mathcal{E}$

$$H^2 N_{U^{238}}^2 \, \delta\tau = \mathcal{E} \left[ \frac{N_{U^{238}}^2}{V^2} \right] \delta V$$

Hence $\quad H^2 = \dfrac{\mathcal{E}}{V^2} \dfrac{\delta V}{\delta \tau}$

Crust "1"
Mantle "2" ← melt $\delta V$
$U^{238}$ concentration $\quad U^{238}$ concentration
$N_{U^{238}}^2 / V^2 \qquad \mathcal{E} N_{U^{238}}^2 / V^2$

$$V^1 = V^{20}(1 - e^{-H^2 \tau / \mathcal{E}})$$

Fig. 10. Relationship between transport constant, chemical fractionation factor, and rate of growth of crust.

taking place in this field is represented by the symposium on lead isotopes at the National American Geophysical Union meeting of 1964.

The evolution of strontium may also be applied to the study of crustal evolution [Gast, 1960; Hurley et al., 1962]. Rb-Sr age determinations on meteorites [Schumacher, 1956; Herzog and Pinson, 1956; Webster, Morgan, and Smales, 1957; Gast, 1962] have shown that there is considerable enrichment in $Sr^{87}$ in chondrites due to $Rb^{87}$ decay. The enrichment observed in primary strontium in terrestrial igneous rocks is much lower as shown in Figure 12. This means that the ratio of Rb/Sr for the source of these materials is considerably lower than for chondritic meteorites as emphasized by Gast [1960]. In terms of the evolution of strontium in the mantle as determined on rocks of supposed mantle origin, this means that the total enrichment is at most only 1.4 per cent (and most probably between 0.7 and 1 per cent) over a period of about 4.5 AE or about 0.3 per cent per AE. The present precision of good quality solid source mass

# GEOCHRONOLOGY

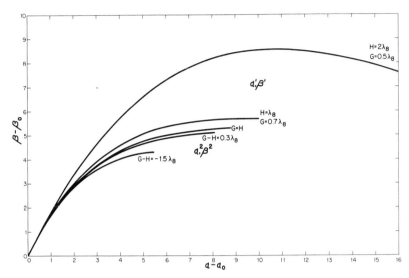

FIG. 11(a). Lead evolution curves for crust (1) and mantle (2) for two-layer model with transport constants $G$ and $H$. It is assumed (without justification) that the crust is enriched in uranium with respect to lead. The composition of lead added to the crust at a given time is read off of the mantle curve for the appropriate time.

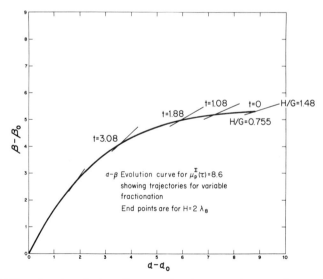

FIG. 11(b). Lead evolution diagram showing growth curve for total system and trajectories for variable U-Pb fractionation (H/G) at different times.

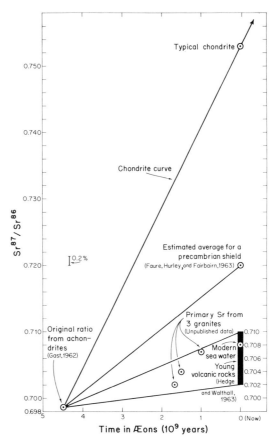

Fig. 12. Schematic diagram for $Sr^{87}/Sr^{86}$ growth with time for chondrites showing the comparison with modern sea water, young volcanic rocks (continental and oceanic) and a present-day shield area.

spectrometric measurements is about 0.1 to 0.2 per cent and hence the detailed exploitation of Sr isotopic evolution is dependent on a refinement in technique. The enrichment in crustal Rb/Sr over that found in presumed mantle material can permit the distinction between remelted crustal material and the addition of new mantle material. The distinction between primary mantle material and refused crustal material can be very positive in the case that the refused crustal material has a long enough history in a Rb rich environment. However, in the case that this ratio is not high enough, the distinction cannot be as strong since the great heterogeneity of crustal materials can certainly provide sources with a variety of Rb/Sr ratios.

With regard to the question of normal strontium, primary $Sr^{87}/Sr^{86}$ values for some granites are typically from 0.702 to 0.708, whereas young basic volcanic rocks have typical ratios of from 0.702 to 0.707. Estimates of the mean $Sr^{87}/Sr^{86}$ ratio in the Precambrian shield of N. America (Gast, Faure, Hurley, and Fairbairn) indicate this to be $\approx 0.72$. The refusion of such shield materials today would produce magmas with an original strontium quite different from the above-mentioned range of values and would indicate a crustal origin. This has already been used on metasedimentary rocks and is clearly an indication of crustal history. The clarification of the time evolution of Sr and the Rb/Sr ratio is being pursued by investigations on Sr rich-Rb poor minerals from rocks with simple histories and of known age.

As mentioned earlier, the assumption of a simple homogeneous mantle reservoir for continental volcanic rocks is uncertain because of the possibility of crustal contamination, and it would appear that the oceanic volcanic materials would yield better samples of upper crustal material. A very recent study by Lessing and Catanzaro [1964] on Hawaiian trachytes, and a study by Gast, Tilton, and Hedge [1964] on the isotopic composition of lead and strontium in several volcanic rocks from Ascension and Gough Islands on the mid-Atlantic ridge showed that this simple assumption is invalid. They reported that $Sr^{87}/Sr^{86}$ varied for different samples from the same island and between two islands. The magnitude of this effect is about 0.7 per cent and would constitute half of the total effect due to mantle evolution as estimated earlier. They also reported large isotopic variations of $\alpha$ and $\beta$ for samples from the same island and between the islands. The samples from Gough Island fall very close to the modern model isochron but have differences of isotopic composition

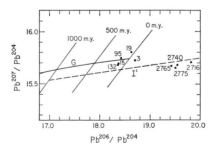

Fig. 13. Part of a lead evolution diagram [parameters after Murthy and Patterson, 1962], showing the experimental points for Gough and Ascension Islands [Gast, Tilton and Hedge, 1964].

which would correspond to model age differences of almost 250 million years. The latter effect is present in the Ascension Island samples, but they would yield negative model ages of up to 700 million years as shown in Figure 13. Anomalous leads and strontium abundances have been observed previously in continental areas, but this result emphasizes that the assumption of a homogeneous upper mantle in oceanic areas is not justified and that some of the anomalies observed in continental areas may not so obviously be used as an indication of crustal contamination. Petrologic investigations have indicated that it is necessary to assume two magma sources in order to explain the extrusives in Hawaii. The isotopic data confirm this minimal degree of complication and demand isotopic as well as chemical heterogeniety on what appears to be a local scale for significant time periods. Such different sources are apparently mechanically available for extrusion over extremely short time periods.

The results on oceanic leads and Sr indicate Pb-U-Th and Sr-Rb fractionation during crustal evolution and suggest more complexities than are even hinted at by a simple fractionating two layer crustal-mantle model. Clarification of such phenomena will demand the interest and activity of all the branches of the earth sciences.

## Acknowledgments

The author wishes to acknowledge valuable discussions with both friendly and hostile colleagues. This work was aided by an understanding wife and a helpful secretary and was disrupted by exciting data and mechanical breakdowns produced in a laboratory supported by a National Science Foundation grant and an Atomic Energy Commission contract.

## References

Aldrich, L. T., and G. W. Wetherill, Geochronology by radioactive decay, *Ann. Rev. Nucl. Sci.*, *8*, 257–298, 1958.

Aldrich, L. T., G. W. Wetherill, G. L. Davis, and G. R. Tilton, Radioactive ages of micas from granite rocks by Rb-Sr and K-Ar methods, *Trans. Am. Geophys. Union*, *39*, 1124–1134, 1958.

Aston, F. W., Isotopic constitution and atomic weight of lead from different sources, *Proc. Roy. Soc. London*, *A*, 535–543, 1933.

Catanzaro, E. J., and P. W. Gast, Isotopic composition of lead in pegmatitic feldspars, *Geochim. Cosmochim. Acta*, *19*, 113–129, 1960.

Compston, W., P. M. Jeffery, and G. H. Riley, Age of emplacement of granites, *Nature*, *186*, 702, 1960.

Eaton, J. P., and K. J. Murata, How volcanoes grow, *Science*, *132*, 925–938, 1960.

Fairbairn, H. W., P. M. Hurley, and W. H. Pinson, The relation of discordant Rb-Sr mineral and whole rock ages in an igneous rock to its time of subsequent metamorphism, *Geochim. Cosmochim. Acta*, *23*, 135–137, 1961.

Gast, P. W., The isotopic composition of Sr and the age of stone meteorites, *Geochim. Cosmochim. Acta*, *26*, 927–943, 1962.

Gast, P. W., Limitations on the composition of the Upper Mantle, *J. Geophys. Res.*, *65*, 1287–1297, 1960.

Gast, P. W., G. R. Tilton, and C. Hedge, Isotopic composition of lead and strontium from Ascension and Gough Islands, *Science*, *145*, 1181–1185, 1964.

Gilluly, J., Distribution of mountain building in geologic time, *Bull. Geol. Soc. Am.*, *60*, 561–590, 1949.

Gilluly, J., The tectonic evolution of the western United States, *Quart. J. Geol. Soc. London*, *119*, 133–174, 1963.

Grant, James, A., A rubidium-strontium isochron study of the Grenville Front near Lake Timagami, Ontario, *Science*, in press, 1964.

Hall, Sir James, Account of a series of experiments showing the effects of compression in modifying the action of heat, *Trans. Roy. Soc. Edinburgh*, *6*, 71 et seq., 1812.

Herzog, L. F., and W. H. Pinson, Rb/Sr age, elemental and isotopic abundance studies of stony meteorites, *Am. J. Sci.*, *254*, 555–566, 1956.

Hills, A., and P. W. Gast, Age of pyroxene-hornblende granitic gneiss of the eastern Adirondacks by the Rb-Sr whole rock method, Abstract, Geol. Soc. Am., 1963 Annual Meetings.

Houtermans, F. G., Die isotopen häufigkeiten in natürlichen blei und das alter des urans, *Naturwissenschaften*, *33*, 185, 1946.

Houtermans, F. G., Das alter des urans, *Z. Naturforsch.*, *2A*, 322–328, 1947.

Hurley, P. M., H. Hughes, G. Faure, H. W. Fairbairn, and W. H. Pinson, Radiogenic $Sr^{87}$ model of continent formation, *J. Geophys. Res.*, *67*, 5315–5334, 1962.

Krogh, T. E., Whole rock isochron tests on the genesis of granitic bodies in the Grenville of Ontario, Abstract, *Trans. Am. Geoph. Union*, *45*, 114, 1964.

Kulp, J. L., and H. Neumann, Some potassium-argon ages on rocks from the Norwegian Basement, *Ann. N.Y. Acad. Sci.*, *91*, 469–473, 1961.

Lanphere, M. A., G. J. Wasserburg, A. L. Albee, and G. R. Tilton, Redistribution of strontium and rubidium isotopes during metamorphism, *Isotopic and Cosmic Chemistry*, North-Holland Publishing Company, Amsterdam, 269–320, 1964.

Leech, G. B., J. A. Lowdon, C. H. Stockwell, and R. K. Wanless, Age determinations and geological studies, *Geol. Surv. Canada Paper*, 63–117, 1963.

Lepp, H., S. S. Goldich, and R. W. Kistler, A Grenville cross section from Port Cartier to Mt. Reed, Quebec, Canada, *Am. J. Sci.*, *261*, 693–712, 1963.

Lessing, P., and E. J. Catanzaro, $Sr^{87}/Sr^{86}$ ratios in Hawaiian lavas, *J. Geophys. Res.*, *69*, 1599–1601, 1964.

Leutz, H., H. Wenninger, and K. Ziegler, Die halbwertszeit des $Rb^{87}$, *Z. Physik*, *169*, 409–416, 1962.

Long, L., Isotopic ages from northern New Jersey and southeastern New York, *Ann. N.Y. Acad. Sci.*, *91*, 400–407, 1961.

Mackin, J. H., Rational and empirical methods of investigation in geology, *The Fabric of Geology*, edited by C. C. Albritton, Jr., p. 135, Addison-Wesley Publishing Company, Reading, Mass., 1963.

Menard, H. W., *Marine Geology of the Pacific*, McGraw-Hill Book Company, Inc., New York, 1964.

Murthy, V. R., and C. C. Patterson, Primary isochron of zero age for meteorites and the earth, *J. Geophys. Res.*, 67, 1161–1167, 1962.

Nier, A. O., Isotope abundance of common lead, *J. Am. Chem. Soc.*, 60, 1571–1576, 1938.

Patterson, C. C., Characteristics of lead isotope evolution on a continental scale in the earth, *Isotopic and Cosmic Chemistry*, North-Holland Publishing Company, Amsterdam, 244–268, 1964.

Polkanov, A. A., and E. K. Gerling, The Pre-Cambrian geochronology of the Baltic Shield, *Ann. N.Y. Acad. of Sci.*, 91, 492–499, 1961.

Reed, G., K. Kigoshi, A. Turkevich, Determinations of concentrations of heavy elements in meteorites, *Geochim. Cosmochim. Acta*, 20, 122–140, 1960.

Russell, R. D., Some recent researches on lead isotope abundances, *Earth Science and Meteoritics*, North-Holland Publishing Company, Amsterdam, 44–73, 1963.

Russell, R. D., and R. M. Farquhar, *Lead Isotopes in Geology*, Interscience Publishers, Inc., New York, 1960.

Schumacher, E., Alter bestimmung von steinmeteoriten mit der Rb-Sr methode, *Z. Naturforsch.*, 11a, 206, 1956.

Silver, L. T., The use of cogenetic uranium-lead isotope systems in geochronology, in *Radioactive Dating*, International Atomic Energy Commission, Vienna, 1963a.

Silver, L. T., and S. Deutsch, Uranium-Lead isotopic variations in Zircons: a case study, *J. Geol.*, 71, 721–758, 1963b.

Stockwell, C. H., A tectonic map of the Canadian Shield, in *The Tectonics of the Canadian Shield*, edited by John S. Stevenson, *Roy. Soc. Canada Spec. Publ.*, no. 4, 1962.

Tilton, G. R., Volume diffusion as a mechanism for discordant lead ages, *J. Geophys. Res.*, 65, 2933–2945, 1960.

Tilton, G. R., G. W. Wetherill, G. L. Davis, M. N. Bass, 1000 m.y. old minerals from the Eastern United States and Canada, *J. Geophys. Res.*, 65, 4173–4179, 1960.

Tilton, G. R., G. W. Wetherill, G. L. Davis, and C. A. Hopson, Ages of minerals from the Baltimore gneiss, *Bull. Geol. Soc. Am.*, 67, 1469–1474, 1958.

Wasserburg, G. J., Crustal history and the Precambrian time scale, *Ann. N.Y. Acad. Sci.*, 91, 583–594, 1961.

Wasserburg, G. J., Diffusion processes in Pb-U systems, *J. Geophys. Res.*, 68, 4823–4846, 1963.

Wasserburg, G. J., Pb-U-Th evolution models for homogeneous systems with transport, *Trans. Am. Geophys. Union*, 45, Abstract, 111, 1964.

Wasserburg, G. J., D. Towell, and R. Steiger, A study of Rb-Sr systematics in some Precambrian granites of New Mexico, *Trans. Am. Geophys. Union.*, 46, Abstract, 173, 1965.

Wasserburg, G. J., G. W. Wetherill, L. T. Silver, and P. T. Flawn, A study of the ages of the Precambrian of Texas, *J. Geophys. Res.*, 67, 4021–4047, 1962.

Webster, R. K., J. W. Morgan, and A. A. Smales, Some recent Harwell analytical work on geochronology, *Trans. Am. Geophys. Union*, 38, 543–545, 1957.

Wetherill, G. W., Discordant uranium-lead ages, *Trans. Am. Geophys. Union*, *37*, 320–326, 1956.
Wetherill, G. W., O. Kouvo, G. R. Tilton, and P. W. Gast, Age measurements on rocks from the Finnish Precambrian, *J. Geol.*, *70*, 74–88, 1962.
Zartman, R. E., Thesis, California Institute of Technology, 1963.
Zartman, R. E., The isotopic composition of lead in microclines from the Llano uplift, Texas, *J. Geophys. Res.*, in press.

## General References

Faure, G., P. M. Hurley, and H. W. Fairbairn, An estimate of the isotopic composition of strontium in rocks of the Precambrian Shield in North America, *J. Geophys. Res.*, *68*, 2323–2329, 1963.
Hedge, C. E., and F. G. Walthall, Radiogenic $Sr^{87}$ as an index of geologic processes, *Science*, *140*, 1214–1219, 1963.
Holmes, A., An estimate of the age of the earth, *Nature*, *157*, 680, 1946.
Nicolaysen, L. O., Graphic interpretation of discordant age measurements on metamorphic rocks, *Ann. N.Y. Acad. Sci.*, *91*, 198–206, 1961.
Nier, A. O., The isotopic constitution of radiogenic leads and the measurement of geological time (II), *Phys. Rev.*, *55*, 153–163, 1939a.
Nier, A. O., The isotopic constitution of uranium and the half life of the uranium isotopes (I), *Phys. Rev.*, *55*, 150–153, 1939b.
Nier, A. O., R. W. Thompson, and B. F. Murphey, The isotopic constitution of lead and the measurement of geological time III, *Phys. Rev.*, *60*, 112–116, 1941.
Patterson, C. C., The isotopic composition of meteoritic, basaltic and oceanic leads and the age of the earth, *Proc. Conf. Nucl. Processes Geol. Settings, Williams Bay, Wisconsin*, N.A.S.-N.R.C., 36–40, 1953.
Patterson, C. C., H. Brown, G. Tilton, and M. Inghram, Concentration of uranium and lead and the isotopic composition of lead in meteoritic material, *Phys. Rev.*, *92*, 1234–1235, 1953.
Silver, L. T., Age determinations on Precambrian diabase differentiates in the Sierra Ancha, Gila County, Arizona, Abstract, *Bull. Geol. Soc. Am.*, *71*, (**12**), 1973, 1960.
Silver, L. T., C. R. McKinney, S. Deutsch, and J. Bolinger, Precambrian age determinations in the western San Gabriel mountains, California, *J. Geol.*, *71*, 196–214, 1963.
Wasserburg, G. J., and R. J. Hayden, Age of meteorites by the $A^{40}$-$K^{40}$ method, *Phys. Rev.*, *97*, 86–87, 1955.
Yoder, H. S., Jr., C. E. Tilley, Origin of basaltic magmas, *J. Petrol. Spec.*, *3*, 342–5320, 1962.

# THERMAL STRUCTURE OF THE UPPER MANTLE AND CONVECTION

*Walter M. Elsasser*

*Department of Geology, Princeton University, Princeton, New Jersey*

## 1. Introduction

When continental drift was first proposed by Wegener 50 years ago it was in a purely descriptive framework, allowing the geologist to organize many separate observations into a coherent picture. It was, correspondingly, not dynamical, and even the descriptive features could only be inferred from the continents, not from the geology of the ocean bottoms, which was then virtually unknown. It seems that the geology of the Southern Hemisphere offers particularly favorable conditions for the empirical study of continental drift; the idea attained there a life of its own and was vigorously developed. Du Toit [1937] replaced Wegener's original concept of a single protocontinent from which the present continents were split off and moved away, by a model of two protocontinents called Laurasia and Gondwana and located in the Northern and Southern Hemisphere, respectively. The idea that these two original continents might have been centered about the poles in the early period of the earth's history is interesting, but there seem to be so far no compelling empirical arguments for such a hypothesis.

Gondwana is supposed to have comprised the present continents of South America, Africa (including Arabia), Antarctica, Australia, and the subcontinent of India. It is well known that this scheme explains the appearance of remnants of the Permian glaciation in all of the present fragments, at places that in some cases are now separated from each other by more than 90° on a great circle. Further evidence for such a motion apart is found in the way in which these assumed fragments can be fitted together in a natural way after the manner of a picture puzzle, the best known case being that of the coastlines of South America and Africa which can be fitted almost perfectly, to within a few per cent of their present distance, if one of the two continents is translated and rotated bodily (and much more recent formations such as river deltas are omitted). The protocontinent of Laurasia is thought to have split into North America and Eurasia, the respective coastlines fitting only slightly less perfectly than those of the Southern Hemisphere. The Appalachian mountain system becomes in this picture a southwesterly extension of the Hercynian orogeny. Originally, the two protocontinents were separated by a wide ocean designated as Tethys.

These highly imaginative ideas are bolstered by a considerable number of tectonic and stratigraphic arguments, of more or less plausibility. Since, however, it has seemed impossible to provide conclusive proof on a geological basis, the idea of continental drift has for a long time led a marginal existence in the limbo of geological speculation. Only in recent years has it reentered the stage, largely as a result of the remarkable confirmation of the hypothesis provided by paleomagnetic methods [see Cox and Doell, 1960; and Runcorn's paper in Runcorn, ed., 1962]. At the same time, in connection with this revival, the question of the *dynamics* of continental drift, and of motions in the mantle in general, began to be the subject of serious inquiry [Ewing and Ewing, 1959]. The advent of paleomagnetic methods was only one of the developments that brought speculations on continental drift as well as on mantle convection to the fore again; the other, equally important, is the exploration of the ocean bottoms which took place in the last few decades. Since the ocean bottoms are separated from the mantle proper only by a relatively tenuous crustal layer, more direct inferences about motions in the mantle should be possible on the basis of ocean bottom topography and geology than on the basis of the crustal geology of the continents. Quite recently, Menard* [1964] has systematically surveyed all the features of the ocean bottom in view of interpreting them as expressions of motions in the mantle. The most conspicuous phenomenon is of course the

* The writer is grateful for the opportunity he had of studying Professor Menard's paper in preprint form.

vast system of rises or ridges which is world-wide and has a total length of nearly twice the circumference of the earth. Many of the ridges are rather accurately centered on medians between continental margins, as for instance in the Atlantic. But others are not. Thus the East Pacific Rise continues into the Gulf of California (which may indeed seem a product of the Rise) whence it extends farther under the western part of the United States; traces of even farther extensions reach into Alaska and the Arctic Sea. Other extensions of ridges passing from the oceans into the continents are found under Siberia.

Whatever the detailed topography and structure of these ridges, one thing seems to be clear, namely, that they are produced by rising mantle material. This is evidenced by their volcanism, seismicity, and the greatly enhanced heat flow found near the middle of several ridges. Further evidence is given by the many transcurrent faults crossing the ridges almost perpendicularly. These can be interpreted as produced by blocks that slide laterally off the center part of the ridges. Menard gives a detailed analysis and discussion of all these features especially insofar as they are taken to be indicative of convective motions in the mantle. One of his main conclusions is that the ridges seem to be transient features on a sufficiently long-time geological scale; this in turn enhances the probability that they are closely related to more widespread mass motions in the mantle.

As Menard points out, all surficial features, whether continental or oceanic, are limited in their diagnostic value with regard to motions in the mantle because they can give only a two-dimensional projection of a basically three-dimensional pattern. The question then arises as to what can be said about possible mass motions in the mantle from a study of the physical properties of the mantle in depth. An inquiry of this type presents a challenge to the geophysicist. Although our knowledge of the physics of the mantle is admittedly poor, it may nevertheless be put to some use. In the present paper we have tried to bring together and analyze, on the one hand, known data about the structure and physics of the upper mantle, and on the other hand, such general principles pertaining to the mechanics of deformable bodies as might be applicable to the kind of motions to be expected. It appears then that from the confrontation of these pertinent data and principles certain conclusions can be drawn which, while sometimes a trifle too general, are nevertheless not entirely indefinite. On the whole, by comparing these arguments with the evidence drawn from surficial observations, a somewhat stronger case can be made in favor of the occurrence of certain types of mass motions in the mantle. Moreover, and this seems almost equally important, on applying known physical principles to the mantle, one might be able to narrow down substantially the range of possible models for mechanical processes which have occurred in the course of the earth's geological history.

We shall make the point, however, that one needs to distinguish clearly between mass motions in the mantle on the one hand and continental drift on the other; the two types of processes cannot be taken as equivalent. There exist good reasons to believe that the top layers of the mantle are "harder" than the material a little farther down (for definitions and details see later). Thus motions somewhat farther down in the mantle might not at all be faithfully reflected in horizontal displacements at or very near the surface. If grounds can be given to show that large-scale mass motions might occur at some depth in the mantle, it does not follow that any of them, or more than a fraction of them, give rise to corresponding surface motion. Observational evidence, especially from the ocean bottom, leaves little doubt that horizontal surficial displacements of the order of a few hundred to over a thousand kilometers have occurred during the not too remote geological history. This does not constitute unambiguous evidence for the widespread occurrence of displacements amounting to many thousands of kilometers as is implied in any of the proposed schemes of continental drift. Menard is rather clear about the fact that the interpretation of rises as motions in the mantle implies surficial horizontal displacements but that it does not in itself confirm hypothetical schemes of continental drift. In the present paper we are concerned with the mechanics of the upper mantle and the possibility of mass motions which may be derived from it, but not necessarily with a confirmation of the drift hypothesis. The tie-up of such conclusions as may be derived from the study of the physics of the mantle with any scheme of continental drift constitutes rather a separate problem; we shall touch on it briefly toward the end.

## 2. Instability. Symmetry Considerations

We shall find that the concept of mechanical instability plays a fundamental role in any theory of motions in the earth's mantle. It seems advantageous, before dealing with the general physics of the mantle, to discuss instability as a problem of mechanics. We begin with the simplest possible cases and proceed from there to more complex situations closer to those that will be significant in tectonics.

The simplest case is that of one-dimensional motion as exemplified by the textbook case of a small ball which can slide or roll on a curve (constraint) under the influence of gravity. In $(a)$, $(b)$, $(c)$, Figure 1,

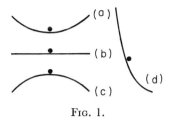

Fig. 1.

are shown the familiar cases of stable, indifferent, and unstable equilibrium. In $(d)$, on the other hand, we may imagine the ball released from an elevated point of the curve, whereupon it begins to fall downward. Note the characteristic difference between $(c)$ and $(d)$. In case $(c)$ we have a true equilibrium point in a mathematical sense, but the equilibrium is not stable; there are two orbits issuing from this point, to the left or right. Which one of these orbits will be taken depends on minor and usually impredictable perturbations or inhomogeneities of the system. Case $(d)$ is quite different: Here the ball once released describes an orbit which as function of space coordinates and of time is uniquely determined. We shall refer to case $(c)$ as an *instability* whereas case $(d)$ represents merely an ordinary *orbital point* of the system.

The general mathematical analysis of such problems takes into account friction; it is due to Poincaré and may be found in textbooks of nonlinear mechanics. We omit the discussion of ordinary orbital points and concentrate on points of equilibrium. In the case of one-dimensional motion, there are always at least two orbits issuing from any point of unstable equilibrium. Poincaré further considers the curvature of the constraint at the point of equilibrium. This curvature will be taken as a variable that can be changed in a continuous fashion; the curvature may be varied from positive values in $(a)$, over zero value in $(b)$, to negative values in $(c)$. Consider now the case where the ball moves with friction. In case $(a)$, if the ball is released somewhere along the curve, it will oscillate about the equilibrium point with decreasing amplitude until it comes to rest at the bottom. In case $(c)$, the ball moves away from the unstable equilibrium point either to the right or to the left and does not return; friction will merely diminish the velocity. Thus the cases $(a)$ and $(c)$ have quite different geometrical properties so far as the orbits of the ball are concerned. If now we consider with Poincaré the changes in the orbits as the parameter (curvature) is varied, we note that condition $(b)$ of vanishing curvature constitutes a singularity of this variation. At this value of the curvature the equilibrium changes from stable to unstable, and the topological nature of the orbits in the neighborhood of the equilibrium point changes. Such a singular point in the continuum of the variable parameter is called a *bifurcation point*. We need not enter here into the general mathematical definition of this concept but can confine ourselves to that case where an originally stable system changes its parameters or variables in such a way that an instability appears as soon as a bifurcation point is passed. It is characteristic of such systems that there are at least two equivalent orbits away from the unstable configuration, and that there exist no simple dynamical criteria to distinguish which one of these orbits the system will take; this is in general determined by minor local perturbations or inhomogeneities.

Next, consider motion in two dimensions. This case may again be illustrated by Figure 1, if now (*a*) and (*c*) represent cross sections of a spherical bowl. In place of the two orbits of the previous case (*c*), there is now an infinity of orbits away from the point of unstable equilibrium; the ball can move in any direction of a 360° angle. Assume now that we start from case (*a*) and decrease the curvature of the constraining surface very slowly. By this we mean so slowly that the ball, under friction, comes to rest at the equilibrium position in a time short compared to the time during which the curvature changes appreciably. Thus most of the time the ball will just be at rest in the equilibrium position. When bifurcation occurs, at (*b*), the equilibrium turns from stable to unstable; thereafter the ball moves away from equilibrium, without return, in a direction which may be determined by minor local disturbances.

In our studies we shall be concerned with motions of a continuum, namely, the material of the mantle. We then speak of the motion of the body as its deformation, which now involves an infinite number of dynamical variables. Naturally, the analysis of instabilities now becomes quite difficult mathematically, and we must confine ourselves to some very simple example which can then be generalized qualitatively. Figure 2 shows a vessel containing two fluids of different densities on top of each other separated by a plane horizontal boundary. For simplicity assume the two fluids as incompressible so that temperature effects can be ignored. Also assume the viscosity of the fluids so high that any disturbance or wave at the boundary is damped aperiodically. Let $\Delta\rho$ be the difference in density between the upper and lower fluids. So long as $\Delta\rho$ is negative, i.e., the upper fluid lighter than the lower, the configuration is completely stable. As $\Delta\rho$ increases, we arrive at a bifurcation point for $\Delta\rho = 0$. For positive $\Delta\rho$, that is, when the upper fluid is denser than the lower, complete instability

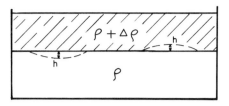

FIG. 2.

obtains. The theory of this type of instability is well worked out [Chandrasekhar, 1961]. The depth or height, say $h$, of any bulge in the boundary increases exponentially in time, as

$$h = h_0 \exp\left[ga\,\Delta\rho t/\eta\right] \quad (1)$$

where $g$ is gravity, $a$ is a length proportional to the breadth of the bulge, and $\eta$ is the viscosity. The rate of increase is therefore dependent on the breadth of the bulge; wide bulges will grow more rapidly than narrow ones. However, any bulge will grow exponentially; the process will clearly stop only when the stratification has been turned upside down and the lighter fluid rests on top of the heavier one. It is clear that, on changing $\Delta\rho$, as soon as the bifurcation point is passed and the system develops instabilities, there is an infinite set of possible deformations. Which deformation actually occurs will depend on such things as minor local inhomogeneities.

Next, consider some types of instabilities that can occur in solids under external stress, as exemplified in Figure 3. We shall consider the applied load as the parameter to be varied. If a thin circular rod is subjected to a longitudinal compression as in $(a)$, it will for small forces simply shorten, and for larger forces it will buckle. The critical force at the bifurcation point where buckling just sets in is given by Euler's formula,

$$F_b = \pi^3\,Er^4/4L^2 \quad (2)$$

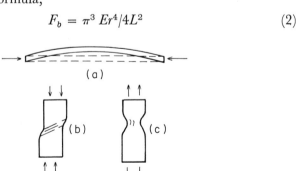

FIG. 3.

where $E$ is the elastic modulus and $r$ and $L$ are the radius and length of the rod. For a longitudinal compressive force in excess of $F_b$, the rod will buckle, if we assume a curved shape which may be located in any place going through the axis of the undeflected rod. Again, we have an infinite set of mathematically equivalent solutions as soon as a bifurcation point is passed and instability sets in.

The case of buckling represents a purely geometrical type of instability; the strains will usually remain within the elastic limits. Examples (*b*) and (*c*) of Figure 3 represent cases where the instability occurs as the result of essentially plastic behavior of the material. If we assume this transition to be sharp, the transition point where stable elastic deformation goes over into unstable plastic behavior will be a bifurcation point. Case (*b*) represents a cylinder under compression in which plastic gliding has occurred; case (*c*) shows a cylindrical specimen under tension in which "necking" has taken place. Again here, the exact place along the cylinder axis where the instability occurs is more or less accidental and is not determined by the gross mechanical properties of the system.

If we look back over the various examples discussed here, we soon notice a common feature: We start out with a body and forces acting upon it that possess a certain symmetry. As a bifurcation point is passed and instability sets in, the total *set* of orbits or deformed configurations may retain the same symmetry as before, but as the system can move along only a single path, the configurations which occur after the onset of instability tend to have a lower degree of symmetry than those of the stable regime. We consider this conclusion of sufficient importance to warrant its formulation as a separate principle: We shall call it the *principle of diminishing symmetry*. To repeat once more: If a body changes in such a way that instabilities of deformation appear, the later stages of the deformation tend to have a lower degree of symmetry than the initial stages, and often also a lower degree of symmetry than the initial forces acting on the system.*

---

\* *Note added in proof*: The foregoing arguments differ radically from those reached by Pierre Curie (*ca.* 1890) in his study of the physical effects of symmetry, as was recently pointed out to me by Professor W. Yourgrau. The latter may be summarized as follows [Curie, 1963]:

"When certain causes produce certain effects, the elements of symmetry in the causes ought to reappear in the effects produced.

When certain effects reveal a certain dissymmetry, this dissymmetry should be apparent in the causes which have given them birth.

The converse of these two statements does not hold, at least practically; that is to say, the effects produced can be more symmetrical than their causes."

In tectonic analysis we are confronted with the inverse problem, the problem of retrodiction. The empirical situations found are usually very complex and highly asymmetrical. The judicious application of the preceding principle can then often save us from specious inferences as to the nature and the hypothetical asymmetry of such physical processes and forces as have brought about the observed structures. We shall discuss some examples.

As the author has pointed out [Elsasser, 1963], the descent of iron from the upper mantle, which must have taken place early in the earth's history, led first to the formation of iron pools of considerable horizontal extent which then drained downward as a result of instabilities of the type shown in Figure 2. While the end product, the core, is highly symmetrical, the intermediate stages of the process are characterized by a rather unexpectedly large asymmetry. One result of this type of mechanical behavior is that the formation of the core must have occurred earlier and far more rapidly than one would otherwise have reason to believe.

Consider next the question of stresses exerted from the oceans upon the continents, or vice versa. Since the continents are less dense than is mantle material under oceans at the same depth, the continents tend to spread laterally; the ultimate stable configuration would be one where the lighter continental material covers the earth uniformly. Correspondingly, there should be a "spreading stress" which the continents exert laterally upon the oceans, as has been particularly emphasized by Orowan [1958]. This stress increases linearly with depth and should amount to roughly 600 bars at a depth of 35 km. However, geological observations do not bear out the existence of a stress which, over sufficiently long periods of geological history, should have led to a spreading, thinning, and possibly breaking up of the continental plates. Instead of this we see that these plates are mainly above sea level, although it is generally agreed that the volume of the oceans has gradually increased over geological times [Rubey, 1951]. Moreover, there is more direct evidence for a pressure exerted from the oceans toward the continents rather than in the opposite direction. This is found in the statistical distribution of heights of the earth's surface, the so-called hypsometric curve [see for instance Scheidegger, 1963]. This curve has two distinct and sharp maxima, one nearly at sea level corresponding to the continental plates and one at a depth of about 5 km, corresponding to the ocean bottoms. These two peaks are separated by a wide and deep minimum of the curve; that is, intermediate elevations are rare. It seems difficult to explain this otherwise than by the assumption that there exists a pressure keeping the continents together, which pressure is far more symmetrically distributed about the coastlines than the extremely asymmetrical location of orogenic belts would lead one to believe.

Survey of the ocean bottoms indicates that the material of the ocean floor often tends to slide away from the center of the ridges [Hess, 1962; Menard, 1964]. Often this sliding is irregular, leading to transcurrent faults perpendicular to the ridges. A conspicuous case is the Mendocino fault off the coast of California where there is a large east-westerly displacement of the surface features to both sides of the fault. The displacements have apparently occurred quite some time ago, as the region is now geologically inactive. But the shift can be seen directly from the offset of the contour lines of the topography* and is verified alternatively by the offset of the contour lines of magnetic anomalies, as determined by Vacquier [see Vacquier's article in Runcorn, editor, 1962]. The relative displacement along the Mendocino fault is 1150 km; similar, somewhat smaller displacements were found along the neighboring Murray and Pioneer faults.

## 3. Temperatures in the Earth

All analysis of mass motions in the upper mantle requires a knowledge of the temperature distribution, if not everywhere in the earth, then at least in the upper mantle itself. We now discuss the temperature variation in the earth's interior in general terms, and subsequently concentrate our attention on the upper mantle. Information about the thermal state of the earth's body is vastly harder to come by than knowledge of its mechanical properties. The latter properties are well established in the realm of seismic frequencies (not for long-term creep of course). Clark [1963] estimates that the errors in density as function of depth are of the order of 2 per cent except in the inner core where they may be as large as 10 per cent. Clark and Ringwood [1964] give two density distributions corresponding to two different, rather extreme sets of assumptions regarding the lithology of the mantle. The two density curves obtained do not differ from each other by more than 5 per cent anywhere, and at most places by much less. A recent, independent determination of the density distribution [McQueen, Fritz, and March, 1964] falls entirely between the limits of the two last-mentioned models. Once the seismic velocities and the density are known, the other mechanical properties pertaining to the

* The writer is indebted to Professor Hess for access to his unpublished map of the area.

seismic region of the spectrum may readily be calculated, the principal ones being pressure, compressibility, and the elastic shear modulus. [See also Birch, 1964.]

In contradistinction, extremely little is known about the earth's temperature. The most significant datum is the average heat flow at the surface, amounting to $1.2 \times 10^{-6}$ cal/cm² sec which, with the usual value of $6 \times 10^{-3}$ cal/cm sec deg for an average heat conductivity of near-surface rocks, gives a temperature gradient below the surface of 20°/km. This implies such a steep rise that the temperature gradient must of necessity decrease as we go deeper into the mantle. This flattening out of the temperature curve is by far its most characteristic feature, whatever the details of the curve.

The temperature of the earth's deep interior is not known. The only "fix" is found in the now rather thoroughly established result [Bullen, 1964] that the inner core is solid and (on the basis of known abundance ratios of elements) consists almost certainly of iron. The temperature at the boundary of the inner core must then be that of the melting point of iron at the pressure prevailing there (about $3.3 \times 10^6$ bar). There will be some melting-point depression owing to appreciable amounts of impurities dissolved in the liquid outer core; Clark [1963] estimates this latter effect as of the order of 300°. The melting point relation $T(p)$ for iron was experimentally determined by Strong [1962] up to 60 kb. We note especially that earlier results of the same author have been retracted, as they were beset by very large calibration errors, and *a fortiori* they cannot be used for extrapolation. The latest curve of Strong's is linear in the interval mentioned; on linear extrapolation to the boundary of the outer core ($1.36 \times 10^6$ bar) it would give a melting point of about 5000°C there. Actually, this extrapolation is of limited value since it is for $\alpha$-iron, which is not the form of iron likely to be stable in the core. It also follows from general thermodynamical arguments that the temperature of the core boundary must be appreciably above the melting point at the pressure prevailing at that boundary: The fluid core, being in convective motion, must be very nearly adiabatically stratified, and the slope of the adiabat is as a rule very much less than the slope of the melting-point curve. Thus the temperature at the outer boundary of the liquid layer, while considerably less than the temperature at the inner boundary, would still be much in excess of the local melting point. Moreover, iron undergoes a number of solid-solid phase transformations. At room temperature and a pressure of 130 kb, it changes from a cubic to a hexagonal closest-packed structure [Takahashi and Bassett, 1964]. As these authors indicate, shock-wave data make it plausible that this is the phase stable in the inner core. A reliable determination of the thermal state of the core should be possible after more detailed shock-wave data at megabar pressures have become available.

Now if we assume, for example, the temperature at a depth of 100 km to be 1300°C and at the bottom of the mantle (2900 km) to be 5000°C, the average temperature gradient in the mantle below 100 km would be 1.3°/km, only a small fraction of the gradient near the surface. This last number will not, however, be in error by more than 30 to 40 per cent, as the central temperature is almost certainly between say 3000 and 7000°C. The physical reasons for this low gradient lie partly in the high efficiency of radiative transfer, which we shall discuss somewhat later, partly in the fact that at greater depth most of the mantle's mass and hence of its radioactive heat sources lies above, leading to a corresponding reduction of the the flux. The flattening out of the temperature curve is thus clearly a result of great generality and in its over-all characteristics subject to little doubt.

It is widely admitted that most if not all of the heat flowing out of the earth is due to the decay of long-lived radioelements (U, Th, K). In fact, no other major source of heat is known or has been proposed. As is well known, the heat flow of the earth is very closely equal to the total heat that would be generated if the over-all concentration of radioactive material in the earth were that of average chondritic meteorites. There are, however, a number of ambiguities in any such model of radioactive heat sources. One of these concerns the K/U ratio; other indeterminacies lie in the unknown increase of concentration of the radioelements toward the upper part of the mantle. Together, these unknown data are the equivalent of several adjustable parameters in thermal models of the earth. The very extensive calculations of MacDonald [1959, 1964] show that a variety of models is compatible with the observed surface heat flow. Of particular interest, therefore, are those properties of thermal models which do not depend too critically on the magnitude and distribution of the radioactive sources assumed.

Consider now the relationship of the surface heat flow to the total radioactive heat output inside the earth; in other words, let us inquire whether the earth as a whole might be heating or cooling. There is also the third and rather likely possibility that in the early days of the earth, when radioactive heat output was larger than it is now, the earth was heating up gradually, while at present it is cooling. We may estimate the heat capacity of the earth's body as follows. We express it in terms of the capacity of a cone having unit area at the earth's surface, and having its apex at the earth's center,

$$C = [\rho c_p]_{\text{av}} \, a/3 \qquad (3)$$

where $C$ is the heat capacity per cm² of earth surface, $a$ is the earth's radius, and the bracket designates an average over the heat capacity per cm³ of volume. To evaluate this last expression, we note that for solids at high-temperature kinetic theory gives

$$c_p = 3nR/M \qquad (4)$$

where $R$ ($= 2$ cal/mole) is the gas constant, $M$ the molecular weight, and $n$ the number of atoms per molecule. For simplicity, assume a mantle consisting of forsterite, $Mg_2SiO_4$, the main constituent of olivine, with a mean mantle density of 4.5. Let the iron core have a mean density of 10.5. From Equations 3 and 4 we obtain numerically, $C = 2.8 \times 10^8$ cal/deg. If all the surface heat flow, of 38 cal/cm² yr, would be used up to heat the earth uniformly, the resulting rate of heating would be 140° per billion years. It follows from this that a moderate deviation from the chondritic heat output, say by a factor of 2, would not lead to a thermal catastrophe in the earth's life, but a very much larger deviation might do so. Let us remark here on the indications so often quoted by geologists, that the earth has been very active tectonically during the last several hundred million years, apparently as active as in its earlier life. This is somewhat surprising in view of the decrease with time of the radioactive heat output. No explanation for this effect has as yet been given.

Before we go on to the temperature distribution in the upper mantle, we digress briefly to consider some thermal problems of the core. The dynamo theory of the main geomagnetic field [Elsasser, 1955] requires that the outer, liquid core be in convective motion. It is true that the magnetohydrodynamic theory can say little about the thermal conditions conducive to convection; merely the existence of convection of sufficient strength to generate and maintain magnetic fields is required. If the core as a whole cools slowly, much of the heat required for convection may be the heat of solidification of iron which would be provided by the gradually growing inner core. This is a view rather widely held. Again, the author has suggested earlier that if there is an exchange of very small amounts of material between mantle and core at their boundary, this might lead to small but geometrically extensive inhomogeneities in density which would generate convection.

It is readily understood how the core can cool down: Heat transport in the mantle at high temperatures is by radiation; this process is many times more effective than conventional heat conduction would be, even in a metallic body having the presumed physical properties of the core. On the other hand, radiative transport in the core must be negligible since all metals are exceedingly opaque in the visible and near-infrared region of the spectrum. Thus apart from convection the mantle is in effect a much better heat conductor than the core; it is not a thermal insulator as it would be at room temperature.

On the other hand, if the core does not cool down but heats up in the course of time, the situation would be very difficult to understand. A convecting fluid remains well mixed, and in a gravitational field its temperature distribution must be very near the adiabat. If heat would flow into the core from the mantle, the stratification would soon become stable and convection would stop. Thus to account for convection with a rising or even with a completely stationary temperature would require sufficiently strong heat sources inside the core to enforce outward flow of heat without a fall in temperature. Such sources could hardly be anything but radioactive, but the existence of appreciable amounts of radioactivity in a metallic core is considered unlikely by geochemists. There is therefore a considerable presumption that the core and the lower mantle are indeed gradually cooling. This conclusion is somewhat hard to reconcile with the observation quoted earlier, that tectonic activity has been rather intense in the most recent periods of the earth's life. Such activity could hardly occur, had the temperature in the *upper* mantle steadily decreased during the lifetime of the earth. It is possible that the temperature of the earth's central parts increased during the earlier periods of the earth's life, owing to the more intense radioactive decay then prevailing, and is now falling gradually. The magnetic field would then have been present only during some fraction of the earth's life. Again, the fact that the temperature of the upper mantle apparently has not decreased during the later parts of the earth's life, or perhaps not during its entire life, may be attributed to the insulating effect of the top 50 to 100 km of the mantle which will be discussed presently. On the whole, however, speculations of the kind just set forth seem to be rather premature.

In what precedes it is silently assumed that no major convection-producing effects have been overlooked. This is somewhat reminiscent of Kelvin's statement that, since the sun will lose all its heat by radiation in about a million years, it cannot be much older. This was said, of course, before the discovery of radioactivity. Years ago Dirac pointed out that over sufficiently long periods the ratios of so-called physical constants may change, that in particular the gravitational constant may have decreased by several per cent during the lifetime of the earth relative to the constants of atomic physics. Dicke [1962; Murphy and Dicke, 1964] has studied extensively the geophysical implications of this assumption. If gravity decreases, so does pressure inside the earth; hence the interior will cool adiabatically by slow expansion. This enhances any existing tendency toward convection. The effects are, however, not of such magnitude that they could readily be disentangled from the effects of radioactive heating.

In the present analysis we have a more limited aim. We are interested in the physical basis, to be found in the upper mantle, of the macrotectonic processes that took place during the more accessible period of geological history, say the last billion years. Since the half-life of potassium ($4.5 \times 10^9$ years) is much longer than this period, and the half-life of U and Th is even longer, the rate of radioactive heat generation during the period considered may be taken as sensibly constant.

## 4. Temperatures of the Upper Mantle

We shall continue to assume that the temperature distribution is horizontally uniform, a function of depth only. In later sections this restriction will be relaxed. We next make use of the fact mentioned before, that any appreciable heating or cooling of the upper mantle ought to have observable geological consequences. In the case of significant heating the tectonic activity, magmatic outpourings, etc., should have drastically increased with time; in the case of significant cooling the same activities should have equally drastically slowed down. There is no evidence for either of these extreme alternatives. We therefore analyze the temperature distribution in the mantle under the assumption that the thermal regime is *stationary*, in a sufficient approximation, with the added assumption that convective heat transport is negligible. The latter assumption is of extremely doubtful validity and will only be made to obtain a consistent model whose consequences can be studied. All conclusions reached in this section are contingent upon the absence of major convective transport. They will later be modified as we take convective heat transport into account. Then, in the stationary case the temperature obeys the differential equation

$$\frac{d}{dz}\left(\kappa(z)\frac{dT}{dz}\right) = -A(z) \tag{5}$$

where $z$ is the depth. Earth curvature has been neglected in Equation 5; $\kappa$ is the thermal conductivity, and $A$ the heat production per unit time and unit volume. Both quantities are in general functions of $z$. The boundary condition on Equation 5 is

$$\kappa_0 \left(\frac{dT}{dz}\right)_0 = F_0 \qquad (6)$$

where the subscript 0 designates conditions at the surface; in particular, $F_0$ is the measured surface heat flow. With these assumptions, the temperature in the upper mantle depends on two unknown functions. Of these, the variation, $A(z)$, of radioactive heat production with depth, is far harder to estimate than the variation of thermal conductivity. We shall find that the critically important region of the upper mantle is quite shallow, some 250 to 300 km deep. If radioactivity were uniformly distributed in the mantle, the heat sources in this layer (10 to 12 per cent of the mantle by volume) could be neglected as an approximation. Actually, there might well be a concentration of lighter elements as well as of radioelements in the top layers of the mantle. The fact that this concentration is far from complete may be put in evidence as follows [for numbers quoted see Mason, 1962]: The crust comprises about 0.6 per cent of the mantle by mass. We might increase this to, say 1 per cent, on taking account of the light material in and under the oceanic rises. On the other hand, stony meteorites contain vastly larger proportions of light materials: The mean ratio of Ca + Al + Na + K by weight to Mg in meteoritic silicates is 0.30. The numerical discrepancy is such that just one interpretation seems possible, namely, that only a fraction of the potential sialic material of the mantle can have become concentrated in its top layers. Radioelements, being alkalis and alkaline earths, are so closely associated with the sialic, crustal material that the same can be said of them. Clark and Ringwood [1964] conclude from an extensive discussion of geochemical data that at least 40 per cent of the surface heat flow must originate at depths below 400 km. Numerical estimates carried out by us have shown that if about a third of the radioactive heat sources lie above 250 to 300 km, the results of the following discussion will be modified quantitatively but the general and qualitative results remain unchanged. We therefore set $A = 0$ for the upper mantle, in Equation 5. The equation can then at once be integrated, and with the boundary condition 6 yields

$$\kappa(z)\, dT/dz = F_0 \qquad (7)$$

This equation of course says simply that the heat flux is independent of depth. To integrate again, we must estimate the variation of $\kappa$ with depth. This has been discussed by Lubimova [1958] who gives the following formula:

$$\kappa = \kappa_1 + \kappa_2 = B\frac{\rho^{2/3}u^{7/2}}{T^{5/4}} + \frac{16n^2\sigma}{3\epsilon}T^3 \qquad (8)$$

We discuss first $\kappa_1$ which is the conventional, or lattice conductivity; $u$ is a suitable mean of the longitudinal and transverse sound (seismic) velocities, and $B$ is a numerical parameter adjusted so that near the surface of the earth $\kappa_1 = \kappa_0$. The value $\kappa_0 = 6 \times 10^{-3}$ cal/cm sec is widely accepted. The formula for $\kappa_1$ is derived from theories of lattice conductivity; references may be found in Lubimova's paper. It is necessary to know $\kappa_1$ in only the topmost 100 km or so; farther down the radiative conductivity, $\kappa_2$, becomes rapidly much larger than $\kappa_1$. In this top layer the most important variation of $\kappa_1$ is its decrease with increasing temperature. The decrease of the heat conductivity on heating of silicates and similar ionic crystals is experimentally well established [Birch et al., editors, 1942; Kingery, 1960]. In the mantle, this leads according to Lubimova to a minimum of the over-all conductivity, which falls from 0.006 at the surface to 0.004 somewhere between 50 and 100 km and thereafter rises again due to the radiative term. Lubimova, who was the first to point out the importance of this effect, speaks of the top layer of the mantle as an efficient *thermal insulator* of the inside of the earth.

We come next to the radiative conductivity. In ionic crystals which are relatively transparent in the near infrared and visible spectrum, radiative heat transfer becomes significant between 500 and 1000°C, and at still higher temperatures predominates completely. This has been known from laboratory experiments for years; it was first systematically applied to the earth's mantle by Clark [1957a]. In order to make the problem mathematically tractable, all authors dealing with it have assumed that the material of the mantle is a "gray" body, that is, a substance in which the coefficient of extinction is independent of wavelength. This coefficient $\epsilon$ is designated as the opacity. We note that the mean free path of a quantum of radiation (the "transparency") is, for any wavelength

$$\lambda = \int_0^\infty e^{-\epsilon x}\,dx = \frac{1}{\epsilon}$$

that is, the reciprocal of the opacity. In Equation 8, $\sigma$ ($= 1.36 \times 10^{-12}$ cal/cm² sec deg) designates the Stefan-Boltzmann constant, $\sigma T^4$ being the total radiation emitted by cm² of a blackbody at temperature $T$. Again, $n$ is the refractive index of the material; from laboratory observations one estimates $n^2 = 3$ in a sufficient approximation for the mantle, a value accepted by most authors. The pronounced flattening of the temperature curve with depth is clearly due to the $T^3$ dependence of the radiative conductivity.

Figure 4, comprising depths from 0 to 400 km, shows three pairs of temperature curves. They correspond to the opacities, $\epsilon = 10$, $\epsilon = 3$, and $\epsilon = 1$ cm$^{-1}$. The lower curve of each pair is computed under the assumption $\kappa_1 = \kappa_0 = $ const. This last assumption is often made but is unjustified on grounds explained previously. The upper curves are calculated on assuming $\rho$ and $u$ constant in Equation 8, equal to their surface values, but taking account of the relationship $\kappa_1 \sim T^{-5/4}$. This tends to enhance slightly the minimum of the conductivity mentioned earlier and hence makes the temperature curves a little steeper than they would be otherwise. (A slight correction

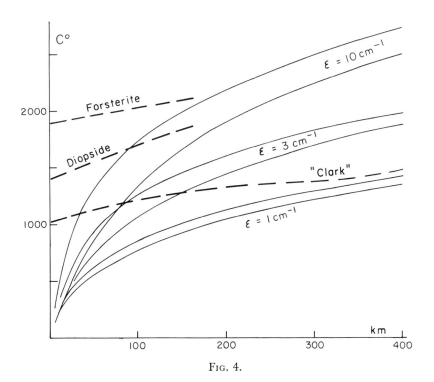

Fig. 4.

for earth curvature, omitted for simplicity's sake from the formulas, has been included in Figure 4.) As mentioned, zero heat production is postulated throughout; if we assume some heat production in the top layers, the curves become correspondingly depressed. If, for instance, it is assumed that 30 per cent of the surface flux originates in the top 100 km, the resulting temperature is lowered by an amount comparable to the distance between the two curves of the middle pair in Figure 4. It follows that those conclusions, which rest essentially on the steepness of the temperature gradient in the top 1 to 200 km, are not critically dependent on assumptions about the distribution of heat sources, unless these assumptions are extreme.

Three melting-point curves are also shown in Figure 4. The uppermost one is for forsterite, probably the most common silicate in the upper mantle, the next one is for diopside, $CaMgSi_2O_6$ [Davis and England, 1964; Boyd and England, 1963]; finally the bottom curve may be considered as a sort of lower bound to any sort of melting occurring in the mantle. It was given by Clark [1963] on the basis of some rather involved thermodynamical estimates. The mantle is a highly complex aggregate chemically speaking, and mixtures tend in general to have a lower melting point than pure substances. In a mixture as complex as the material of the mantle, a liquidus is ill-defined; on the other hand if the material contains a few per cent of a liquid phase, its mechanical resistance to sufficiently slow deformations will no doubt be reduced by many orders of magnitude. We therefore take Clark's curve as indicating roughly the temperature for a given depth where the mantle begins to change its mechanical properties rapidly toward much greater ductility.

The main feature to notice in Figure 4 is the remarkably low opacity that must exist in the upper mantle if it is to be kept from melting. In case of the topmost pair of curves, where the mean free path is 1 mm, a layer of the upper mantle would eventually melt completely. For free paths of 2 to 3 mm the mantle material may become "softened" but would not entirely melt; for still more transparent material even the lowest-melting components would stay solid. These values of opacity are quite remarkably low, that is, the material is surprisingly transparent. Clark [1957b] has measured the absorption spectrum of single crystals of olivine and diopside and on averaging over the various absorption bands for a spectral distribution corresponding to 1000°K arrives at $\epsilon = 0.31$ for olivine and $\epsilon = 0.14$ for diopside. He thinks that the mantle material at depth must be more opaque owing to graininess, irregularities, and impurities. We feel, however, that this conclusion may not be fully warranted.

We come here to a point of great importance for the physics of the upper mantle. Since the upper mantle seems to be close to partial melting at least, and since by the evidence of volcanism partial melting has indeed occurred throughout geological history but only locally and intermittently, there must be mechanisms which, in the likely case of a heat accumulation in the upper mantle, enhance the heat flow and therefore act as *stabilizers* for the thermal condition of the layers involved. If such stabilization did not exist, it would be very difficult indeed to understand the occurrence of magmatic processes through geological history in an intermittent, localized, irregular, but on the average far from completely unsteady manner. Two heat flow-enhancing processes come to mind immediately; namely, volcanism itself in its various forms, and mass motions, that is, convection.

With the frequently quoted value of 1 km$^3$ of basalt appearing near the earth's surface per year, the heat-transporting capacity of volcanic processes is small even if the material comes from a depth of, say, 100 km. So far as convection is concerned, we shall deal with it in more detail later on. Here, we wish to point out that in addition to the large-scale processes there might be processes on a local, microscopic scale which also can enhance the heat transport.

Let us note, first, that melting of the upper mantle would take a long time. Forsterite and diopside have heats of fusion of about 360 cal/cm$^3$ at room pressure and might have slightly smaller but comparable values at some depth. If we now assume that 10 per cent of the surface heat flux of 38 cal/cm$^2$ year be used to melt a given layer, melting would proceed at a rate of about 10 km per hundred million years. If only a minor component melts, this rate will be correspondingly slower. Now suppose a layer, say 2 to 300 km below the surface, to be at the melting point of diopside and therefore only 2 to 300° below the melting point of pure forsterite, and probably closer to the lower melting point of typical olivine. The material may remain in this state for periods that are long even by geological standards. We may then inquire into the localized physicochemical and atomistic processes that will occur under these conditions. In the main, there will be two different but interrelated processes on the molecular scale: *(1)* Diffusion of those impurity atoms which have low solubility will be greatly enhanced. A configuration of lower free energy will eventually be reached where the impurities form thin films covering the grains of the basic silicate matrix. *(2)* Diffusion, and ultimately disappearance at the surface, of dislocations will also be greatly enhanced. This process will in the first place facilitate the diffusion of impurities;

second, it will lead to the migration of grain boundaries or, what is the same thing, to growth of larger crystallites at the expense of smaller ones, with the impurities being squeezed out where grain boundaries grow together. The last-named process tends to increase the thickness of the impurity surface layer. To illustrate this quantitatively, let the material consist of a number of adjacent cubes of side $a$, and let it contain a fraction $\gamma$, by volume of impurities, where $\gamma$ is a small number. If all the impurities become collected in surface layers between the cubes, the average thickness of these layers is $\gamma a/3$. Again, if the material consists of separate spheres of radius $a$ and impurity content $\gamma$, the thickness of the layer surrounding each sphere when the impurities are extruded is given by the same formula.

Large crystallites, fist-size or larger, are often found in extrusions from great depth, of which the kimberlite pipes are a well-known example. We may assume that even more often the large crystallites get sheared while they rise to the surface, especially when not surrounded by a liquid medium. It is known that a sufficient shear strain will break a large single crystal into many small grains; given enough time some of the impurities can diffuse back into the newly formed grain boundaries. There seems therefore no valid objection to the idea that below, say, about 75 to 100 km depth there is a layer of the mantle consisting largely of rather big and relatively pure crystallites of olivine and other basic materials surrounded by thin films of less abundant constituents. Now Al and Ca can form crystalline phases with Mg in rather dense silicate lattices which will tend to form separate phases; the principal relatively abundant elements which will be expelled from close-packed lattices are Na, K, and H, as water.

The physical consequences of the transformations suggested are not hard to assay. They are twofold. In the first place, the appearance of large, relatively homogeneous, and pure crystallites tends to *increase the transparency* of the material. Radiative mean free paths of the order of 2 to 3 mm, as required by Figure 4 to keep the bulk of the material from melting, are not now so unreasonable as they would seem for a fine-grained and mechanically well-mixed material. Clark's value of $\epsilon = 0.3$ cm$^{-1}$ measured on olivine crystals now takes on added significance.

We come now to a second effect of the assumed growth of large, relatively pure crystallites in the upper mantle. The material which does not fit into the crystal structure, especially alkali compounds and water, will be extruded into thin layers between the grains or into small interstitial pools. These substances have a much lower melting point than the main body of mantle material; moreover, general experience shows that they are rather ductile. Their volume fraction must be rather large. In terms of weight, for stony meteorites, $(Na + K)/Mg = 0.06$; furthermore, if we assume that the upper mantle down to a depth of 300 km contains as much water again as the present oceans, we find for water $\gamma = 0.01$. With values of $\gamma$ for the substances mentioned of the order of one to several volume per cent, one may safely conclude that these minor constituents must have a most important influence on the mechanical properties of the mantle as a whole. It is known in the mechanics of materials* that grain-boundary sliding is one of the most common mechanisms for the deformation of solids. Hence it may readily be presumed that the extrusion of light materials such as alkali compounds and water from the main body of the silicate grains of the mantle is conducive to *a pronounced enhancement of the ductility of the material*, quite possibly by many orders of magnitude. Now while some of the arguments of this type can be taken over more or less bodily from engineering data, the establishment of this presumption for the upper mantle beyond reasonable doubt will require much additional research. It seems therefore prudent at this time to state only the bare proposition while leaving elaborations to the future.

We shall finally say a few words about the properties of the lower mantle. Clark [1957a] has remarked that at sufficiently high temperatures the opacity must increase again, owing to the appearance of thermally excited states which act as absorption centers. This conclusion seems incontestable. Thus one may expect that in the lower parts of the mantle, where the temperature is high enough, there will be a continued rise of the temperature as we go down. There will not be the pronounced decrease of the gradient which would result from a fixed value of $\epsilon$ applied to the radiation of a gray body.

On the other hand, there might also occur a rise of the melting point in the lower mantle. Clark's [1963] graph shows a strong upward curvature of his "melting-point curve of the mantle" around 6 to 700 km depth; thereafter the curve rises steeply, to attain about 8000° at the core boundary. While the curve of the actual temperature will also rise, one may seriously doubt whether it rises as steeply as the melting-point curve of silicates, since in this case the existence of an inner core of solid iron would be very hard to understand. It is therefore possible that in the lower part of the mantle the actual temperature remains substantially below the melting point, perhaps by one or two thousand degrees, and that this state of affairs may have existed since

---

* We are grateful for specific information on this subject to Professor E. Orowan of the Massachusetts Institute of Technology.

the beginning of the earth. The lower mantle may then not yet have undergone the complete transformation from the fine-grained dust from which the earth may have accreted, to the more coarse-grained and far more ductile material presumed to exist now in the upper mantle.

## 5. Horizontal Temperature Variation

It appears from the foregoing that we might distinguish two main divisions in the upper mantle. Following a suggestion by E. Orowan, we shall call them the *outer shell* and the *soft layer*. The outer shell is characterized by a very steep rise of the temperature, though the temperature stays well below the melting point throughout. The latter condition expresses itself in a certain hardness of the layer, and in fact, at the very top this becomes genuine brittleness. We may estimate the thickness of this layer as between 50 and 100 km, say roughly, 75 km. This is followed by the soft layer where the temperature gradient flattens out rapidly and where the temperature is not too far below the melting point of the chief constituents, and possibly at the melting point of some of the minor constituents. The soft layer is of course nothing but the well-known "asthenosphere" [Daly, 1938]. We are using the term soft layer here as a strict synonym of the established geological term asthenosphere, no more; this does not mean of course that the earlier estimates of geologists as to the depths and physical properties of this layer need altogether agree with our own. We may estimate that the soft layer reaches down to a depth of, say, 300 km, although this number may well require modification. The region below the soft layer will here summarily be designated as the lower mantle. Seismologists usually recognize a transition zone between the upper and the lower mantle which zone may reach down to 4 to 600 km depth. With respect to the thermal and convective properties under discussion we find our simple terminology convenient in spite of the fact that these layers are not sharply defined strata separated from each other by surfaces of discontinuity but that they must be thought of as shading into each other.

Some curious phenomena and some interesting theoretical conclusions are related to the initial sudden rise followed by a gradual leveling off of the temperature. This rise occurs mainly in the top 5 per cent or so of the depth of the mantle. Perhaps the oldest and best known of these phenomena is the so-called low-velocity layer of seismic waves associated with the name of Gutenberg [1959; Press, 1956]. The minimum of the seismic velocities is rather faint and occurs at a depth of 100 to 150 km; thereafter the velocities increase with increasing depth. This behavior can be satisfactorily understood in terms of the initial rapid rise of the temperature which leads to a decrease of the elastic constants overbalancing their gradual increase with increasing pressure. Somewhat deeper down the temperature curve flattens out and the pressure dependence becomes the predominant effect.

Recently, Anderson and Archambeau [1964] have succeeded in studying the absorption of the seismic waves in the mantle as function of depth. The absorption is very weak in the lower mantle but has a pronounced peak in the upper mantle. Although the resolution in depth is limited, the authors have established that the average absorption coefficient above 400 km is some 25 times larger than below that depth.

A somewhat hypothetical but rather interesting set of ideas has lately been proposed by Takeuchi* and Hasegawa [1965]. These authors have studied the relaxation times required for adjustment to deviations from static equilibrium in the earth. These times are of the order of some thousands of years for limited regional adjustments such as the Fennoscandian uplift, whereas the adjustment of the earth's equatorial bulge to its diminishing rate of rotation seems to require much larger times, of the order of some millions of years. The authors, using a model of a viscous "channel" surrounding a rigid lower mantle, show that these different numbers can be reconciled on this model, provided the channel is shallow enough, not much more than 200 km deep. The use of Newtonian viscosity in this theory confers a somewhat preliminary character upon the model.

To return now to the Gutenberg low-velocity layer of seismic waves, perhaps its most distinct characteristic is its regional variation [summarized in Clark and Ringwood, 1964]. Under oceans, the low-velocity layer has its upper limit at a shallower depth (about 50 km) than under continents and especially under ancient shields (about 120 km). The lower limit of the low-velocity layer is around a depth of 250 km, somewhat deeper in continental regions. More detailed regional differences are tabulated by Clark and Ringwood. While the resolution of the measurements is limited and does not admit of too much quantitative detail, the existence of regional differences, and especially a distinction between suboceanic and subcontinental upper mantle, seems established beyond reasonable doubt. Thus the conspicuous division of the earth's surface into continental and oceanic

* The author is indebted to Professor Takeuchi for the opportunity to see this paper in preprint form.

areas does in one form or the other continue down to a depth of 200 to 250 km. Such a basic fact requires explanation.

A widespread school of thought claims that regional variability is essentially due to differences in vertical chemical differentiation. This school, then, denies that large-scale horizontal displacements have taken place in the past. The chemical composition of the material under any part of the earth's surface is assumed to be the same everywhere, only at some places the lighter constituents have come up to form the crust, and at others they are still buried in the upper mantle. This model can also explain the fact that the mean heat flows in continental and in oceanic areas are very nearly the same in spite

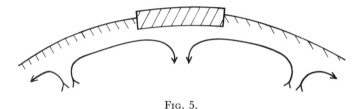

Fig. 5.

of the concentration of radioactivity in the continental plates. On this view, the same amount of crustal material and of radioactivity per unit surface exists everywhere, but in oceanic regions it is distributed throughout much of the upper mantle.

An alternate view is to assume that the development of the earth's surface features and upper layers involves mechanical processes of a far more extensive kind, in particular large-scale horizontal displacements of matter. The observed regional variations, both in the outer shell as shown by the difference between continental crust and ocean bottom, and in the soft layer as evidenced by the regional variation of seismic velocities, should then be explicable in terms of such large-scale dynamical processes. The comparative uniformity of the heat flow might similarly be the result of dynamical agencies tending to equalize the flow in spite of regional variations of the chemical composition. Now the physics and history of the earth's upper layers are so complex that it is rarely possible to adopt a single, monolithic theory at the expense of all other theoretical attempts. The theory must ultimately reach the complexity of the facts, but it seems well worth the effort of elaborating the dynamical aspects of regional variability on their own.

Consider the simple model (Figure 5) of a circulation in the upper mantle rising under the oceans and sinking under the continents. This is not meant to be a realistic model so far as any details are concerned. In the first place the mantle is not a viscous fluid but a crystalline solid, and its deformation will be much closer to plastic than to viscous flow. In the second place, these motions are subject to the principle of diminishing symmetry and will certainly not have the simplistic form shown in Figure 5. One virtue of this model, however, is that it reproduced a basic fact of observation, namely, the tendency toward compression of the continents of which we spoke earlier. We shall now describe the manner in which the circulatory pattern typified in Figure 5 changes the temperature in regions of vertical descent or ascent.

If the speed of motion is fast compared to the rate of heat conduction, a particle of the moving mass changes its temperature adiabatically. The rate of adiabatic heating or cooling on a change of ambient pressure can be estimated from thermodynamical formulas; here we can confine ourselves to noting the fact that the adiabatic temperature gradient in the upper mantle is relatively small. It is usually given as lying between 0.25 and 0.40 deg/km, much smaller than the actual gradient in the soft layer. The latter is smaller than in the outer shell but still of the order of several degrees per km. We introduce the abbreviation

$$\tau_z = \frac{dT}{dz} - \left(\frac{\partial T}{\partial z}\right)_S \approx \frac{dT}{dz} \qquad (9)$$

to designate the difference between the actual and the adiabatic temperature gradient ($z$ is again counted positive downward). If now a particle of the mantle is displaced vertically by a distance $\Delta z$, it will differ in temperature from the surrounding material at the *new* level. If at the initial level it had the temperature of the surrounding material, the temperature difference at the final level will be

$$\Delta_1 T = -\tau_z \Delta z \qquad (10)$$

A somewhat more intricate temperature variation results from horizontal convergence (divergence) of flow, which of course is accompanied by a compensating vertical extension (contraction). The thermal effects of such mechanical processes have been analyzed in detail in meteorology and oceanography. From the mechanical viewpoint the problem is a purely kinematic one, that is, the motions are assumed given and are not considered in relation to the forces generating them. Let the horizontal flow be only in the $x$ direction and consider a prism (extending perpendicular to the plane of the paper, Figure 6) of rectangular cross section, width $L$ and height $H$.

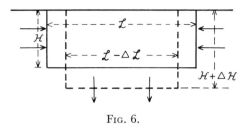

Fig. 6.

Assume that the upper boundary of the prism is fixed so that on lateral contraction it will expand by motion of the lower boundary. Let the contraction be homogeneous, that is, independent of $z$ within the layer. If after the motion the width is $L - \Delta L$, the height will be $H + \Delta H$. From continuity we have, to the first order

$$L \Delta H - H \Delta L = 0$$

Let $z = 0$ be the fixed upper boundary; a particle at depth $z$ will sink by

$$\Delta z = (\Delta H/H)z = (z/L) \Delta L$$

Thus the temperature change relative to the unmodified medium at level $z$ will be

$$\Delta_2 T = -\tau_z z \Delta L/L \tag{11}$$

where $\tau_z$ is defined in Equation 9.

Any actual process of convergence, sinking, and cooling will be described by a linear combination of the cooling rates (10) and (11), say

$$T = a \Delta_1 T + b \Delta_2 T \tag{12}$$

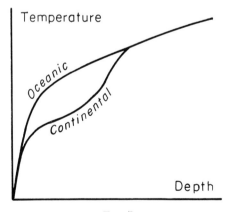

Fig. 7.

with $a + b = 1$, for convenience. The same applies of course to horizontal divergence, rising, and heating where $\Delta T$ is positive. We shall not try to pursue the mathematics further at this place, given the great variety of possible combinations. We may draw a qualitative and general conclusion: If, as in Figure 5, there is sinking matter under the continents and rising matter under the oceans, we can expect that the temperature will be modified as shown schematically in Figure 7, there being colder material under the continents and hotter material under the oceans. This should lead to a regional variation of seismic velocities which qualitatively agrees with what is observed. Temperature variations of this type have of course been assumed by previous authors, but their arguments were based on differences in radioactivity and heat conductivity, not on the effects of motions.

## 6. Thermal Instability and Circulation

So far as we know the mantle is a crystalline solid. Such a body can undergo elastic deformation and hence is capable of a rather large variety of equilibrium configurations. On the other hand, we also know from experience that some elastic stresses can become relaxed as time goes on. One may estimate that at reasonably high temperatures in the earth such stresses should become relaxed within some thousands, or within a few tens of thousands of years (certain stresses are not relaxable, especially those which generate closed circulations in a body).

We are in particular interested here in shear stresses. A substance which cannot sustain shear stresses in equilibrium is a fluid. Leaving out mathematical details we can state that, as elastic stresses become relaxed, the equilibrium configuration of a solid will tend toward that of a fluid.

The equilibrium conditions for a fluid as found in any textbook on fluid mechanics are very simple and also very restrictive: Given a fluid in an external gravitational field of potential $\phi$, say, the fluid is in equilibrium *if* and *only if* any equipotential surface, $\phi = $ const., is at the same time a surface of constant density, $\rho = $ const. It follows then readily that any such surface is also an isobaric surface, $p = $ const. Thus in equilibrium we have a functional relationship between the three quantities, $\phi$, $\rho$, $p$, such that this relationship is independent of the coordinates $(x, y, z)$. It is convenient to choose $\rho$ as the independent variable on stating that in equilibrium $p$ is a function of $\rho$ only, not of the coordinates. Next, it follows that deviations from equilibrium are produced by local variations of the density; call $\Delta \rho$ the deviation of the density from its standard value along an equipotential surface. This deviation engenders forces of buoyancy. Since we are counting $z$ positive downward, the force of bouyancy is counted as positive (directed downward) when $\Delta\rho$ is positive.

Now a variation in density may be produced in two different ways (and no others so far as we know), namely, thermally or chemically. The meaning of thermal expansion or contraction with varying temperature needs no elaboration, nor does the meaning of the separation of a given material into a lighter fraction which rises and a heavier fraction which sinks. We do not treat phase transformations as a separate effect here but shall consider them as special cases of thermal expansion or contraction. This cavalier treatment would be awkward if phase transformations were of the major importance in the upper mantle claimed for them by some authors; but a detailed analysis of these phenomena will fortunately not be needed here.

To determine the thermal force of buoyancy, we imagine a particle displaced vertically by an amount $\Delta z$. In this displacement the particle changes its temperature adiabatically, but the layer itself is not adiabatically stratified. For simplicity, we assume the layer to be homogeneous, meaning that its chemical composition is independent of depth. Assume further, as before, that the adiabatic temperature gradient is small compared to the actual temperature gradient in the layer. Some simple thermodynamical manipulations show that the difference in density between the displaced particle and the material at the *new* level is very approximately,

$$\Delta\rho = \rho\alpha\frac{dT}{dz}\Delta z \tag{13}$$

where $\alpha$ is the volume coefficient of thermal expansion. Note that $\Delta\rho$ has the same sign as $\Delta z$. A particle moving away from its equilibrium therefore generates a force of buoyancy which drives it *farther away* from its equilibrium position. Since this holds for any particle whatever in the layer, we see that the layer is radically unstable gravitationally. The well-known, more detailed mathematical analysis shows that a stratified fluid is unstable whenever the actual temperature gradient exceeds the adiabatic one and is stable whenever the actual gradient is less than the adiabatic one. In the case of the upper mantle, the gravitational instability is extreme since the actual gradient is very large compared to the adiabatic one.

Although the fact that the upper mantle is unstable has been known for many years, it seems that its implications have not yet been fully exploited. Instability must no doubt also have an effect upon the possibility of buckling the outer shell, but whether and when this effect is quantitatively appreciable has not, to our knowledge, been investigated. Sometimes one finds even an actual density inversion (that is, a decrease of density with increasing depth) as shown for instance in the density curves of Clark and Ringwood [1964] for the outer shell. Ultimately, this effect is due to the exceptionally large thermal expansion of olivine. We note that a density inversion, while of course indicative of instability, is not in itself dynamically meaningful. The only quantity which measures stability or instability is the difference between the actual and the adiabatic density gradient.

Assume now a column of height $h$ subsiding (or rising) by the amount $\Delta z$. From Equation 13 we can readily compute the pressure excess at the bottom (pressure deficiency at the top) of the displaced column, assuming that the column forms part of a layer that is homogeneous and of uniform temperature gradient:

$$\Delta p = g\rho\alpha \frac{dT}{dz} \cdot h \, \Delta z \qquad (14)$$

where $g$ is gravity. We choose conditions appropriate for the soft layer, say $dT/dz = 3.3°/\text{km}$ (see middle curves of Figure 4) and $\alpha = 3.0 \times 10^{-5}$, say [see Birch, 1952; Lubimova, 1955]. Using practical units we find then, numerically,

$$\Delta p \text{ (bar)} = 3.4h \text{ (km)} \, \Delta z \text{ (km)} \qquad (15)$$

On scaling, the right-hand side of Equation 15 is proportional to the square of the linear dimensions; this indicates that large-scale phenomena engender much larger forces of buoyancy and therefore have a strong mechanical advantage. This is corroborated by the general observation that the oceanic ridges are the main regions of rising and the orogenic belts the main regions of sinking in the mantle. These phenomena are indeed of the largest scale known on the earth.

A rather conspicuous effect of the thermal instability of the mantle is the long-term persistence of orogenic belts. While such a persistence has not been established in detail for oceanic rises, it is clear that many regions of orogenic activity on continents have lasted for 1 to 200 million years, and in this time the belts have been the stage for lengthy sequences of orogenic events, each lasting perhaps only a few millions of years. This becomes now more readily understandable if we remember that once a vertical displacement of some magnitude has been started, the buoyancy forces tend to continue it in the same direction. Thus if orogenic processes are related to sinking motions in the soft layer it is plausible that they occur at the same localities where orogenies had already taken place before. This may go on until for some as yet unknown reason the source of potential energy for these vertical displacements has been substantially reduced in magnitude. Then, by virtue of the principle of diminishing symmetry, an orogenic movement taking place elsewhere might reduce the local stresses to a point where the old belt stops its activity altogether.

We now come to chemically induced motions in the mantle. Separation of lighter, sialic material from the mantle matrix mainly takes place in the outer shell owing to the greater ease of diffusion at the lower pressures. Basaltic lava appears as a result near the surface. Menard [1964] comments on vast outpourings of basalt at the ocean bottom. Fyfe and Verhoogen [1958] have demonstrated that the

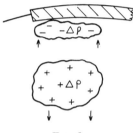

Fig. 8.

continental roots in orogenic regions must grow at least in part by accretion of material from below. The evidence is based on the fact that the basements of orogenic provinces have very commonly undergone widespread metamorphosis. This requires large quantities of heat. The energy of mechanical stresses is negligible for the purpose; hence there seems to be no other explanation than that hot sialic material and water have been brought up from the depths.

Now if light material segregates and rises, there remains a specifically heavier residue which we may call a "sinker" (Figure 8; it seems that this concept has occurred independently to Menard, Orowan, and this writer). An elementary calculation shows that if the density is diminished by 0.01 (or 0.3 per cent of the total density), the excess pressure at the bottom of a 100-km column would be 100 bars, probably enough to cause slow sinking; the rising sialic material above would be about 300 meters thick. The example shows that the assumptions underlying this argument are quantitatively realistic. Now the material of the sinkers may spread laterally in the soft layer; again, one may assume that it falls farther down into the mantle. The existence of deep-focus earthquakes can be interpreted tentatively as evidence for the latter. Below a depth of 300 km, earthquakes occur almost entirely in two regions [Gutenberg and Richter, 1954]. One of these is under the Andes, an orogenic strip having but recently been rising rather rapidly. The other region is the western border of the Pacific, an extremely active orogenic belt, reaching from Japan over the Philippines and Celebes down to New Zealand.

If the sinkers fall deeper down into the mantle, they will encounter the transition layer located at a depth of 400 km or below, whose existence has been made virtually certain by Birch [1961] and which is interpreted as representing the phase transformation from silicates to oxides. As Verhoogen* [1965] has shown, such a phase transition layer tends to smooth out small disturbances but does not act as an obstacle to convective motions on a larger scale. Hence there seems so far no radical objection to interpreting these deep-seated seismic activities as resulting from the sinking of heavier remnants of earlier chemical segregations. As the sinkers fall farther down, they will generate horizontal motions in the soft layer converging toward the region of sinking. These, transmitted by shear stresses to the outer shell, might well contribute to the familiar compressive stresses near the surface, so characteristic of orogenic belts.

Perhaps the most significant conclusion which results from the subdivision of convection-generating forces into thermal and chemical ones, concerns the mutual relationship of these two mechanisms. It seems clear that actual aggregation of sialic material takes place in the outer shell, some of it perhaps just below it. Farther down, therefore, we have only chemically heavier "sinkers"; the dynamical effect of the corresponding rising material on top of the sinker will be negligible at depth. Hence in the orogenic belts, both the chemical sinker and the thermal instability of downgoing material *pull in the same direction.* Under the oceanic rises, on the other hand, the two effects are in opposite directions at depth; they are in the same direction only in a thin layer of the order of the outer shell, where there is some rising sial or water supporting the thermal pull of the rising mantle material. The geological consequences of these facts can undoubtedly be disentangled only slowly. Let us estimate, however, the depth which a purely thermal root of an oceanic rise would have; this is given by the relation of isostasy,

$$h = \alpha H \Delta T \tag{16}$$

where $h$ and $H$ are the heights of the rise and its root, respectively, and

---

* The author is indebted to Professor Verhoogen for the opportunity of seeing this paper in manuscript form.

$\Delta T$ the mean excess temperature in the root. After correcting for the buoyancy in the surrounding water, the mean height of an oceanic rise is about 2 km; letting $\alpha = 3 \times 10^{-5}$ as before and $\Delta T = 200°$, we find $H = 330$ km. This indicates that if one attributes a primarily thermal root to the oceanic ridges, the root would extend, in order of magnitude, to the bottom of the soft layer. It does not prove of course that the root is purely thermal. Hess [1954, 1955] has given strong arguments for the view that much, or most, of the contour of the ocean bottom is due to serpentinization, that is, to addition of water which is chemically bound to mantle material and diminishes the density of the latter. It would therefore seem that the upper part of rising columns in the oceans represents a combination of thermal and chemical density changes. While in this way they are analogous to the sinking columns under orogenic belts, the relationships of buoyancy are different in the two cases, as already explained. In this context, the hypothesis is intriguing that in the sense of a uniformitarian scheme, most if not all orogenic processes of the past have been mixtures, in various proportions, of chemical as well as thermal effects.

## 7. The Anisometry of Convection*

We have on the whole now completed the task we set ourselves in the beginning, namely, to supplement the two-dimensional picture resulting from surface observations by a survey of such properties of the upper mantle in depth as may be derived from general geophysical arguments. The question as to the geometrical nature of the convection pattern or patterns is an intriguing one. It can be answered by observation only. A deformable body is capable of an infinity of patterns of deformation and, as every student of fluid mechanics

---

\* Notes on terminology:

(a) The term "anisotropic turbulence" is commonly used in geophysical hydrodynamics. Anisotropy in crystal physics designates the dependence of physical properties on direction in an otherwise homogeneous material. This applies to turbulence in that the rate of turbulent diffusion is anisotropic in this strict sense. To avoid confusion, we designate upper-mantle convection which has long horizontal and short vertical legs as "anisometric." The term isometry is applied to an object having about the same linear dimensions in all directions (e.g., an isometric grain) and correspondingly for anisometry.

(b) Convection is motion in a gravitational field driven by internal density differences. In the geophysical literature, convection is sometimes associated with the concept of "cells," that is, with comparatively simple, closed systems of circulations. Such cells, however, are primarily mathematical tools (they are orthogonal vector fields in the language of the mathematician) and the actual motions occurring are likely to be much more complicated. It appears to us impossible, from the present surface observations, to form a model of the topological connectivity of motions a little farther down in the mantle.

knows only too well, there is nothing simple about these shifting patterns. The task is yet different when the material considered is not a fluid but a solid with an extremely nonlinear relationship between shear stress and shear velocity. Nevertheless, it is possible to draw a few general conclusions. The following sketch is based upon a mixture of geological generalizations, laboratory results, and geophysical arguments, to be summarized under three heads:

(a) The analysis of submarine geological observations has led Hess [1962] to an empirical theory of mantle convection. In this view, mantle material rises under the oceanic ridges, thence spreads laterally in directions away from the ridge to both sides of it. Hess points out that nowhere in the oceans have sediments older than Cretaceous been found; he attributes this to replacement of ocean bottom by fresh mantle material spreading out from the ridges and estimates that in 3 to 400 million years most of the existing ocean bottom may have been replaced by new material.

Let us note here that previous assumptions about continental drift suffered from the fatal flaw of not providing a mechanical explanation for the shifting geometrical relationships of continents and oceans. The older idea of continents plowing through the oceans somewhat after the manner of ships seems incompatible with what we know about the mechanical properties of crust and mantle. If the American continents and Eurafrica have drifted apart, it becomes almost axiomatic that the bottom of the Atlantic be a newly generated surface. The consistent extension of this point of view lies in the working hypothesis that the same holds, or has held at some time or other, for most ocean bottoms.

(b) Mantle material has frequently been treated as a viscous fluid. Actually, crystalline solids at high temperatures show a behavior designated as "high-temperature creep" in which the relationship between shear stress and the rate of shear deformation is highly nonlinear (Figure 9). This behavior resembles far more the plasticity of solids as commonly observed under high stresses, although it should be said at once that the stress, or stress interval, which separates the region of small deformations from that of large deformations (the "yield stress," or stress interval, $Y$ of Figure 9) might be quite different in absolute magnitude from the yield stress for plasticity. For plastic flow in the laboratory, $Y$ is of the order of some thousands of bars. We know nothing about the yield stress in the earth, but it cannot be so large if convection is to occur: Simple calculations based on thermal or chemical variations of a reasonable magnitude show that shear stresses in the upper mantle can rarely exceed a few hundred bars, under the sensible assumption that relaxable stresses older than some millions of years have been relaxed.

Recently, E. Orowan* [1964, 1965] has for the first time systematically applied the concepts of solid-state creep to the earth's mantle. Since the creep curve rises so steeply past the yield region, one may in a first approximation use solutions of the equations of plasticity. In models of idealized plasticity, there is a sharply defined yield point; for lower stresses there is no motion at all, and for higher stresses there

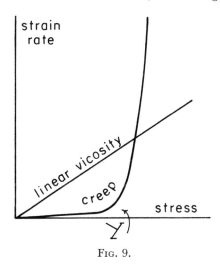

Fig. 9.

is no appreciable resistance to motion. In a plastic body the shear *strains* are sharply concentrated: Very often the material moves as a block. A viscous fluid, as is well known, develops a parabolic velocity profile. In case of a plastic body, on the other hand, the shear stress becomes concentrated on the walls and the material moves through the tube uniformly, like a plug.

As Orowan points out, if convection occurs in the upper mantle, there will be rising columns with relatively well-defined boundaries; he compares the oceanic ridges with dikes which are well-defined plates of rising hot material. It is then required, of course, that the material flows off sideways from the top of the ridges. Orowan further notes that the relationship between stress and strain rate below the yield point is by no means linear as some authors have claimed. From laboratory behavior of high-temperature creep, a fourth-power law seems fairly descriptive of the actual conditions.

* We are greatly indebted to Professor Orowan for the opportunity to study the manuscripts of these forthcoming papers.

(c) Any attempt at devising a scheme of circulation in the upper mantle must take into account the extremely superadiabatic thermal stratification. There are two chief possibilities for convection. The first one consists in a complete, relatively rapid overturn of the upper mantle or a layer of it, not perhaps at the same time all over the earth but regionally, any such overturn to be followed by a period of quiescence. In the approximation in which the adiabatic gradient is small and may roughly be equated to zero, the temperature gradient becomes simply reversed by a sufficiently rapid overturn of a layer. Thereafter, the temperature would decrease with increasing depth in the layer. If this occurred near the surface, in the outer shell say, tremendous surface heat flows would occur on overturn, followed by a period where no outward heat flow would be measurable.

The main argument against a model of this type is found in the rough uniformity of the surface heat flow which is observed at the present time (except for the crests of the oceanic ridges). Provided relatively rapid overturns have occurred in the past, it seems unlikely that there would be no detectable remnants of them. The most conspicuous feature would be a rather wide variation in the surface heat flow, depending on whether the region in question has or has not turned over in more recent geological times. Nothing of the sort seems observed.

We therefore shall adopt the alternative assumption, *that extremely superadiabatic thermal gradients are a steady-state feature of the upper mantle.* This is tantamount to a severe limitation on the convective pattern: The space-time configuration of convection must be so that large deviations from the average temperature gradient do not as a rule occur; if they occur they must be confined to narrow regions (e.g., the crests of oceanic ridges). In order to satisfy this last condition we must take into account the fact that motions confined to the upper mantle are highly *anisometric*. Since the lower mantle has a much more stable thermal stratification than the upper one, we shall assume, as has already been implied above, that the lower mantle does not participate appreciably in the convective motions considered here. We estimated the depth of the upper mantle as about 300 km but this might conceivably be too low, and 400 km or 500 km could also be a realistic value. We simply do not know this at present. The depth of the upper mantle is then only on the order of 1 per cent of the earth's circumference. This is a measure of the anisometry.

The condition of anisometry can be satisfied by a variety of flow patterns which may be schematized here in the form of two limiting cases. First we may assume that convective motions are as a rule sloping, with a very small slant angle. The other alternative consists in assuming that a fraction of the material moves vertically up or down at any one time, whereas the remainder of the mass, most of it in fact, moves very nearly horizontally. The observations clearly indicate that the truth is much closer to the second than to the first alternative. The oceanic rises are rather symmetrical, and we have so far little reason to doubt that the material comes up nearly vertically. For downgoing material, the deep-focus earthquakes, under the Andes for instance, indicate slant angles near 45°. Compared to the over-all anisometry of the upper mantle this is so steep that the idealization of localized purely vertical motions seems sensible.

We next compute the mean vertical heat transport of a motion schematized by assuming that each parcel of the upper mantle which moves appreciably at all moves either strictly vertically or strictly horizontally. Let $\eta$ be the fraction of volume (and since the upper mantle is thin, also of the surface) in which material moves vertically upwards. We shall make the approximations which correspond to assuming $\eta$ to be a small number. To estimate its magnitude, we start from the statement of Menard [1964] that the aggregated length of the ridges is almost twice the circumference of the earth. Dropping the "almost," we set the surface area occupied by the ridges equal to $4\pi ba$, where $b$ is the mean breadth of a ridge and $a$ the earth's radius. Comparing this to the earth's surface area, we see at once that, approximately,

$$\eta = b/a \tag{17}$$

The same fraction $\eta$ of the surface must be covered by downgoing material on the simple assumption that the downward and upward velocities are equal. A sinking particle carries "cold" down; so far as its effect on net transport of heat is concerned, it is equivalent to a rising particle. The *average* heat delivered upward per cm² of the earth's surface, per second, is now, using Equation 17,

$$W = (2b/a)(\rho c_p) w \Delta T \tag{18}$$

where $w$ is the vertical velocity and $\Delta T$ the temperature difference in the undisturbed earth between the mean end points of a vertical path.

The oceanic rises are often about 1000 km wide. It is likely, however, that part of this width is due to sideways outflow near the top and that somewhat farther down the column is narrower, say $b = 500$ km. This gives by Equation 17 $\eta = 0.078$. Let us further assume that the average travel of a particle is from a depth of about 260 to 40 km. This gives, using the middle curves of Figure 4, $\Delta T = 900°$. We have, approximately, $\rho c_p = 1.0$. We now assume in order of magnitude that convection carries half the surface flux, that is $W = F_0/2 = 19$ cal/cm$^2$ year. Solving Equation 18, we obtain $w = 0.14$ cm/year. Contrary to what one may think, the corresponding horizontal velocities will be larger than the vertical velocities, namely,

$$v = (b/2H)w \qquad (19)$$

Physically, this difference in velocities can be explained by remembering that the vertical motion is limited by the extremely severe constraint of isostasy: material must be removed sideways before the vertical motion can continue. There is no similar constraint on the horizontal motion. If we set $H = 75$ km, the assumed height of the outer shell, we find $v = 0.45$ km/year, or a distance of travel of 1000 km in 220 million years. We note that we are likely to have underestimated the velocities since the concurrent action of conduction and radiation tends to reduce the effective temperature difference $\Delta T$ available for convective transport. Hess [1962], on an empirical basis, estimates a horizontal velocity of about 1 cm/year and, in view of the crudeness of the scheme used here, the agreement is satisfactory.

Elementary as this analysis is, it would make no sense unless it can be shown that the conductive-radiative transport processes are efficient enough to smooth out the radical changes in temperature produced by convection and thus to maintain a superadiabatic vertical temperature gradient. We shall use the outer shell as an example; since radiative transport in the soft layer is much more rapid than conduction in the outer shell, the temperature distribution in the soft layer will be smoothed out much more rapidly than that in the outer shell.

Consider a flat plate of heat conductivity $\kappa$ whose bottom is kept at temperature $T$, while the top is kept at zero temperature. The stationary temperature gradient will be $T/H$, where $H$ is the thickness of the plate. Now let at time $t = 0$ the temperature at the bottom be raised to $T + \Delta T$ and let the temperature be kept at this value thereafter. We inquire into the time that elapses until this temperature change affects the top of the plate. The general theory of heat conduction [Carslaw and Jaeger, 1959] gives quantitative answers to this, as well as to a host of similar questions. We define a "penetration time,"

$$t_p = H^2(\rho c_p/4\kappa) \qquad (20)$$

with the following property: When $t/t_p$ is small there is no appreciable penetration to the other side; when $t/t_p$ is large there is complete penetration, meaning that a new stationary gradient has been established which is larger than the original gradient by the amount $\Delta T/H$. Similar relationships, also governed by the parameter 20, hold if we replace the "step function" in time just referred to by other, more complicated time functions. Since the functional dependence of the penetration on $t/t_p$ is essentially exponential, the transition from negligible penetration to practically complete penetration takes place within a limited interval of the variable $t/t_p$ (within a factor 3 to 4 say). Numerically, for the outer shell, with $H = 75$ km and a mean value of $\kappa = 0.005$, we obtain $t_p = 90$ million years. If replacement of the ocean bottoms by convection takes several hundred million years, it appears now altogether possible that conduction in the outer shell can maintain the typically observed, superadiabatic temperature distribution, apart from limited local variations. Conversely, if hot material reaches the surface, the top will cool and the temperature gradient at the top will asymptotically tend toward the average value for the earth. Since material rises primarily under the oceans, convective heat will be brought up to the ocean bottoms; this should tend to compensate largely for the otherwise bigger heat flow that issues from the continents on account of the concentration of radioactivity in the continental plates. Moreover, if part of the heat flow from the soft layer is carried by convection, the values of transparency given in connection with Figure 4 can be lowered, although this will hardly invalidate the arguments given in Section 4 regarding the thermal mechanisms that tend to increase transparency.

We have so far spoken only of upwellings occurring on the ocean bottoms. It seems, however, that ridges can develop almost anywhere on the earth; if they appear under a continental plate, they split the latter apart and the fragments are pushed away from the ridge. The continental plates therefore tend to be found in the regions where mantle material goes down, analogous to the way a piece of wood moves toward a region where there is a downdraft in the water. Thus the association of downdraft with continental plates is essentially a *secondary* effect caused by the relatively passive way in which these plates are subjected to the convective motions of the upper mantle. The actual pattern seems to be quite far from the symmetry indicated in Figure 5. Most of the orogenic regions where we assume downdrafts to occur or to have occurred are at the edges of continental blocks. Sometimes these downdrafts seem to be relatively independent of continental margins, as exemplified by the Western border of the Pacific, the archipelago between Asia and Australia with its shallow seas. This is perhaps an early stage of an orogenic development resembling the conditions that may have prevailed in the Western part of North America during the Paleozoic. According to Hess, the basaltic cover of the ocean floor resists being pulled down into the mantle owing to its buoyancy; after a slight initial downdraft it rises and is welded onto the continental plates.

This entire scheme is clearly too rough to contain more than the sketch of a theory of convection and of orogeny. The tremendous variability of the observed phenomena will not come as a surprise to any student of solid-state physics. There are, however, three rather definite empirical regularities which, while not perhaps rigorous laws, are still clearcut enough to require at least a semiquantitative explanation. These are (*1*) the hypsometric distribution showing a double maximum of elevations, one for the continents near sea level and one for the ocean bottoms near $-5$ km; (*2*) the rough equality of the average heat flow over continents on the one hand and over ocean bottoms on the other; (*3*) the rather exact centering of certain oceanic ridges, especially the Atlantic one, relative to the edges of the continents between which they lie. While a qualitative type of explanation for these three effects may be gathered from the preceding discussion, a true theory, which is still in the future, must no doubt be precise and quantitative on these points.

In this paper we have not even mentioned an observed phenomenon which is obviously closely related to the soft layer, namely, pole migration. Given the existence of the soft layer, one may be almost sure that pole migration represents a sliding of the outer shell over the remainder of the mantle with the soft layer as the principal shear zone. Any such motion must be intimately connected with the anisotropic convection pattern described, but the nature of this relationship remains to be worked out.

## Acknowledgments

The field covered here is a complex one, and I owe a very great deal to a number of my colleagues. I am particularly indebted for assistance and much valuable criticism to Professor E. Orowan of the Massachusetts Institute of Technology and Professor H. H. Hess of Princeton University. Several of the members of the Geology Department at Princeton have shown great patience in explaining to me numerous pertinent geological facts, especially H. Greenwood, R. Hargraves, and J. Maxwell. This work has been supported by a grant from the National Science Foundation.

## References

Anderson, D. L., and C. B. Archambeau, The anelasticity of the earth, *J. Geophys. Res.*, *69*, 2071–2084, 1964; see also *70*, 1441-1448, 1965.

Birch, F., Elasticity and the constitution of the earth's interior, *J. Geophys. Res.*, *57*, 227–286, 1952.
Birch, F., Composition of the earth's mantle, *Geophys. J.*, *4*, 295–311, 1961.
Birch, F., Density and composition of mantle and core, *J. Geophys. Res.*, *69*, 4377–4388, 1964.
Birch, F., J. F. Schairer, and H. C. Spicer, editors, *Handbook of Physical Constants*, Geol. Soc. Amer. Special Papers, No. 36, 1942.
Boyd, F. R., and J. L. England, Effect of pressure on the melting of diopside and albite in the range up to 50 kilobars, *J. Geophys. Res.*, *68*, 311–324, 1963.
Bullen, K. E., New evidence on rigidity in the earth's core, *Proc. Natl. Acad. Sci. U.S.*, *52*, 38–42, 1964.
Carslaw, M. S., and J. C. Jaeger, *Conduction of Heat in Solids*, Clarendon Press, Oxford, 510 pp., 1959.
Chandrasekhar, S., *Hydrodynamic and Hydromagnetic Stability*, Clarendon Press, Oxford, 654 pp., 1961.
Clark, S. P., Radiative transfer in the earth's mantle, *Trans. Am. Geophys. Union 38*, 931–938, 1957a.
Clark, S. P., Absorption spectra of some silicates in the visible and near infrared, *Am. Mineralogist*, *42*, 732–742, 1957b.
Clark, S. P., Variation of density in the earth and the melting curve in the mantle, in *The Earth Sciences*, edited by T. W. Donelly, pp. 5–42, University of Chicago Press, 195 pp., 1963.
Clark, S. P., and A. E. Ringwood, Density distribution and constitution of the mantle, *Rev. Geophys.*, *2*, 35–88, 1964.
Cox, A., and R. R. Doell, Review of paleomagnetism, *Bull. Geol. Soc. Am.*, *71*, 645–768, 1960.
Curie, Marie, *Pierre Curie*, Dover Publications, New York, 118 pp., reprint, 1963; see p. 24.
Daly, R. A., *Architecture of the Earth*, Appleton, New York, 211 pp., 1938.
Davis, T. C., and J. L. England, The melting of forsterite up to 50 kilobars, *J. Geophys. Res.*, *69*, 1113–1116, 1964.
Dicke, R. H., The earth and cosmology, *Science*, *138*, 653–664, 1962.
Du Toit, A. L., *Our Wandering Continents*, Oliver and Boyd, Edinburgh, 366 pp., 1937.
Elsasser, W. M., Hydromagnetic dynamo theory, *Rev. Mod. Phys.*, *28*, 135–163, 1955.
Elsasser, W. M., Early history of the earth, in *Earth Science and Meteoritics*, dedicated to F. G. Houtermans, pp. 1–30, North-Holland Publishing Company, Amsterdam, 312 pp., 1963.
Ewing, J., and M. Ewing, Seismic refraction measurements in the Atlantic Ocean basins, in the Mediterranean Sea, on the Midatlantic Ridge, and in the Norwegian Sea, *Bull. Geol. Soc. Amer.*, *70*, 291–318, 1959.
Fyfe, W. S., and J. Verhoogen, Water and heat in metamorphism, in *Metamorphic Reactions and Metamorphic Facies*, edited by W. S. Fyfe, Turner, and J. Verhoogen, pp. 187–198, Geol. Soc. Amer. Memoirs, No. 73, 259 pp., 1958.
Gutenberg, B., *Physics of the Earth's Interior*, Academic Press, Inc., New York, 240 pp., 1959.
Gutenberg, B., and C. F. Richter, *Seismicity of the Earth*, Princeton University Press, Princeton, N. J., 310 pp., 1954.

Hess, H. H., Geological hypotheses and the earth's crust under oceans, *Proc. Roy. Soc. London, A, 222*, 341–348, 1954.
Hess, H. H., The oceanic crust, *J. Marine Res., 14*, 423–439, 1955.
Hess, H. H., History of ocean basins, in *Petrological Studies: a Volume to honor A. F. Buddington*, pp. 599–620, Geol. Soc. Amer., 1962.
Kingery, W. D., *Introduction to Ceramics*, John Wiley & Sons, New York, 781 pp., 1960.
Lubimova, H. A., On the heating of the earth's interior in the process of its formation, *Izv. Akad. Nauk SSSR, Ser. Geofiz.*, No. 5, 1955.
Lubimova, H. A., Thermal history of the earth with consideration of the variable thermal conductivity of the mantle, *Geophys. J., 1*, 115–134, 1958.
Mason, Brian, *Principles of Geochemistry*, 2nd ed., John Wiley & Sons, New York, 310 pp., 1962.
MacDonald, G. J. F., Calculations on the thermal history of the earth, *J. Geophys. Res., 64*, 1967–2000, 1959.
MacDonald, G. J. F., Dependence of the surface heat flow on the radioactivity of the earth, *J. Geophys. Res., 69*, 2933–2946, 1964.
McQueen, R. G., and G. N. Fritz, and S. P. March, On the composition of the earth's interior, *J. Geophys. Res., 69*, 2947–2966, 1964.
Menard, H. W., Sea floor and mantle convection, to appear in *Physics and Chemistry of the Earth*, edited by L. H. Ahrens, McGraw-Hill Book Company, Inc., New York, 1964.
Murphy, C. T., and R. H. Dicke, The effect of a decreasing gravitational constant in the interior of the earth, *Proc. Am. Phil. Soc., 108*, 224–246, 1964.
Orowan, E., Mechanical problems of geology, unpublished lecture notes, California Institute of Technology, 1958.
Orowan, E., Continental drift, and the origin of mountains, *Science, 146*, 1003-1010, 1964.
Orowan, E., Non Newtonian viscosity, continental drift and mountain building, to appear in *Proc. Roy. Soc. London, A*, 1965.
Press, F., Some implications on mantle and crustal structure from G-waves, *J. Geophys. Res., 64*, 567–568, 1956.
Rubey, W. W., Geological history of sea water, *Bull. Geol. Soc. Am., 62*, 1111–1147, 1951.
Runcom, S. K., editor, *Continental Drift*, Academic Press, New York, 338 pp., 1963.
Scheidegger, A. E., *Principles of Geodynamics*, Academic Press, New York, 362 pp., 1963.
Strong, H. M., Melting and other phase transformations at high pressure, in *Progress in Very High Pressure Research*, edited by Bondi, Hibbard, and Strong, pp. 182–194, John Wiley & Sons, New York, 314 pp., 1961.
Takahashi, T., and W. A. Bassett, High pressure polymorph of iron, *Science, 145*, 483–485, 1964.
Takeuchi, H., and Y. Hasegawa, Viscosity distribution within the earth, in print, 1965.
Verhoogen, J., Phase changes and convection in the earth's mantle, to appear in *J. Geophys. Res.*, 1965.